Post-Processing of Parts and Components Fabricated by Fused Deposition Modeling

This book describes several post-processing techniques that can be used to enhance the mechanical strength, isotropy, surface quality, and dimensional accuracy of 3D printed components using the Fused Deposition Modeling (FDM) technique. It also discusses the usage of adhesives, interlocks, fasteners, ultrasonic, frictional, and microwave energy to join FDM-3D printed parts. Furthermore, the book also covers the scope of future research and challenges in the post-processing of FDM parts, as well as some of the most popular approaches in the field, such as Big Area Additive Manufacturing (BAAM), Machine Learning, and Internet of Things (IoT).

Features:

- Covers all necessary details related to post-processing of Fused Deposition Modeling (FDM) parts.
- Provides an overview of various joining techniques for 3D printed FDM parts.
- Focuses on the latest developments related to sustainability and optimization in post-processing of FDM parts.
- Includes microwave joining of 3D printed parts.
- Reviews case studies on cutting edge research, innovation, and development aspects.

This book is aimed at researchers and graduate students in additive manufacturing, materials science, as well as manufacturing engineering.

Advanced Materials Processing and Manufacturing

Series Editor: Kapil Gupta

The CRC Press Series in Advanced Materials Processing and Manufacturing covers the complete spectrum of materials and manufacturing technology, including fundamental principles, theoretical background, and advancements. Considering the accelerated importance of advances for producing quality products for a wide range of applications, the titles in this series reflect the state-of-the-art in understanding and engineering the materials processing and manufacturing operations. Technological advancements for enhancement of product quality, process productivity, and sustainability, are on special focus including processing for all materials and novel processes. This series aims to foster knowledge enrichment on conventional and modern machining processes. Micro-manufacturing technologies such as micro-machining, micro-forming, and micro-joining, and hybrid manufacturing, additive manufacturing, near net shape manufacturing, and ultra-precision finishing techniques are also covered.

Advanced Materials Characterization: Basic Principles, Novel Applications, and Future Directions
Ch Sateesh Kumar, M. Muralidhar Singh and Ram Krishna

Thin-Films for Machining Difficult-to-Cut Materials: Challenges, Applications, and Future Prospects
Ch Sateesh Kumar and Filipe Daniel Fernandes

Advanced Materials Processing and Manufacturing: Research, Technology, and Applications
Amogelang Sylvester Bolokang and Maria Ntsoaki Mathabathe

Advanced Joining Technologies
Edited by Manjaiah M, Shivraman Thapliyal and Adepu Kumar

Nanofinishing of Materials for Advanced Industrial Applications
Edited by Faiz Iqbal, Dilshad Ahmad Khan and Zafar Alam

Environmentally Benign Machining
Edited by Şenol Bayraktar and Sunil Pathak

Post-Processing of Parts and Components Fabricated by Fused Deposition Modeling: Techniques and Advancements
Edited by Vinayak R. Malik, Vivek Kumar Tiwary and Arunkumar Padmakumar

For more information about this series, please visit: www.routledge.com/Advanced-Materials-Processing-and-Manufacturing/book-series/CRCAMPM

Post-Processing of Parts and Components Fabricated by Fused Deposition Modeling

Techniques and Advancements

Edited by
Vinayak R. Malik, Vivek Kumar Tiwary and
Arunkumar Padmakumar

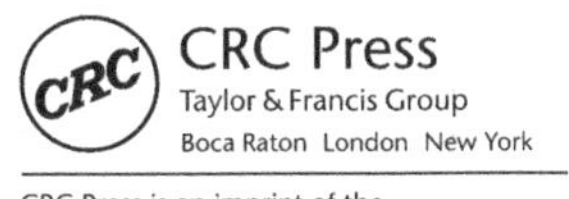

CRC Press
Taylor & Francis Group
Boca Raton London New York

CRC Press is an imprint of the
Taylor & Francis Group, an **informa** business

Designed cover image: Shutterstock

First edition published 2025
by CRC Press
2385 NW Executive Center Drive, Suite 320, Boca Raton FL 33431

and by CRC Press
4 Park Square, Milton Park, Abingdon, Oxon, OX14 4RN

CRC Press is an imprint of Taylor & Francis Group, LLC

© 2025 selection and editorial matter, Vinayak R. Malik, Vivek Kumar Tiwary and Arunkumar Padmakumar; individual chapters, the contributors

ISBN: 9781032527710 (hbk)
ISBN: 9781032665344 (pbk)
ISBN: 9781032665351 (ebk)

DOI: 10.1201/9781032665351

Typeset in Times
by Newgen Publishing UK

Contents

Chapter 1 An Overview of Post-Processing of Fused Deposition Modelling
3D Printed Products 1

Mehmet Şükrü Adin and Menderes Kam

SECTION I Post-Processing Techniques for Improving the Quality of FDM 3D Printed Parts

Chapter 2 Chemical-Based Methods for Polishing Surfaces Produced
by Material Extrusion Process..13

Mohammad Vahid Ehteshamfar

Chapter 3 Surface Roughness Evaluation of Fused Deposition Modeling:
Additive-Manufactured Polylactic Acid Components Affected by
3D-Printing Parameters and Vapor Post Processing Treatments........24

*Mohammad Azadi, Maryam Chiyani, Shokouh Dezianian,
Ali Dadashi, Amir Hasan Rahaei and
Mohammad Kamyab*

Chapter 4 Post-Processing of Additive Manufacturing Functional Polymeric
Parts: Influence on Surface, Dimensional Quality and Mechanical
Performance...43

Giovanni Gómez-Gras and Marco A. Pérez

Chapter 5 Coating Methods and Materials for 3D Printed (FDM) Parts............81

*Sateeshkumar Kanakannavar, Kiran Shahapurkar,
Gangadhar M. Kanaginahal, Rayappa Shrinivas Mahale and
Prashant P. Kakkamari*

SECTION II Joining/Welding as Post-Processing Techniques for FDM 3D Printed Parts

SECTION III Miscellaneous Topics

Preface

Additive Manufacturing or 3D printing is considered an important technological innovation of Industry 4.0 which relies on the benefits like design freedom, less setup cost, and ease of customization. However, one of the important limitations of FDM-3D printed parts is the requirement of several post-processing measures to improve upon their mechanical and aesthetic characteristics. Unfortunately, there isn't enough systematic fundamental understanding to solve this issue. This book *Post-Processing Treatments for Fused Deposition Modeling 3D Printed Parts: Techniques and Advancements* tries to fill in this gap by providing specific information and recent advances from multiple perspectives.

Reasonably wide coverage in sufficient depth has been attempted, giving importance to various post-processing techniques to improve the surface finish quality, mechanical strength, dimensional accuracies and isotropy of FDM-3D printed parts. The highlight of the book is the different welding techniques including Friction Stir, Friction Stir Spot, Spin Friction, Microwave and Ultrasonic welding that could be attempted on the FDM-3D printed parts. For nascent researchers in the field of 3D printing, microwave welding of 3D-printed parts is a new topic that has not been extensively studied. A special emphasis is on the case studies from different companies which are also included in the book. Further, separate chapters are devoted to Big Area Additive Manufacturing (BAAM), IoT, Machine learning, Hybrid Large Format Additive Manufacturing (LFAM) and the future challenges and research scope in post-processing of FDM parts.

The book is well organized into three sections with fourteen chapters. Section 1 is preceded by a general introduction to the various post-processing techniques for FDM parts (Chapter 1). Section 1 provides a detailed discussion of various post-processing techniques to improve the quality of FDM 3D printed parts including Chemical-based methods (Chapter 2), Vapor Smoothing technique (Chapter 3), Post-processing of functional parts (Chapter 4), and Coating methods (Chapter 5). Section 2 tries to discuss various joining/welding methods as post-processing techniques for FDM 3D printed parts which includes an overview of different post-processing techniques to overcome the volume limitation of FDM-3D printers (Chapter 6), Adhesive joining (Chapter 7), Friction-based welding techniques (Chapter 8), and newer welding techniques like Microwave and Ultrasonic Welding (Chapter 9). Section 3 contains separate chapters on recent post-processing technologies including Big Area Additive Manufacturing (BAAM) (Chapter 10), Machine Learning (Chapter 11), a study on additive manufacturing processes, standards, and mechanical properties (Chapter 12), the Role of IoT (Chapter 13) and finally ends with the Future challenges and Research scope in post-processing of FDM parts (Chapter 14). For better comprehension, each chapter is accompanied by relevant diagrams, figures, and actual photos.

The book is expected to benefit undergraduates, postgraduates, individuals working with additive manufacturing in government, industry, academia, 3D printing consultants, researchers, and professionals in mechanical, industrial, civil, and aeronautical engineering. This book is an equal combined effort of Dr. Vinayak R. Malik, Vivek Kumar Tiwary, and Dr. Arunkumar Padmakumar It has original and novel contributions from many different authors from around the world.

To improve the content of this book, we will gladly accept reader recommendations. We will make an effort to include the best ones in subsequent versions.

Vinayak R. Malik
Vivek Kumar Tiwary
Arunkumar Padmakumar

About the Editors

Vinayak R. Malik is working as associate professor in the department of Mechanical Engineering at KLS Gogte Institute of Technology. He holds a Doctorate degree in Material Science from Indian Institute of Science (IISc), Bangalore. His research areas include Friction Stir Welding and Processing, Microwave Sintering, Materials technology, Additive Manufacturing and 3D printing. He has published numerous research papers in reputed journals along with two grants and four patents to his credit. His research contributions in close association with industry partner Enerzi Microwave Systems Pvt. Ltd. have yielded metallurgical solutions to Industrial problems.

Vivek Kumar Tiwary is working as assistant professor in the department of Mechanical Engineering at KLS Gogte Institute of Technology. He holds a Master's degree in Material Science & Metallurgy from National Institute of Technology, Surathkal. His research areas include Materials technology, Additive Manufacturing, Welding/joining of 3D printed parts and Heat treatment of steels. His Ph.D. title is "Experimental Investigations of Joining as Post-Processing Techniques to Circumvent the Bed Size Limitation of FDM-3D Printers" earned from VTU. He has published 15 research papers in reputed SCI journals. He has obtained and executed grants worth Rs. 55 lakhs from KCTU and KSCST. Recently, his final year UG project has also bagged the reputed IMTEX and KSCST awards. His most recent "IOT based 3D printing" event was one of several workshops and training programs he has held for students and industry personnel. He has been actively offering consultancy services to the local MSME's industries in the fields of 3D printing and 3D scanning.

Arunkumar Padmakumar is an associate professor at the KLS Gogte Institute of Technology, Belagavi. He was previously the institute's chief of examiner (COE), and he is currently the Dean of academics. His experience in the field of academics and industry extends over a period of 18 years. His research areas include CAD/CAM, Process Planning and Additive Manufacturing. He has 30 international journal and conference publications to his credit along with two research grants. He has conducted several workshops/FDP for MSME's industries in the fields of 3D printing and 3D scanning. He has received invitations to serve as a resource on reverse engineering and 3D printing from numerous organizations and businesses.

Contributors

Mehmet Şükrü Adin
Batman University
Besiri OSB Vocational School
Turkey

Nergizhan Anaç
Engineering Faculty of Zonguldak Bulent
    Ecevit University
Turkey

Mohammad Azadi
Faculty of Mechanical Engineering
Semnan University
Semnan, Iran

Shreyans Bhandari
Department of Architecture at
KLS Gogte Institute of Technology
Belagavi, India

Maryam Chiyani
Faculty of Mechanical Engineering
Semnan University
Semnan, Iran

Ali Dadashi
Faculty of Mechanical Engineering
Semnan University
Semnan, Iran

Shokouh Dezianian
Faculty of Mechanical Engineering
Semnan University
Semnan, Iran

Mohammad Vahid Ehteshamfar
University of Guelph
Canada

Giovanni Gómez-Gras
IQS School of Engineering
Ramon Llull University
Barcelona, Spain

Prashant P. Kakkamari
Department of Mechanical Engineering
KLS Gogte Institute of Technology
Belagavi, India

Menderes Kam
Duzce University
Dr. Engin Pak Cumayeri Vocational School
Turkey

Mohammad Kamyab
Manufacturing and Design Department
    Irankhodro Powertrain Company
    (IPCO)
Tehran, Iran

Gangadhar M. Kanaginahal
Department of Mechanical Engineering
Jain College of Engineering and Research
Belagavi, Karantaka, India

Sateeshkumar Kanakannavar
Department of Mechatronics and
    Automation Engineering
IIIT Bhagalpur
Bihar, India

Oğuz Koçar
Engineering Faculty of Zonguldak Bulent
    Ecevit University
Turkey

Raman Kumar
Department of Mechanical and
Production Engineering
Guru Nanak Dev Engineering College
    Ludhiana, Punjab, India

Ranvijay Kumar
University Centre for Research and
    Development
Chandigarh University

Rayappa Shrinivas Mahale
Department of Mechanical Engineering
Jain College of Engineering and Research
Belagavi, India

Vinayak R. Malik
Department of Mechanical Engineering
    KLS Gogte Institute of Technology
Belagavi, Karnataka, India

Arunkumar Padmakumar
Department of Mechanical Engineering
    KLS Gogte Institute of Technology
Belagavi, Karnataka, India

Jatinder Pal
Mechanical Engineering at Thapar
    Institute of Engineering and
    Technology Patiala
Faculty at Guru Nanak Dev
Engineering College
Ludhiana, Punjab, India

Marco A. Pérez
IQS School of Engineering
Ramon Llull University
Barcelona, Spain

Amir Hasan Rahaei
Faculty of Mechanical Engineering
    Semnan University
Semnan, Iran

Srikrishna B. Rao
Jain University
Jain Global Campus
Bengaluru, India

Kiran Shahapurkar
School of Mechanical, Chemical and
    Materials Engineering
Adama Science and Technology University
Adama, Ethiopia

Vivek Kumar Tiwary
Department of Mechanical Engineering
    KLS Gogte Institute of Technology
Belagavi, Karnataka, India

Introduction

Welcome to the forefront of additive manufacturing innovation! Fused Deposition Modeling (FDM) has revolutionized the landscape of modern manufacturing, offering unparalleled flexibility, affordability, and accessibility. As the demand for FDM-printed parts and components continues to surge across industries ranging from aerospace and automotive to consumer goods and healthcare, the need for optimizing post-processing techniques becomes increasingly imperative.

This book represents a comprehensive exploration of post-processing methodologies tailored specifically for parts and components fabricated through Fused Deposition Modeling. From foundational principles to cutting-edge advancements, our journey navigates through the intricate realm of additive manufacturing, focusing on refining the quality, functionality, and aesthetics of FDM prints.

Before delving into the intricacies of post-processing, it's essential to grasp the fundamentals of Fused Deposition Modeling. FDM, a subset of additive manufacturing, operates on the principle of layer-by-layer deposition of thermoplastic materials to form three-dimensional objects. This technique offers unparalleled versatility in material selection, ranging from standard polymers like ABS and PLA to advanced composites and engineering-grade thermoplastics.

While FDM presents numerous advantages such as rapid prototyping, design iteration, and cost-effectiveness, the inherent layer-by-layer deposition process often results in surface imperfections, structural weaknesses, and aesthetic limitations. Addressing these shortcomings necessitates a holistic approach encompassing meticulous design considerations, optimized printing parameters, and, crucially, effective post-processing techniques.

Post-processing serves as the bridge between raw FDM prints and refined, application-ready components. By strategically employing an array of techniques, engineers and designers can enhance mechanical properties, improve surface finish, and imbue prints with desirable functional characteristics. Moreover, post-processing enables customization, enabling the tailoring of parts to meet specific performance requirements or aesthetic preferences.

In this book, we elucidate the pivotal role of post-processing in the additive manufacturing workflow, emphasizing its potential to unlock the full capabilities of FDM technology. Through a systematic exploration of techniques and advancements, readers will gain invaluable insights into optimizing their post-processing workflows to achieve superior results efficiently and cost-effectively.

Our journey through post-processing methodologies traverses both traditional techniques and cutting-edge advancements, offering a comprehensive toolkit for optimizing FDM prints. From mechanical finishing and chemical smoothing to thermal treatments and surface coatings, each chapter provides a deep dive into a specific aspect of post-processing, accompanied by practical insights, case studies, and real-world applications.

Furthermore, we spotlight recent advancements in post-processing technology, including emerging innovations such as machine learning-driven optimization, in-situ monitoring, and hybrid additive-subtractive manufacturing approaches. By staying at the forefront of these developments, readers can harness the latest tools and methodologies to stay competitive in an ever-evolving additive manufacturing landscape.

As additive manufacturing continues to proliferate across industries, the importance of post-processing cannot be overstated. *Post-Processing of Parts and Components Fabricated by Fused Deposition Modeling: Techniques and Advancements* represents a definitive guide for engineers, designers, and enthusiasts seeking to unlock the full potential of FDM technology. Whether you're a seasoned professional or a novice exploring the possibilities of additive manufacturing, this book promises to be an indispensable resource in your journey towards manufacturing excellence.

1 An Overview of Post-Processing of Fused Deposition Modelling 3D Printed Products

Mehmet Şükrü Adin and Menderes Kam

1.1 INTRODUCTION

Modern manufacturing industries give more importance to new products and new production processes in parallel with the rapid progress in technology. Therefore, they devote large financial resources to new product development and new production processes. In the manufacturing processes, improving the quality of the products, reducing the costs and shortening the production time have been determined as the main goals. Contemporary additive manufacturing techniques have been created and used in accordance with these objectives before the latter portion of the 1980s. The main characteristic feature of additive manufacturing processes is that materials are usually added layer by layer until the complete desired product is manufactured. Thanks to this developed manufacturing method, it is possible to make geometrically very complex products that are intricate or almost impossible to fabricate with traditional manufacturing processes [1, 2, 3, 4, 5, 6].

1.2 ADDITIVE MANUFACTURING PROCESS

Additive manufacturing (AM) is also called Three-dimensional (3D) printing and 3D printers in different manufacturing industries. 3D printers are very popular equipment used in educational institutions as well as in many manufacturing industries [1, 2, 7, 8].

As a result of all the research on 3D printers, it is stated that this technology is transformative. For this reason, 3D printers will directly affect component design, product cost, product delivery, global business models and logistics. Moreover, it will allow low environmental impact and high energy efficiency [2, 3, 9, 10].

In general, 3D printer processes are classified into several categories. These can be expressed as a set of similar 3D printer processes such as geometric shape, material, machine type, required post-processing, surface finish and process. For many years, 3D printers had to be used without classification due to lack of organization. As a result of this situation, problems were experienced in the transfer of information on this subject both in technical education and in the manufacturing industry. Later,

DOI: 10.1201/9781032665351-1

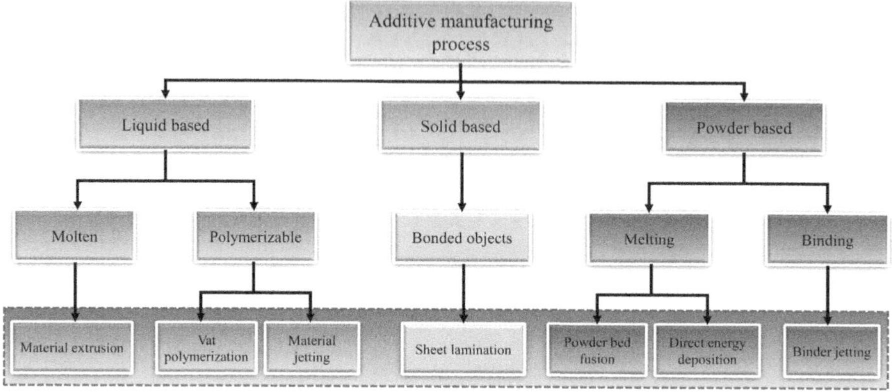

FIGURE 1.1   Additive manufacturing categories according to ISO/ASTM 52900 standard [2, 11]

TABLE 1.1
Classification according to the principle used [12]

Principle used	AM process
VAT Polymerisation (VPP)	–
MaterialTJetting (MJT)	Drop onTDemand (DOD)
BinderTJetting (BJT)	–
MaterialTExtrusion (MEX)	Fused DepositionTModelling (FDM)
Powder Bed Fusion (PBF)	Direct MetalTLaser Sintering (DMLS)
	Electron BeamTMelting (EBM)
	Selective HeatTSintering (SHS)
	Selective LaserTMelting (SLM)
	Selective LaserTSintering (SLS)
	Three-Dimensional Printing (3DP)
Sheet Lamination (SHL)	Laminated ObjectTManufacturing (LOM)
Directed Energy Deposition (DED)	–

thanks to the International Standardization Organization (ISO) and the American Society for Testing and Materials (ASTM), a solution has been found for this situation [2, 9, 10]. Thus, the ISO/ASTM 52900 standard for 3D printers, also known as AM, has emerged. As per the ISO/ASTM 52900 standard, AM is divided into seven broad categories. These classifications are shown in Figure 1.1 [2, 11].

As seen in Figure 1.1, according to the ISO/ASTM 52900 standard, the processes of 3D printers are determined as MaterialTextrusion (MEX), Vat polymerization (VPP), Material jetting (MJT), SheetTlamination (SHL), Powder BedTFusion (PBF), Directed Energy Deposition (DED) and BinderTJetting technology (BJT), respectively [11]. Also included in Table 1.1 are the guiding principles for several AM process types [12].

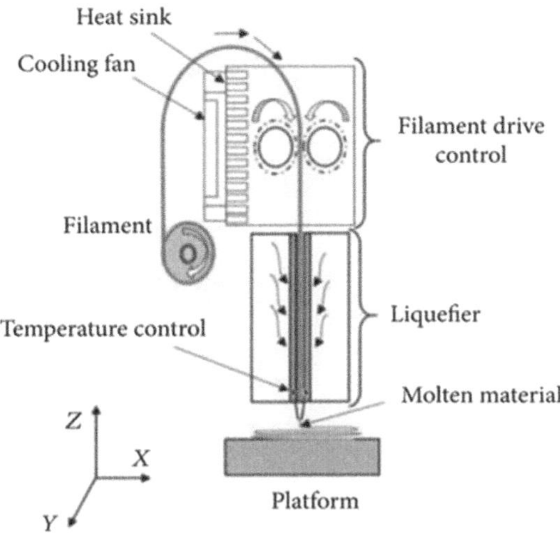

FIGURE 1.2 The process of FDM technology [19]

The 3D printers are seen as a revolution in the manufacturing industry as they allow rapid prototyping. That's why it attracts more attention than usual. Due to this extraordinary interest, many different technologies have been developed for 3D printers, for example Fused Deposition Modelling (FDM), Selective Laser Sintering (SLS) and Stereolithography (SLA) [13, 14]. Among the most popular 3D printing processes is FDM, which is one of these various technologies. This very popular technology is used commercially in most 3D printers. Moreover, it is low-maintenance and highly cost-effective. In the FDM technology, parts are produced by extruding molten material layer by layer as the material hardens [14, 15, 16, 17]. Extruded materials such as ABS, PLA, and PETG [1, 11, 12, 18] are used in FDM technology. These materials make it possible to create pieces with intricate shapes in layers [2]. As seen in Figure 1.2, the process of FDM technology is depicted [19].

1.3 POST-PROCESSING METHODS OF FDM 3D PRINTED PRODUCTS

FDM technology is extensively used in many domains such as aviation, space, automobile and medical because of its many advantages such as ease of application and production of functional parts [20, 21, 22, 23, 24]. As can be seen in Figure 1.3, some application areas of parts produced using FDM technology are shown [24].

Despite all these advantages, there is a need for extensive research on materials produced with FDM technology. The main reason for this is that there are some problems mentioned by the manufacturers. The most frequently mentioned problem in production with FDM technology is poor surface quality or characteristic texture (due to staircase effect) [25, 26, 27, 28]. In the literature, there are studies on this subject. In these studies, it is stated that the parts produced with FDM inevitably have surface roughness. In some studies, researchers using different techniques have

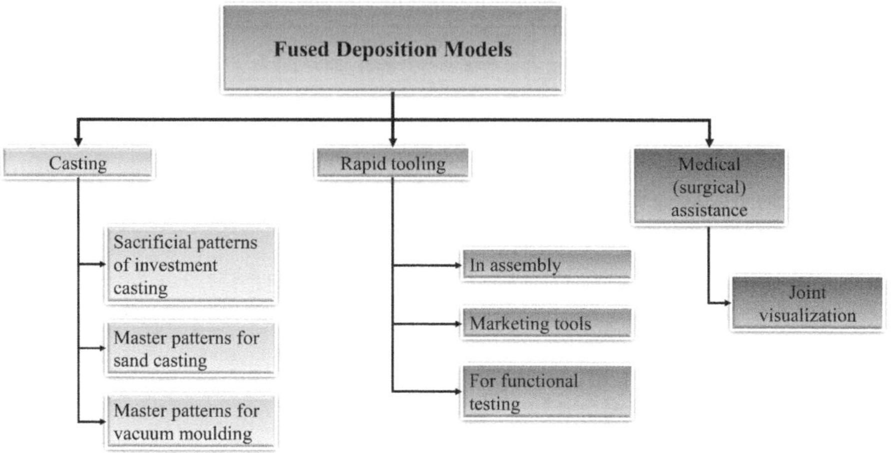

FIGURE 1.3   Applications of FDM products [24]

shown that they are successful in controlling and reducing surface roughness [29]. Even pre-treatment practices have been used to reduce the bad surface quality. As a result of these investigations examined, both pre-processing and post-processing were found to be quite remarkable [30]. Pre-processing technologies are generally expressed as the determination of manufacturing parameters before the production of the part. However, post-processing technologies generally refer to any operation performed after part manufacture. Post-processing is classified as mechanical [31, 32], thermal [33, 34] and chemical [35, 36] treatments. Moreover, it is stated that the fastest result among these treatments is obtained with the second treatment [12]. Summary results of some studies on the post-processing techniques are given in Table 1.2.

In an experimental study supporting this situation, it is reported that the acetone vapour smoothing treatment performed on ABS parts reduces the surface roughness by approximately 90% [24].

Post-processing techniques are divided into two different groups according to the application method. These are expressed as conventional and non-conventional techniques (Figure 1.4) [12].

Also, the categories according to the raw material used in AM are shown in Figure 1.5 [12].

1.4   CONCLUSION

In this chapter, open source literature on post processing treatments of FDM 3D printed products has been reviewed. These conclusions were arrived at based on the findings of the investigations:

- Additive manufacturing is a popular method extensively employed because of its low cost and simplicity.

TABLE 1.2
An overview of some studies on post-processing techniques

	Post-processing techniques		
	Process	**average surface improvement**	**Ref.**
Mechanical finishing techniques	Manual sanding	Not suitable for parts with complex shapes	[29]
	Abrasive Milling	40%–42%	[37, 38]
	Abrasive Flow Machining	65%–70%	[39, 40]
	Sand Blasting	Matte finish problem	[29, 41]
	Vibratory Bowl Finishing	increase in surface roughness	[42]
	Barrel Tumbling	45%–52%	[31, 43]
	Hot Cutter Machining	Unsuitable for components with intricate shapes	[44, 45]
	Ball Burnishing	Not suitable for parts with complex shapes	[46, 47]
Thermal finishing technique	Thermal	75%–85% (fastest result)	[29, 33, 34]
Chemical finishing techniques	Manual Painting	increase in surface quality	[29]
	Acetone Dipping	95%–98%	[48, 49, 50]
	Vapour Smoothing	80%–90%	[51, 52, 41]
	Electroplating	increase in surface roughness (in some tests)	[53, 54, 55]

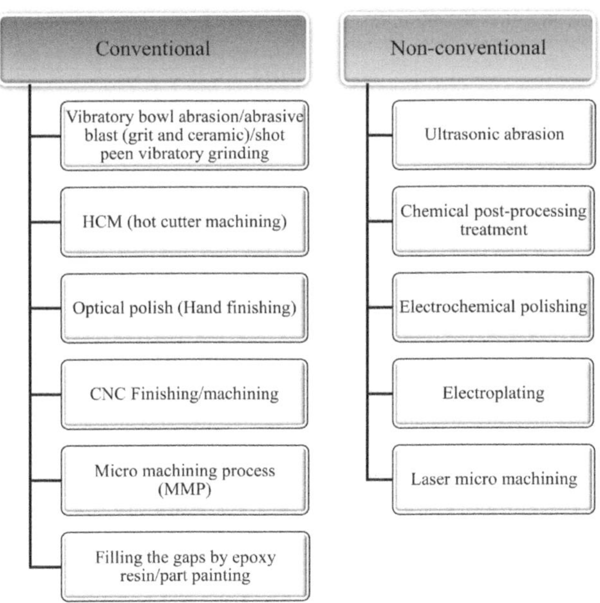

FIGURE 1.4 Conventional and non-conventional techniques [12]

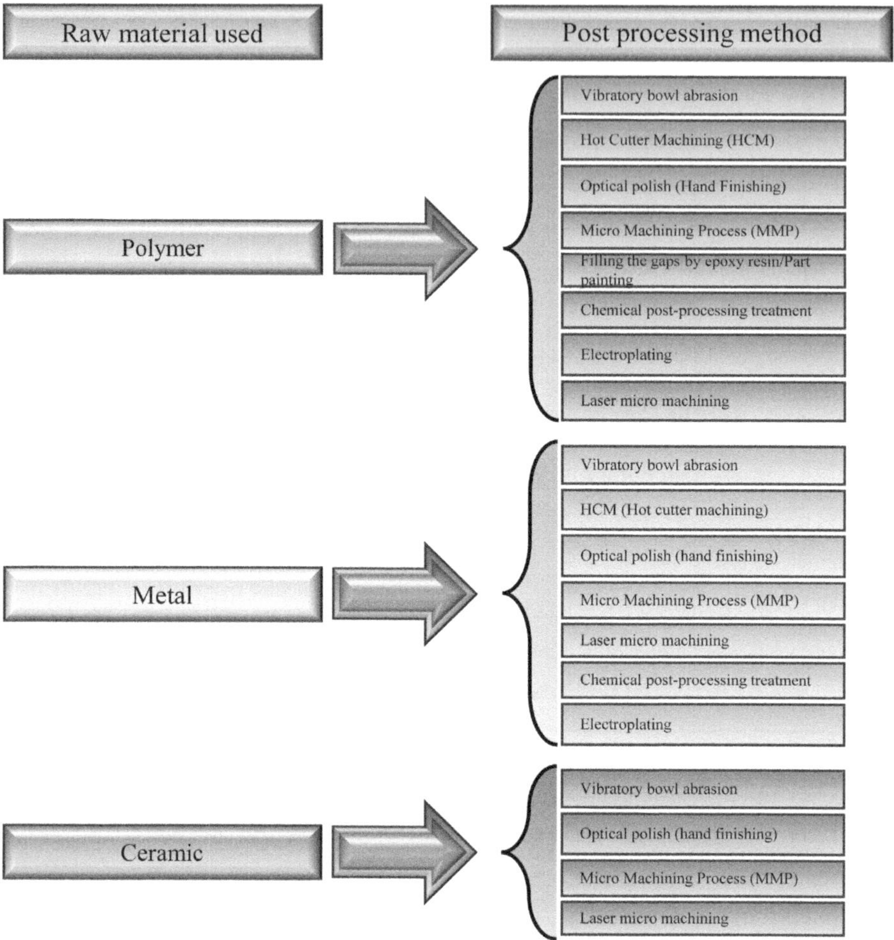

FIGURE 1.5   The categories according to the raw material used in AM [12]

- Due to its many benefits, including simplicity of use and fabrication of useful parts, AM has become popular in a variety of industries, including aerospace, automotive, and biomedicine.
- One benefit of 3D printing/printers is the fact that there are no geometric restrictions.
- FDM has some limitations due to their poor dimensional accuracy and surface quality.
- Academicians have shown that they are successful in regulating and improving surface quality using various methods.
- Post-processing methods are divided into two different groups according to the application method as conventional and non-conventional techniques.
- Post-processing technologies generally refer to any operation performed after part manufacture. Post-processing is classified as mechanical, thermal, and

chemical treatments. Moreover, it is stated that the fastest result among these treatments is obtained with the second treatment.

REFERENCES

[1] Çevik, Ü., and Kam, M., "A review study on mechanical properties of obtained products by FDM method and metal/polymer composite filament production," *Journal of Nanomaterials*, vol. 2020, pp. 1–9, 2020.

[2] D. Godec, T. B., Katalenić, M., Nordin, A., Diegel, O., Kristav, P., Motte, D., and Tavčar, J., Applications of AM, in: D. Godec, J. Gonzalez-Gutierrez, A. Nordin, E. Pei, J. Ureña Alcázar, *A Guide to Additive Manufacturing*, Springer Nature Switzerland AG, Cham, pp. 1–344, 2022.

[3] Frazier, W. E., "Metal additive manufacturing: A review," *Journal of Materials Engineering and Performance*, vol. 23, pp. 1917–1928, 2014.

[4] Mitchell, A., Lafont, U., Hołyńska, M., and Semprimoschnig, C., "Additive manufacturing—A review of 4D printing and future applications," *Additive Manufacturing*, vol. 24, pp. 606–626, 2018.

[5] Tiwary, V. K., Padmakumar, A., and Malik, V., "Adhesive bonding of similar/dissimilar three-dimensional printed parts (ABS/PLA) considering joint design, surface treatments, and adhesive types," *Proceedings of the Institution of Mechanical Engineers, Part C: Journal of Mechanical Engineering Science,* vol. 236, no. 16, pp. 8991–9002, 2022.

[6] Tiwary, V. K., Ravi, N., Arunkumar, P., Shivakumar, S., Deshpande, A. S., and Malik, V. R., "Investigations on friction stir joining of 3D printed parts to overcome bed size limitation and enhance joint quality for unmanned aircraft systems," *Proceedings of the Institution of Mechanical Engineers, Part C: Journal of Mechanical Engineering Science,* vol. 234, no. 24, pp. 4857–4871, 2020.

[7] Tiwary, V. K., Deshpande, A. S., and Rangaswamy, N., "Surface enhancement of FDM patterns to be used in rapid investment casting for making medical implants," *Rapid Prototyping Journal,* vol. 25, no. 5, pp. 904–914, 2019.

[8] Tiwary, V. K., Padmakumar, A., and Malik, V. R., "Investigations on FSW of nylon micro-particle enhanced 3D printed parts applied to a Clark-Y UAV wing," *Welding International,* vol. 36, no. 8, pp. 474–488, 2022.

[9] Gibson, I., Rosen, D., Stucker, B., Khorasani, M., Rosen, D., Stucker, B., and Khorasani, M., *Additive Manufacturing Technologies*, Springer Nature Switzerland AG, 2021.

[10] Hegab, H., Khanna, N., Monib, N., and Salem, A., "Design for sustainable additive manufacturing: A review," *Sustainable Materials and Technologies*, vol. 35, pp. e00576, 2023.

[11] ISO/ASTM, *ISO/ASTM 52900: 2015, Additive Manufacturing–General Principles–Terminology*: ASTM F2792-10e1, West Conshohocken, PA 19428-2959, USA, 2015.

[12] Kumbhar, N. N., and Mulay, A., "Post processing methods used to improve surface finish of products which are manufactured by additive manufacturing technologies: A review," *Journal of The Institution of Engineers (India): Series C*, vol. 99, pp. 481–487, 2018.

[13] Gibson, I., Rosen, D., and Stucker, B., "Additive manufacturing technologies: Rapid prototyping to direct digital manufacturing," in: I. Gibson, D. W. Rosen, B. Stucker, Springer, Heidelberg, Germany, pp. 415–435, 2010.

[14] Hague, R., Dickens, P., and Hopkinson, N., *Rapid Manufacturing: An Industrial Revolution for the Digital Age*, John Wiley & Sons, Chichester, England, 2006.

[15] Kam, M., Saruhan, H., and İpekçi, A., "Experimental investigation of vibration damping capabilities of 3D printed metal/polymer composite sleeve bearings," *Journal of Thermoplastic Composite Materials*, vol. 36, no. 6, pp. 08927057221094984, 2022.

[16] Kam, M., İpekçi, A., and Şengül, Ö., "Investigation of the effect of FDM process parameters on mechanical properties of 3D printed PA12 samples using Taguchi method," *Journal of Thermoplastic Composite Materials,* vol. 36, no. 1, pp. 307–325, 2023.

[17] Onwubolu, G. C., and Rayegani, F., "Characterization and optimization of mechanical properties of ABS parts manufactured by the fused deposition modelling process," *International Journal of Manufacturing Engineering,* vol. 2014, pp. 1–13, 2014.

[18] Atakok, G., Kam, M., and Koc, H. B., "Tensile, three-point bending and impact strength of 3D printed parts using PLA and recycled PLA filaments: A statistical investigation," *Journal of Materials Research and Technology,* vol. 18, pp. 1542–1554, 2022.

[19] Hodzic, D., and Pandzic, A., "Influence of carbon fibers on mechanical properties of materials in FDM technology." pp. 334–342, 2019. DOI: 10.2507/30th.daaam. proceedings.044

[20] Anaç, N., Koçar, O., and Hazer, B., "Katmanlı imalatla üretilen parçaların birleştirilmesinde yapıştırma bağlantı dayanımının incelenmesi," *International Journal of 3D Printing Technologies and Digital Industry,* vol. 6, no. 3, pp. 449–458, October 2022.

[21] Atakok, G., Kam, M., and Koc, H. B., "A Review of mechanical and thermal properties of products printed with recycled filaments for use in 3d printers," *Surface Review and Letters,* vol. 29, no. 2, pp. 2230002, 2022.

[22] Kam, M., İpekçi, A., and Şengül, Ö., "Effect of FDM process parameters on the mechanical properties and production costs of 3D printed PowerABS samples," *International Journal of Analytical, Experimental and Finite Element Analysis, RAME Publishers,* vol. 7, no. 3, pp. 77–90, 2020.

[23] Mahale, R. S., Kanaginahal, G. M., Vasanth, S., Tiwary, V. K., Shashanka, R., Sharath, P., and Patil, A., "Applications of fused deposition modeling in dentistry," in: R. Keshavamurthy, Vijay Tambrallimath, J. Paulo Davim, *Development, Properties, and Industrial Applications of 3D Printed Polymer Composites*, IGI Global, Hershey, Pennsylvania, USA, pp. 211–219, 2023.

[24] Singh, R., Singh, S., Singh, I. P., Fabbrocino, F., and Fraternali, F., "Investigation for surface finish improvement of FDM parts by vapor smoothing process," *Composites Part B: Engineering,* vol. 111, pp. 228–234, 2017.

[25] Kam, M., Saruhan, H., and İpekci, A., "Investigation the effects of 3D printer system vibrations on mechanical properties of the printed products," *Sigma Journal of Engineering and Natural Sciences,* vol. 36, no. 3, pp. 655–666, 2018.

[26] Kam, M., Saruhan, H., and İpekçi, A., "Investigation the effect of 3D printer system vibrations on surface roughness of the printed products," *Düzce Üniversitesi Bilim ve Teknoloji Dergisi,* vol. 7, no. 2, pp. 147–157, 2019.

[27] Rao, A. S., Dharap, M. A., Venkatesh, J. V., and Ojha, D., "Investigation of post processing techniques to reduce the surface roughness of fused deposition modeled parts," *International Journal of Mechanical Engineering and Technology,* vol. 3, no. 3, pp. 531–544, 2012.

[28] Vijay, P., Danaiah, P., and Rajesh, K., "Critical parameters effecting the rapid prototyping surface finish," *Journal of Mechanical Engineering and Automation,* vol. 1, no. 1, pp. 17–20, 2011.

[29] Chohan, J. S., and Singh, R., "Pre and post processing techniques to improve surface characteristics of FDM parts: a state of art review and future applications," *Rapid Prototyping Journal,* vol. 23, no. 3, pp. 495–513, 2017.

[30] Singh, G., and Kumar, P., "Methods to Improve Surface Finish of Parts Produced by Fused Deposition Modeling. Manuf," *Science and Technology,* vol. 2, pp. 51–55, 2014.

[31] Boschetto, A., and Bottini, L., "Surface improvement of fused deposition modeling parts by barrel finishing," *Rapid Prototyping Journal,* vol. 21, no. 6, pp. 686–696, 2015.

[32] Galantucci, L. M., Lavecchia, F., and Percoco, G., "Experimental study aiming to enhance the surface finish of fused deposition modeled parts," *CIRP Annals,* vol. 58, no. 1, pp. 189–192, 2009.

[33] Kumar, R., Singh, R., and Ahuja, I., "Investigations of mechanical, thermal and morphological properties of FDM fabricated parts for friction welding applications," *Measurement,* vol. 120, pp. 11–20, 2018.

[34] Kumar, R., Singh, R., Ahuja, I., Amendola, A., and Penna, R., "Friction welding for the manufacturing of PA6 and ABS structures reinforced with Fe particles," *Composites Part B: Engineering,* vol. 132, pp. 244–257, 2018.

[35] Hambali, R., Cheong, K. M., and Azizan, N., *"Analysis of the influence of chemical treatment to the strength and surface roughness of FDM,"* IOP Conference Series: Materials Science and Engineering, vol. 210, International Technical Postgraduate Conference 5–6 April 2017, University of Malaya, Kuala Lumpur, Malaysia, p. 012063.

[36] Lalehpour, A., and Barari, A., "Post processing for fused deposition modeling parts with acetone vapour bath," *IFAC-PapersOnLine,* vol. 49, no. 31, pp. 42–48, 2016.

[37] Spencer, J. D., "Vibratory finishing of stereolithography parts," Materials Science, Engineering, 1993. DOI:10.15781/T2DJ5906T

[38] Spencer, J. D., Cobb, R., and Dickens, P., "Surface finishing techniques for rapid prototyping," *Technical Papers-Society Of Manufacturing Engineers-All Series,* 1993.

[39] Leong, K. F., Chua, C. K., Chua, G. S., and Tan, C. H., "Abrasive jet deburring of jewellery models built by stereolithography apparatus (SLA)," *Journal of Materials Processing Technology,* vol. 83, no. 1–3, pp. 36–47, 1998.

[40] Williams, R. E., and Melton, V. L., "Abrasive flow finishing of stereolithography prototypes," *Rapid Prototyping Journal,* vol. 4, no. 2, pp. 56–67, 1998.

[41] Zinniel, R. L., "Surface-treatment method for rapid-manufactured three-dimensional objects," Google Patents, 2014.

[42] Trivedi, A., "Experimental investigations for surface quality improvement of abs patterns prepared by FDM using barrel finishing," *M. Tech Thesis GNDEC, Ludhiana,* 2014.

[43] Fischer, M., and Schöppner, V., "Some investigations regarding the surface treatment of Ultem* 9085 parts manufactured with fused deposition modeling," 2013.

[44] Pandey, P. M., Reddy, N. V., and Dhande, S. G., "Improvement of surface finish by staircase machining in fused deposition modeling," *Journal of Materials Processing Technology,* vol. 132, no. 1–3, pp. 323–331, 2003.

[45] Pandey, P. M., Venkata Reddy, N., and Dhande, S. G., "Virtual hybrid-FDM system to enhance surface finish," *Virtual and Physical Prototyping,* vol. 1, no. 2, pp. 101–116, 2006.

[46] Bottini, L., Boschetto, A., and Veniali, F., "Estimation of material removal by profilometer measurements in mass finishing," *Key Engineering Materials*, pp. 615–622, May 2014. https://doi.org/10.4028/www.scientific.net/kem.611-612.615

[47] Vinitha, M., Rao, A., and Mallik, M., "Optimization of speed parameters in burnishing of samples fabricated by fused deposition modeling," *International Journal of Mechanical and Industrial Engineering*, vol. 2, no. 2, pp. 10–12, 2012.

[48] Colpani, A., Fiorentino, A., and Ceretti, E., "Characterization of chemical surface finishing with cold acetone vapours on ABS parts fabricated by FDM," *Production Engineering*, vol. 13, pp. 437–447, 2019.

[49] Garg, A., Bhattacharya, A., and Batish, A., "On surface finish and dimensional accuracy of FDM parts after cold vapor treatment," *Materials and Manufacturing Processes*, vol. 31, no. 4, pp. 522–529, 2016.

[50] Kuo, C.-C., and Mao, R.-C., "Development of a precision surface polishing system for parts fabricated by fused deposition modeling," *Materials and Manufacturing Processes*, vol. 31, no. 8, pp. 1113–1118, 2016.

[51] Anthamatten, M., Letts, S. A., and Cook, R. C., "Controlling surface roughness in vapor-deposited poly (amic acid) films by solvent-vapor exposure," *Langmuir*, vol. 20, no. 15, pp. 6288–6296, 2004.

[52] Priedeman, W., and Smith, D., "Smoothing method for layer manufacturing process," United States Patent No. US8123999B2, USA, 2011.

[53] Daneshmand, S., Aghanajafi, C., and Ahmadi Nadooshan, A., "The effect of chromium coating in RP technology for airfoil manufacturing," *Sadhana*, vol. 35, pp. 569–584, 2010.

[54] Kannan, S., and Senthilkumaran, D., "Assessment of mechanical properties of Ni-coated ABS plastics using FDM process," *International Journal of Mechanical & Mechatronics Engineering*, vol. 14, no. 3, pp. 30–35, 2014.

[55] Raja, K., Naiju, C., Narayanan, S., Jeeva, P., and Karthikeyan, S., "Metallization of ABS plastics prepared by FDM-RP process and evaluation of corrosion and hardness characteristics," *International Journal of ChemTech Research*, vol. 6, no. 14, pp. 5490–5493, 2014.

Section I

Post-Processing Techniques for Improving the Quality of FDM 3D Printed Parts

2 Chemical-Based Methods for Polishing Surfaces Produced by Material Extrusion Process

Mohammad Vahid Ehteshamfar

2.1 INTRODUCTION

Material Extrusion or MEX, an alternate term for Fused Deposition Modelling (FDM), ISO 52900, is an extremely well-known Additive Manufacturing (AM) process and is also the most economical [1], [2]. Unfortunately, it is impossible to completely remove the stair case effect of produced components, which might restrict their use, even by optimizing production settings [3]. Figure 2.1 illustrates this staircase effect graphically. Therefore, it seems that the surface texture of parts fabricated by the cheapest AM technology has to be enhanced since it is necessary for a variety of applications, including customized biomedical implants, which demand great surface finish, mechanical characteristics, and dimensional accuracy [4]. There are several ways to modify this situation, the most common of which is chemical treatment.

In this chapter, various methods for chemically modifying the surfaces of components will be discussed in the first stage. Following this, articles that employed these methods will be examined along with their advantages and disadvantages. These approaches will be contrasted in the final phase. Figure 2.2 illustrates various chemical treatment types.

2.2 TYPES OF CHEMICAL TREATMENT

2.2.1 Dips in Acetone

A simple technique to change the surface of manufactured components is to dip them in acetone or other chemicals, a process known as chemical bath treatment. This technique produces components with remarkable surface peeling capabilities by immersing them in an acetone aqueous solution for a set period of time [5]. What is needed for an acetone dipping machine is not really complicated as depicted in

DOI: 10.1201/9781032665351-3

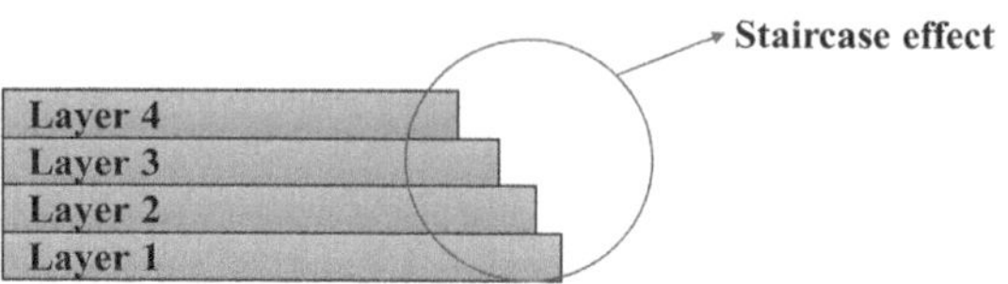

FIGURE 2.1   Staircase effect of parts manufactured by MEX

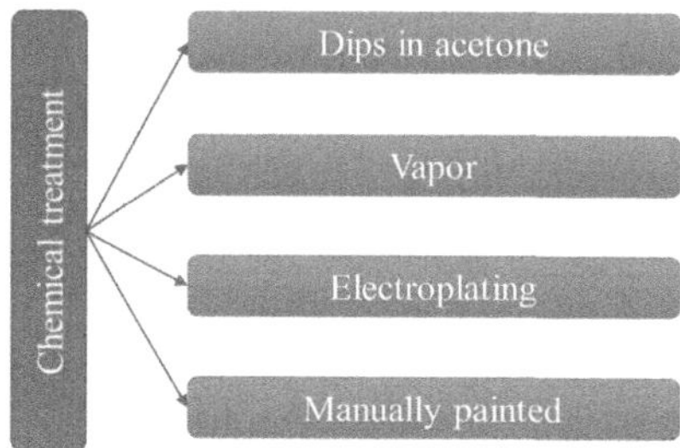

FIGURE 2.2   Various chemical treatment methods

FIGURE 2.3   Acetone dipping machine [15]

Figure 2.3. [6]–[8] proposed an acetone-based chemical finishing technique to completely transform the surface quality of MEX objects. The scientists initially produced the MEX component under controlled process conditions using a range of solvents as a surface finishing agent. They discovered that this magic liquid, acetone (C_3H_6O), works best for cleaning ABS parts that have been MEX produced. However, there is a chance that the fragments will disintegrate if they are left in the solution for a long time. If they need to be submerged longer, the solution should be weakened by the addition of water or another solvent [9]. Rao et al. looked at variables such as the concentration of different chemicals, the starting roughness, and the merge time [10], while Lalehpour et al. [11] looked at the length of time and the quantity of smoothing procedures used in this finishing technique. Furthermore, the significance of acetone on the exterior roughness of ABS components, which were made by MEX technology, was evaluated. In this technique, the components were submerged in an diluted acetone solution containing 40–60% acetone between 1 to 8 hours, according to Mccullough et al. [9], and this procedure yielded the best outcomes for the acetone dipping approach. The surfaces of MEX-printed ABS components were finished

FIGURE 2.4 Vapor treatment [15]

using an acetone-based surface finishing process by Chen et al., [12] Havenga et al. [13] and Nguyen and Lee [14]. According to the authors, acetone is successfully used to polish the freeform surfaces of parts, produced by MEX.

It should be noticed that during the dipping procedure, acetone seeps into the cracks on the ABS pieces. Due to acetone's rapid evaporation, the component's outer layers may close and acetone may become trapped in the cracks, causing the part to become distorted. Additionally, employing a diluted solution can prolong immersion times, and prolonged exposure to a component may be harmful. For a more exact and organized finishing, the process has to be controlled, automated, and mechanized [16].

2.2.2 VAPOR TREATMENT

The Vapor Smoothing Technique is a state-of-the-art method to make the dimensional precision, surface quality, and aesthetic appeal of items produced using MEX better. Figure 2.4 displays the process' schematic. It is plainly obvious from the tools used in Figure 2.4 that this is also a low-cost method of changing the surface quality of components.

Both Colpani et al. [17] and Garg et al. [18] examined cold acetone vapor treatment to dramatically alter the surface roughness of specimens fabricated by MEX technology. The results of the former research revealed that the surface quality has experienced a significant improvement (98%) while the latter reached Ra = 0.63μm. In addition to cold acetone working, Kalyan et al. [19] finished the MEX-printed ABS pieces using hot acetone vapors. They found that the surface finish was greatly enhanced by post-processing with hot acetone vapors compared to preprocessing. Singh et al. [20] proved that the benefit of applying vapor of acetone for improving the exterior roughness of ABS parts, which were additively produced. The average S/N ratio and dimensional deviation for surface roughness are 39.8050 and 10.98, respectively. Additionally, Jayanth et al. [21] looked into how vapor treatment influences the surface roughness and tensile strength. The numerous surface enhancement methods for MEX-printed patterns for the IC of bio implants have been evaluated by Singh et al. [22]. The significance of surface finish for bio implant made employing IC application has been determined by the authors. They also talked about the widespread technique of producing implants by combining IC and AM procedures. The in vitro test for bio implants has also been covered by the authors. Using in vitro biocompatibility studies, the adaptation of produced bio implants to the human body has been studied. The capacity of a material to adapt to the human body is the most crucial factor to take into account when selecting it for biomedical implants. A biomaterial should be durable and able to carry out its functions till the person's death. Some of the causes of implant failure include mechanical, tribological, chemical, manufacturing, and biocompatibility concerns. One of them is oxidation or corrosion, which is a major clinical problem for implant failure. The in-vitro investigation of bio implants has led authors to the conclusion that employing vapor smoothing to improve the surface quality of the MEX printed pattern of a bio implant is a safe and healthy approach in terms of biocompatibility concerns and corrosion behavior. The combination of three processes—fused deposition modelling, vapor smoothing, and investment casting—was employed to create implants composed of stainless steel 316L in the Singh et al. [22] study. The implants have also been put through in-vitro testing to check for biocompatibility and corrosion behavior, as well as to decide on their use and establish the effect of manufacturing processes on their corrosion properties. Khan and Mishra [23] created an original, chemical-based strategy to enrich the surface quality of ABS parts fabricated by the MEX process. These AM components were completed by the authors using acetone vapor. They claimed that compared to the other characteristics, the air gap has more sensible effect on the roughness. Sekhar et al. [24] has examined the MEX machine settings and how they affect the surface roughness of PLA components that have been produced using the MEX process. These MEX parts were finished by the authors using dichloromethane vapor. Dichloromethane is shown to be the greatest chemical for breaking down PLA material and eradicating layer lines to lessen surface roughness. A vapor smoothing procedure was employed by Chohan et al. [16] to complete the MEX-printed ABS imprints for IC application. According to the researchers, these MEX parts' surface roughness was greatly decreased, making them acceptable for producing biological implants. Chemical post treatment was employed by Kuo et al. [25] to improve the surface quality of the MEX printed pieces. The surface roughness of these MEX components has significantly improved, according to the authors, however following chemical treatment; these parts lose their height and form. Hand finishing, spray painting, computer aided machining, shot peening, barrel finishing, and chemical treatment were compared by Nsengimana et al. [26]. The scientists claimed that of the aforementioned post-processing procedures, chemical treatment is considered the most significant method. In order to make bio implants utilizing IC, Tiwary et al. [27] experimentally carried out an examination of surface quality enhancement of MEX imprints. With pre-processing treatments, the surface roughness of the MEX-based ABS imprints decreased from 22.63 to 14.3 μm. The best method for further lowering the surface roughness to 0.30 μm was chemical treatment. The average surface roughness of medical implants that underwent various pre and post-processing procedures was 0.68 mm. Rapid investment casting process might be a preferable approach for limited volume, tailored products with special specifications, according to cost and lead time comparisons. Heshmat and Abdelrhman [28] have suggested a unique finishing method that makes use of a silica-water combination slurry to augment the exterior quality of PLA-based MEX products. The experimental examination revealed that the stair-casing effect in these PLA components had been significantly reduced. The same authors [28] have also enhanced the surface smoothness of ABS-based MEX components using vs. post-processing. The ranking of process parameters was examined by the authors using a multi-criteria decision-making method. A chemical-based finishing method employing ethyl acetate vapors for PLA-based MEX components has been proposed by Lavecchia et al. [29]. The largest roughness reduction, according to the scientists, was observed using 5.5 ml of ethyl acetate ($C_4H_8O_2$) solvent for 360s, or around 90%. Hot acetone vapor has been employed by Riva et al. [30]. They use this technique to enhance the 3D-printed products' surfaces. Dimethyl ketone has also been utilized by authors to complete MEX components. Their analysis of the data led them to the conclusion that the average surface roughness had been decreased by 98%. Nevertheless, PC, PMMA, PET, etc. are mentioned in the literature. It has been noted that alternative thermoplastic polymers manufactured using MEX do not need vapor smoothing. It will be essential to test the vapor smoothing technique on various thermoplastic materials in the future and to work with various chemical vapors at various speeds.

2.2.3 ELECTROPLATING

The electroplating process is schematically depicted in Figure 2.5. Palladium serves as a bonding agent in this procedure by first cleaning the surface and then coating it with it. By employing this technique, a thin coating of chromium is deposited, making the product more resilient and improving its wear assistance, mechanical strength, and surface polish [31]. The materials that are most frequently employed as coatings in this approach are gold, copper, silver, chrome, nickel, zinc, brass, etc. [25]. Ni electroplating has been suggested by Eßbach et al. [32] as a way to improve the surface polish of ABS items that have been printed using MEX. By employing electroplated material to patch any surface faults or gaps in stair casing, the exterior smoothness of MEX products was enhanced (Ni). Original surface roughness of the parts being 13.46µm was decreased to 1.02µm after electroplating. Sathishkumar et al. [33] used

FIGURE 2.5 Electroplating [15]

this technique because of its superior coating qualities, straightforward process, low cost, and outstanding surface polish make it superior to other techniques. Five different orientations were used to create components, and the outcomes of both electroplated and non-electroplated components were contrasted and verified to use in functional end tooling applications. The results show that the electroplating process significantly improved the tensile, flexural, and surface roughness properties of 3D-printed HIPS parts when compared to non-electroplated components. Based on Kannan et al. [34] investigation, while samples with a thickness of 70 and 80 µm significantly improved in ductility, the electroplating procedure used on the ABS components with a thickness of 60 µm exhibited lesser ductility. Sugavaneswaran et al. [35] investigated how different build orientations affected the surface roughness and dimensions of MEX parts. The surface roughness and change in size of electroplated MEX pieces at various orientations are predicted using an advanced interpolation approach. Kumar et al. [36] investigated how this procedure affected the mechanical features of MEX-3D-printed PLA pieces. They found that Young's modulus, UTS, and impact strength improved when electro-deposited nickel is applied to MEX prototypes. Kannan et al. [37] found that copper, nickel, and chrome coatings improve the impact strength and hardness of ABS specimens. Electroplating ABS components increases the impact strength and hardness, which results in the creation of useful prototypes and finished goods. A smooth surface is another benefit of electroplating for MEX products, according to their research. In Equbal et al. [38] compares contrasts two methods of copper electroplating on MEX components. The two techniques are as follows: (I) Surface preparation for electroplating using chromic acid (ii) an electro-plating technique that uses Al-C paste to prepare the surface. The conclusion suggests that employing Al-C paste instead of chromic acid for surface preparation results in improved copper deposition. This method has various drawbacks, including the high cost of processing it and the industry in which it is most often used. More hardness and mechanical strength are provided by this approach, although the size of the pieces is altered [39].

2.2.4 Manual Painting

Thinner, acetone, toluene, and heptane are just a few of the chemicals which may be utilized to smooth down ABS components. Using this technique, the layers are connected while the extra edges are deleted [40]. However, human painting won't be as consistent as automated painting, leading to uneven surface finishing, even if it can save time and money.

2.3 COMPARISON AND CONCLUSION

Figure 2.6 summarizes the surface roughness range attained using several chemical-based surface treatment approaches. As could be observed, dips in acetone came in second with almost 0.4µ*m*, whereas vapour post-processing results in the least amount of surface roughness. Table 2.1 displays the advantages and limitations of these three techniques. Table 2.2 contrasts these technologies' post-processed components' mechanical characteristics and microstructure as well.

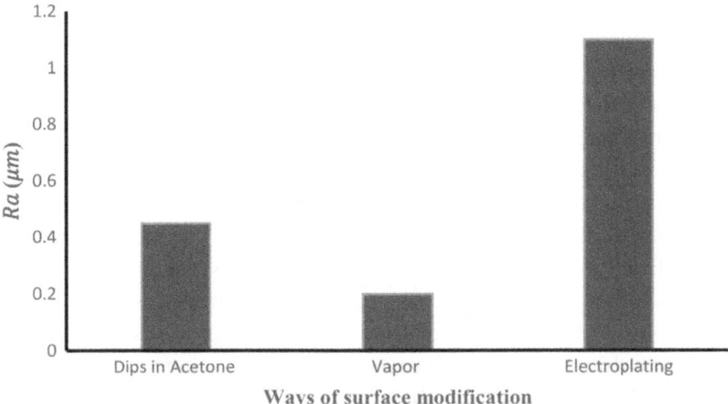

FIGURE 2.6 Comparison of achieved surface roughness via three main chemical treatment methods

TABLE 2.1

Benefits and drawbacks of using bath, vapor smoothing, and electroplating

Type of finishing method	Benefits	Drawback
Bath	• Even application • Can be used for complex parts • Cheap and fast	• More acetone is used than is necessary • Least secure approach • Probability of damaging parts
Vapor smoothing	• Minimal quantity of acetone is used • Safer than bath method	• Even smoothness with bigger components is challenging to maintain.
Electroplating	• completing complicated shapes and surfaces and generating improved surface quality	• Increased processing time and expense

TABLE 2.2

The influence of using bath, vapor smoothing, and electroplating on mechanical properties and micro structure

Type of finishing method	Enhanced mechanical characteristics	Microstructure
Bath	Low	Average
Vapor smoothing	Medium	Average
Electroplating	Low	Poor

To produce parts with complex geometry with the minimum amount of time, AM is an excellent alternative to traditional manufacturing such as milling and turning. Although it is challenging to produce any object utilizing the AM method without flaws, including somehow unacceptable surface quality, anisotropy, and porosities, or flaws that are either more visible than those found in conventional production techniques. Researchers have investigated a number of post-processing methods, enhancing the surface quality of the AM components in an effort to overcome these worries. The potential for the final roughness, the time allotted for achieving the anticipated surface quality, and the related costs all have an impact on the decision of a surface finishing method. The four primary chemical methods for modifying components produced by additive manufacturing were discussed in this section. Each methodology was described in the initial section of this review, which was followed by a summary of the findings from the works of other authors who looked into how various techniques affected surface roughness, dimensional correctness, etc. Each section's last paragraph either discussed the method's limitations or potential future research projects. Finally, a brief comparison of various approaches was made in terms of surface roughness, mechanical characteristics, and microstructure.

REFERENCES

[1] M. S. Javadi, M. V. Ehteshamfar, and H. Adibi, "A comprehensive analysis and prediction of the effect of groove shape and volume fraction of multi-walled carbon nanotubes on the polymer 3D-printed parts in the friction stir welding process," *Polym. Test.*, vol. 117, no. July 2022, p. 107844, 2023, doi: 10.1016/j.polymertesting. 2022.107844.

[2] M. H. Bahrami, M. V. Ehteshamfar, and H. Adibi, "The effect, prediction, and optimization of Fe particles on wear behavior of Fe–ABS composites fabricated by Fused Deposition Modeling," *Arab. J. Sci. Eng.*, vol. 49, no. 7, pp. 2001–2016, 2023.

[3] M. V. Ehteshamfar, M. S. Javadi, and H. Adibi, "Surface modification of prototypes in fused deposition modelling using lapping process," *Rapid Prototyp. J.*, vol. 28, no. 7, pp. 1382–1393, 2022, doi: 10.1108/RPJ-06-2021-0148.

[4] J. Singh, R. Singh, and H. Singh, "Dimensional accuracy and surface finish of biomedical implant fabricated as rapid investment casting for small to medium quantity production," *J. Manuf. Process.*, vol. 25, pp. 201–211, 2017, doi: 10.1016/j.jmapro.2016.11.012.

[5] G. Percoco, F. Lavecchia, and L. M. Galantucci, "Compressive properties of MEX rapid prototypes treated with a low cost chemical finishing," *Res. J. Appl. Sci. Eng. Technol.*, vol. 4, no. 19, pp. 3838–3842, 2012.

[6] N. B. Crane, Q. Ni, A. Ellis, and N. Hopkinson, "Impact of chemical finishing on laser-sintered nylon 12 materials," *Addit. Manuf.*, vol. 13, pp. 149–155, 2017, doi: 10.1016/j.addma.2016.10.001.

[7] L. M. Galantucci, F. Lavecchia, and G. Percoco, "Experimental study aiming to enhance the surface finish of fused deposition modeled parts," *CIRP Ann. – Manuf. Technol.*, vol. 58, no. 1, pp. 189–192, 2009, doi: 10.1016/j.cirp.2009.03.071.

[8] E. Lyczkowska, P. Szymczyk, B. Dybała, and E. Chlebus, "Chemical polishing of scaffolds made of Ti-6Al-7Nb alloy by additive manufacturing," *Arch. Civ. Mech. Eng.*, vol. 14, no. 4. pp. 586–594, 2014. doi: 10.1016/j.acme.2014.03.001.

[9] E. J. McCullough and V. K. Yadavalli, "Surface modification of fused deposition modeling ABS to enable rapid prototyping of biomedical microdevices," *J. Mater. Process. Technol.*, vol. 213, no. 6, pp. 947–954, 2013, doi: 10.1016/j.jmatprotec.2012.12.015.

[10] A. S. Rao, M. Dharap, J. Venkatesh, and D. Ojha, "Investigation of post processing techniques to reduce the surface roughness of fused deposition modeled parts," *IJMET*, vol. 3, no. 3. pp. 531–544, 2012.

[11] A. Lalehpour and A. Barari, "Post processing for Fused Deposition Modeling Parts with Acetone Vapour Bath," *IFAC-PapersOnLine*, vol. 49, no. 31, pp. 42–48, 2016, doi: 10.1016/j.ifacol.2016.12.159.

[12] Y. F. Chen, Y. H. Wang, and J. che Tsai, "Enhancement of surface reflectivity of fused deposition modeling parts by post-processing," *Opt. Commun.*, vol. 430, no. January, pp. 479–485, 2019, doi: 10.1016/j.optcom.2018.07.011.

[13] S. P. Havenga, D. J. de Beer, P. J. M. van Tonder, and R. I. Campbell, "The effect of acetone as a post-production finishing technique on entry-level material extrusion part quality," *South African J. Ind. Eng.*, vol. 29, no. 4, pp. 53–64, 2018, doi: 10.7166/29-4-1934.

[14] T. K. Nguyen and B. K. Lee, "Post-processing of MEX parts to improve surface and thermal properties," *Rapid Prototyp. J.*, vol. 24, no. 7, pp. 1091–1100, 2018, doi: 10.1108/RPJ-12-2016-0207.

[15] N. Sunay, M. Kaya, and Y. Kaynak, "Properties of parts fabricated by additive manufacturing," *Sigma J. Eng. Nat. Sci.*, vol. 38, no. 4, pp. 2027–2042, 2020.

[16] J. S. Chohan and R. Singh, "Pre and post processing techniques to improve surface characteristics of MEX parts: A state of art review and future applications," *Rapid Prototyping Journal*, vol. 23, no. 3. pp. 495–513, 2017. doi: 10.1108/RPJ-05-2015-0059.

[17] A. Colpani, A. Fiorentino, and E. Ceretti, "Characterization of chemical surface finishing with cold acetone vapours on ABS parts fabricated by MEX," *Prod. Eng.*, vol. 13, no. January 2022, pp. 437–447, 2019, doi: 10.1007/s11740-019-00894-3.

[18] A. Garg, A. Bhattacharya, and A. Batish, "Chemical vapor treatment of ABS parts built by MEX: Analysis of surface finish and mechanical strength," *Int. J. Adv. Manuf. Technol.*, vol. 89, no. 5–8, pp. 2175–2191, 2017, doi: 10.1007/s00170-016-9257-1.

[19] K. Kalyan, J. Singh, G. S. Phull, S. Soni, H. Singh, and G. Kaur, "Integration of MEX and vapor smoothing process: Analyzing properties of fabricated ABS replicas," *Mater. Today: Proc.*, 2018, vol. 5, no. 14, pp. 27902–27911, doi: 10.1016/j.matpr.2018.10.029.

[20] R. Singh, S. Singh, I. P. Singh, F. Fabbrocino, and F. Fraternali, "Investigation for surface finish improvement of MEX parts by vapor smoothing process," *Compos. Part B Eng.*, vol. 111, pp. 228–234, 2017, doi: 10.1016/j.compositesb.2016.11.062.

[21] N. Jayanth, P. Senthil, and C. Prakash, "Effect of chemical treatment on tensile strength and surface roughness of 3D-printed ABS using the MEX process," *Virtual Phys. Prototyp.*, vol. 13, no. 3, pp. 155–163, 2018, doi: 10.1080/17452759.2018.1449565.

[22] D. Singh, R. Singh, K. S. Boparai, I. Farina, L. Feo, and A. K. Verma, "In-vitro studies of SS 316 L biomedical implants prepared by MEX, vapor smoothing and investment casting," *Compos. Part B Eng.*, vol. 132, pp. 107–114, 2018, doi: 10.1016/j.compositesb.2017.08.019.

[23] M. S. Khan and S. B. Mishra, "Minimizing surface roughness of ABS-MEX build parts: An experimental approach," in *Mater. Today: Proc.*, vol. 26, no. March 2020, pp. 1557–1566, 2019, doi: 10.1016/j.matpr.2020.02.320.

[24] S. S. Panda, R. Chabra, S. Kapil, and V. Patel, "Chemical vapour treatment for enhancing the surface finish of PLA object produced by fused deposition method using the Taguchi optimization method," *SN Appl. Sci.*, vol. 2, no. 5, p. 916, 2020, doi: 10.1007/s42452-020-2740-1.

[25] C. C. Kuo, C. W. Wang, Y. F. Lee, Y. L. Liu, and Q. Y. Qiu, "A surface quality improvement apparatus for ABS parts fabricated by additive manufacturing," *Int. J. Adv. Manuf. Technol.*, vol. 89, no. 1–4, pp. 635–642, 2017, doi: 10.1007/s00170-016-9129-8.

[26] J. Nsengimana, J. Van der Walt, E. Pei, and M. Miah, "Effect of post-processing on the dimensional accuracy of small plastic additive manufactured parts," *Rapid Prototyp. J.*, vol. 25, no. 1, pp. 1–12, 2019, doi: 10.1108/RPJ-09-2016-0153.

[27] V. K. Tiwary, P. Arunkumar, A. S. Deshpande, and N. Rangaswamy, "Surface enhancement of MEX patterns to be used in rapid investment casting for making medical implants," *Rapid Prototyp. J.*, vol. 25, no. 5, pp. 904–914, 2019, doi: 10.1108/RPJ-07-2018-0176.

[28] M. Heshmat and Y. Abdelrhman, "Improving surface roughness of polylactic acid (PLA) products manufactured by 3D printing using a novel slurry impact technique," *Rapid Prototyp. J.*, vol. 27, no. 10, pp. 1791–1800, 2021, doi: 10.1108/RPJ-09-2020-0227.

[29] F. Lavecchia, M. G. Guerra, and L. M. Galantucci, "Chemical vapor treatment to improve surface finish of 3D printed polylactic acid (PLA) parts realized by fused filament fabrication," *Prog. Addit. Manuf.*, vol. 7, no. 1, pp. 65–75, 2022, doi: 10.1007/s40964-021-00213-2.

[30] L. Riva, A. Fiorentino, and E. Ceretti, "Characterization of chemical surface finishing with hot acetone vapours on ABS parts fabricated by FFF," *Prog. Addit. Manuf.*, vol. 7, no. 4, pp. 785–796, 2022, doi: 10.1007/s40964-022-00265-y.

[31] S. Daneshmand, C. Aghanajafi, and A. Ahmadi Nadooshan, "The effect of chromium coating in RP technology for airfoil manufacturing," *Sadhana – Acad. Proc. Eng. Sci.*, vol. 35, no. 5, pp. 569–584, 2010, doi: 10.1007/s12046-010-0036-7.

[32] C. Eßbach, D. Fischer, and D. Nickel, "Challenges in electroplating of additive manufactured ABS plastics," *J. Manuf. Process.*, vol. 68, pp. 1378–1386, 2021, doi: 10.1016/j.jmapro.2021.06.037.

[33] N. Sathishkumar, N. Arunkumar, L. Balamurugan, L. Sabarish, and A. Samuel Shapiro Joseph, "Investigation of mechanical behaviour and surface roughness properties on copper electroplated MEX high impact polystyrene parts," *Adv. Addit. Manuf.Join.*, 2020, pp. 287–300, doi: 10.1007/978-981-32-9433-2_25.

[34] S. Kannan and D. Senthilkumaran, "Investigating the influence of electroplating layer thickness on the tensile strength for fused deposition processed abs thermoplastics," *Int. J. Eng. Technol.*, vol. 6, no. 2, pp. 1047–1052, 2014.

[35] M. Sugavaneswaran, M. T. Prince, and A. Azad, "Effect of electroplating on surface roughness and dimension of MEX parts at various build orientations," *FME Trans.*, vol. 47, no. 4, pp. 880–886, 2019, doi: 10.5937/fmet1904880S.

[36] K. T. V. Naveen, M. V Kulkarni, M. Ravuri, E. K., and K. Saisha, "Effects of Electroplating on the Mechanical Properties of MEX-PLA Parts," *i-manager's J. Futur. Eng. Technol.*, vol. 10, no. 3, pp. 29–37, 2015, doi: https://doi.org/10.26634/jfet.10.3.3346.

[37] S. Kannan and D. Senthilkumaran, "Assessment of mechanical properties of Ni-coated ABS plastics using MEX process," *Int. J. Mech. Mechatronics Eng.*, vol. 14, no. 3, pp. 30–35, 2014.

[38] A. Equbal, Md. Asif Equbal, and A. Kumar Sood, "Electroplating of MEX Parts: A comparison via two different routes," *Adv. Mater. Proc.*, vol. 2, no. 7, pp. 414–419, 2021, doi: 10.5185/amp.2017/702.

[39] D. Impens and R. J. Urbanic, "A comprehensive assessment on the impact of post-processing variables on tensile, compressive and benDing characteristics for 3D printed components," *Rapid Prototyp. J.*, vol. 22, no. 3, pp. 591–608, 2016, doi: 10.1108/RPJ-02-2015-0018.

[40] G. Pyka, A. Burakowski, G. Kerckhofs, M. Moesen, S. Van Bael, J. Schrooten, and M.Wevers, "Surface modification of Ti6Al4V open porous structures produced by additive manufacturing," *Adv. Eng. Mater.*, vol. 14, no. 6, pp. 363–370, 2012, doi: 10.1002/adem.201100344.

3 Surface Roughness Evaluation of Fused Deposition Modeling

Additive-Manufactured Polylactic Acid Components Affected by 3D-Printing Parameters and Vapor Post Processing Treatments

*Mohammad Azadi, Maryam Chiyani,
Shokouh Dezianian, Ali Dadashi,
Amir Hasan Rahaei and
Mohammad Kamyab*

3.1 INTRODUCTION

Nowadays, design engineers from various industries are taking an interest in additive manufacturing, compared to subtractive manufacturing such as machining. This is due to the reduction of initial materials. The other reason for using additive manufacturing is the capability of fabricating any complex geometries for structures [1–2]. These issues are advantages of additive manufacturing, in contrast to disadvantages, which are the low speed of mass production and the fabrication quality, especially for the surface roughness. Therefore, some complementary processes are required to improve the performance of such components.

One technique for fabricating industrial parts is 3D-printing in fused deposition modeling (FDM) for polymeric parts. The material is extruded from a nozzle and a heater module to fabricate the structure, layer by layer. As a literature review, several works were performed on the 3D-printing parameters to obtain a superior performance (strength or lifetime) of additive-manufactured structures [3–6]. In some case studies, the extruded samples were compared to the 3D-printed specimens [5,7]. In some other cases, hybrid processes were used to achieve such objectives with

DOI: 10.1201/9781032665351-4

different filaments [7–8]. These research jobs are still continued on the effects of 3D-printing parameters.

One notable other issue is the material type that is 3D-printed. Usually, polymeric filaments are utilized for 3D-printing such as polylactic acid (PLA) and acrylonitrile butadiene styrene (ABS). The second one has distortion after 3D-printing, compared to PLA and this issue could be another reason for obtaining a proper flatness (a surface characteristic) of the components. However, the post processing treatments could be a way for all these issues of 3D-printing to improve these characteristics [9–11]. Different 3D-printing variables could impact on the performance of the fabricated structures, including the nozzle diameter, bed temperature, nozzle temperature, layer thickness, printing speed, printing direction or raster orientation, etc. [9]. These parameters could lead to some concerning issues about manufacturing, as shown in Figure 3.1, based on surface characteristics after 3D-printing of polymers [10]. These issues are layer-by-layer surface, internal pores or porosities, rough surface, warping or distortion, geometric deviation, and moisture absorption.

As mentioned, these layer-by-layer surfaces could develop a rough surface. In industrial applications, a flat surface should be provided for some cases, especially when a flow exists on the surface or a contact occurs between two solids. To achieve this objective, post processing works could be used such as sanding, polishing, gap-filling, vapor smoothing, epoxy coating, dipping, metal plating, and painting. Moreover, using superior 3D-printing parameters could be another choice to improve the surface roughness. For better clustering, two different types of techniques (physical and technological) could be used for obtaining superior surface properties. These approaches contain thermal, chemical, optical, and mechanical post processing methods for the physical technique, besides the technological ones as additive, subtractive, and transformative techniques [10]. In another clustering for the surface modification approaches, mechanical, irradiation, thermal, and chemical treatments could be categorized. All these post processing techniques could be seen in Figure 3.2.

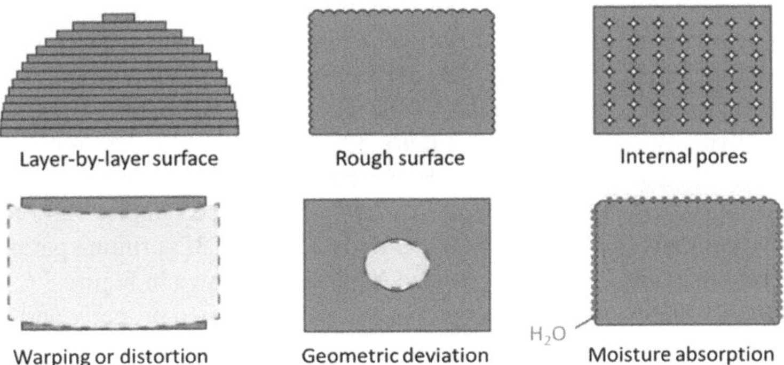

FIGURE 3.1 The surface characteristics after 3D-printing of polymers

Notably, chemical treatments seem to be a cheaper method, as also a simple approach, especially for vapor smoothing. The chemical treatment is done by the action of submerging the component in the vaporized or liquid solution. Then, the specific chemical solutions initiate the reactions on the surface of the 3D-printed polymeric structures. In other words, while the chemical reactions cause melting of the 3D-printed filament, the material reflows filling the valleys and reducing the peaks through the rough surface [11]. This modification approach is considered for this work to check the surface roughness, besides the 3D-printing parameters.

3.2 MATERIALS AND METHODS

The major objective of this research is to evaluate the surface quality of a real component, which is 3D-printed by roughness. This part is a flow-box in the automotive engine, which has been used for combustion developments. Figure 3.3 shows the geometry of this component. According to the function of this part, a smooth surface should be provided for better air movement through the canals. If there is a rough surface, the turbulency leads to perturbation, which is an issue for better combustion performance. Therefore, an important requirement is to have a smooth surface.

One common manufacturing process is a subtractive technique such as machining. However, for the complicated geometry of this component, it is hard to fabricate such a part. Therefore, it is better to use 3D-printing devices for additive manufacturing. However, the issue is the layer-by-layer surfaces in this technique. Moreover, there is a probability of defects on the surface during 3D-printing (Figure 3.4).

To overcome the mentioned issue in 3D-printed parts, post processing treatments could be utilized such as gap-filling, sanding, painting, vapor smoothing, dipping, polishing, metal plating, and epoxy coating. As another technique to find this objective, 3D-printing parameters could be considered to achieve a nice surface finish, such as speed and nozzle temperature.

In this case study, 3D-printing variables were examined besides the chloroform vapor smoothing process. For this objective, a 3D-printing device with the fused deposition modeling (FDM) technique was used in the research laboratory of Advanced Materials Behavior (AMB), located in the Faculty of Mechanical Engineering, at Semnan University (Semnan, Iran). Figure 3.5 depicts the fabrication process of the flow-box. In this case, two different Polylactic acid (PLA) filaments were utilized including transparent and white ones. On the first try, three samples were fabricated. For these specimens (AMB_01, AMB_02, and AMB_03), 3D-printing parameters are also reported in Table 3.1. The mentioned samples are shown in Figure 3.6. Based on the first quality check, two first samples were not 3D-printed properly and they were removed by the eye examinations. The reason was due to the filament type, which was PLA-transparent. Moreover, two other samples were also 3D-printed under new directions of printing, as AMB_04 and AMB_05. Their images are demonstrated in

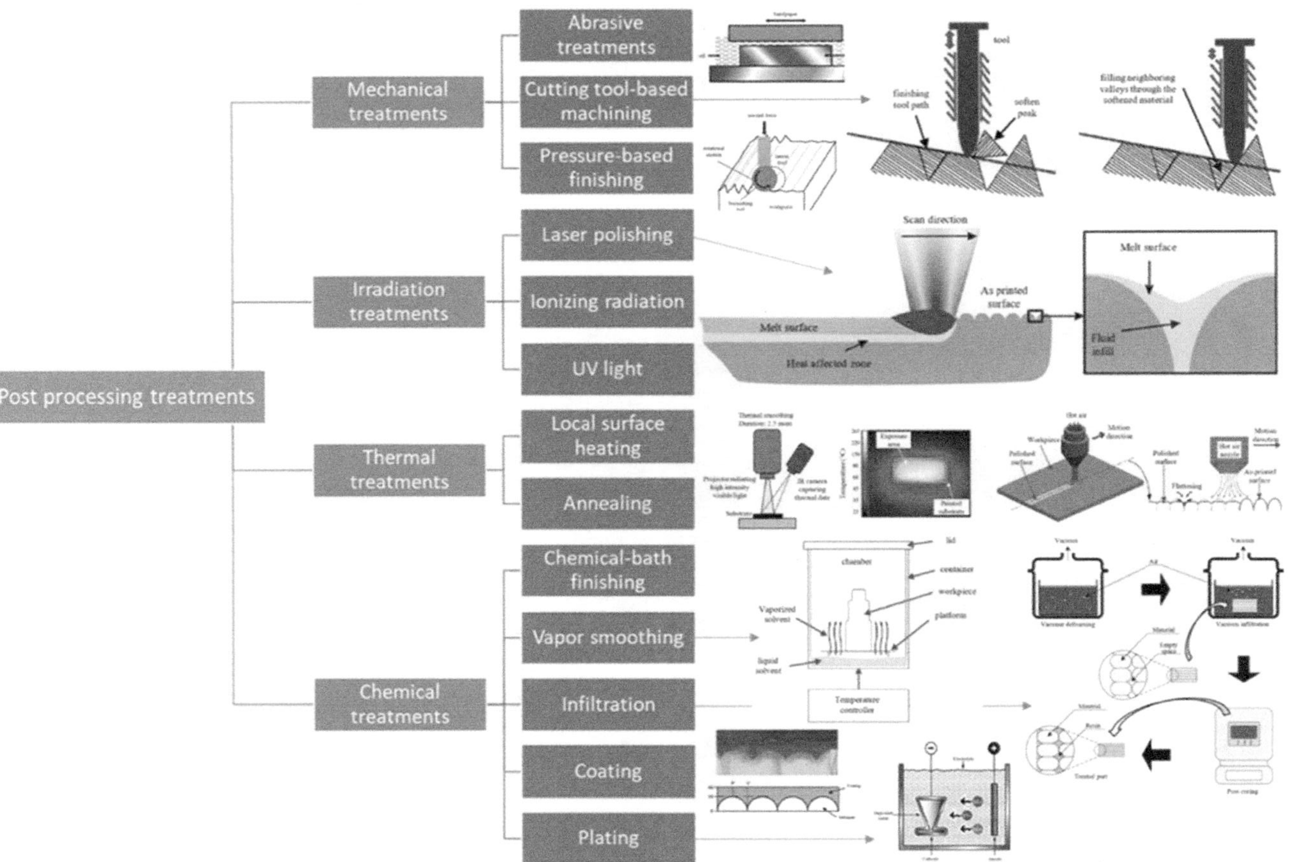

FIGURE 3.2 Clustering of the post processing approaches for 3D-printed structures

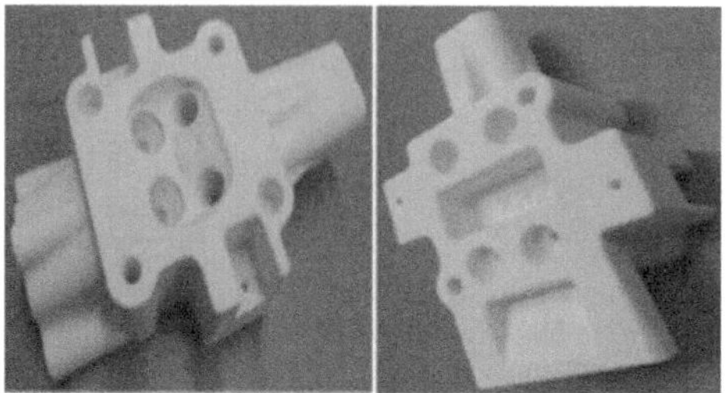

FIGURE 3.3   The flow-box component in the automotive engine

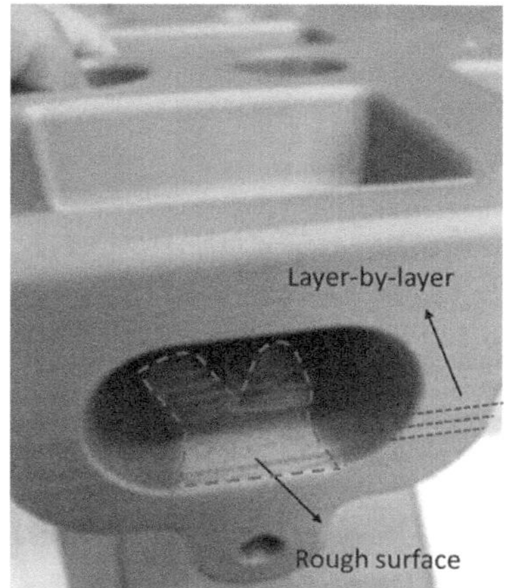

FIGURE 3.4   The defects on 3D-printed components

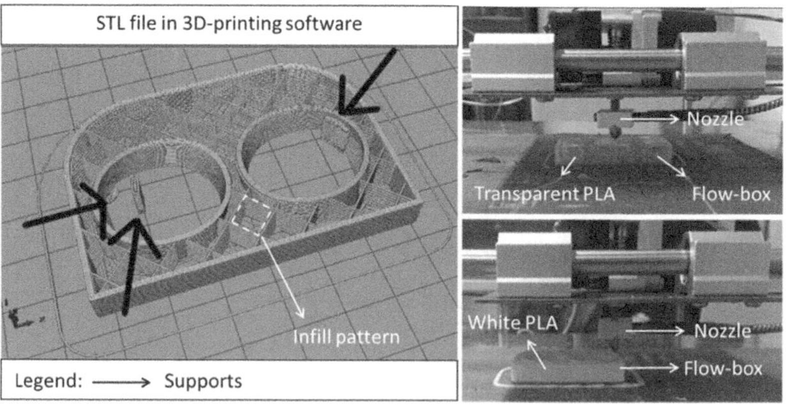

FIGURE 3.5   The fabrication process of the flow-box by the FDM 3D-printer

TABLE 3.1
3D-printing parameters for the first three samples

Input variable	Sample: AMB_01	Sample: AMB_02	Sample: AMB_03
Filament	**PLA-Transparent**	**PLA-Transparent**	**PLA-White**
Filament diameter	1.75 mm	1.75 mm	1.75 mm
Nozzle diameter	0.4 mm	0.4 mm	0.4 mm
Nozzle temperature	**230°C**	**180°C**	**200°C**
Infill	**5%**	**10%**	**10%**
Infill pattern	Rectangular	Rectangular	Rectangular
Infill of first/last layers	100%	100%	100%
Layer thickness	0.4 mm	0.4 mm	0.4 mm
Printing direction	45°	45°	45°
Printing speed (outer layer)	**20 mm/s**	**25 mm/s**	**25 mm/s**
Printing speed (inside)	20 mm/s	20 mm/s	20 mm/s
Support printing speed	5 mm/s	-	-
Brim support	**NO**	**YES**	**YES**

* The bolded text shows the changes.

Figure 3.6 and 3D-printing parameters are also mentioned in Table 3.2. The difference between 3D-printing of the first three samples and the second two specimens could be found in Figure 3.7. Notably, the nozzle diameter and layer thickness were lower, compared to the first three specimens.

A device was used for measuring the roughness, which is shown in Figure 3.8. Notably, the gauge (or the probe) was moved on the surface and the measurement was performed at a distance of 0.5 mm through a line, 1 mm far from the sample edge. The line length for this roughness measurement was 4.0 or 6.5 mm. Then, the average value plus the standard deviation was reported. In some cases, the scattered data were removed to have a reliable measurement.

As mentioned, two first 3D-printed specimens fabricated by PLA-transparent filaments were not considered due to the low initial quality. Therefore, the roughness of the AMB_03 sample was evaluated, which was fabricated by PLA-white filaments. Moreover, for another try, the surface of this specimen was also polished in one case. In another case study, after polishing, the chloroform vapor smoothing process was additionally used, also known as etched. Consequently, three conditions were examined under the roughness evaluation for the components with PLA-white, including 3D-printed, polished, and smoothed conditions. On this specimen, the measurement areas are illustrated in Figure 3.9. It should be noted that one face of the sample was polished with sandpapers (No. 666) and the other face was modified with sandpapers and then using the chloroform solution. Chloroform is a colorless liquid and its chemical formulation is $CHCl_3$ with a purity of 98%. Chloroform is a non-polar solvent and is widely used in the synthesis of organic compounds. Such a process could be also done with acetone.

TABLE 3.2
3D-printing parameters for the second two samples beside the AMB_03 specimen

Input variable	Sample: AMB_03	Sample: AMB_05[**]	Sample: AMB_06[**]
Filament	PLA-White	PLA-White	PLA-White
Filament diameter	1.75 mm	1.75 mm	1.75 mm
Nozzle diameter	**0.4 mm**	**0.2 mm**	**0.2 mm**
Nozzle temperature	200°C	200°C	200°C
Infill	10%	10%	10%
Infill pattern	Rectangular	Rectangular	Rectangular
Infill of first/last layers	100%	100%	100%
Layer thickness	**0.4 mm**	**0.2 mm**	**0.2 mm**
Printing direction	45°	45°	45°
Printing speed (outer layer)	**25 mm/s**	**20 mm/s**	**20 mm/s**
Printing speed (inside)	20 mm/s	20 mm/s	20 mm/s
Support printing speed	-	20 mm/s	20 mm/s

*   The bolded text shows the changes.
** The AMB_05 and AMB_06 samples were 3D-printed under a diagonal angle.

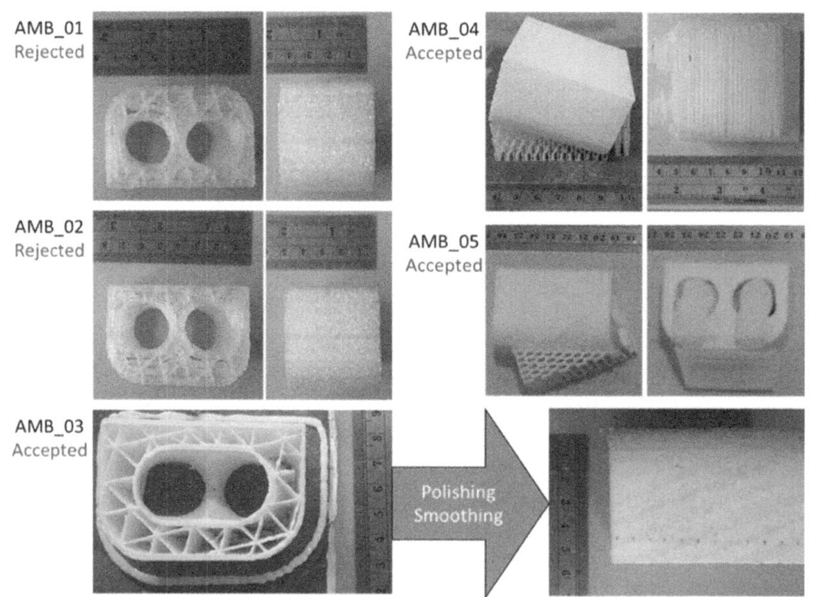

FIGURE 3.6   The first three 3D-printed samples and two other specimens

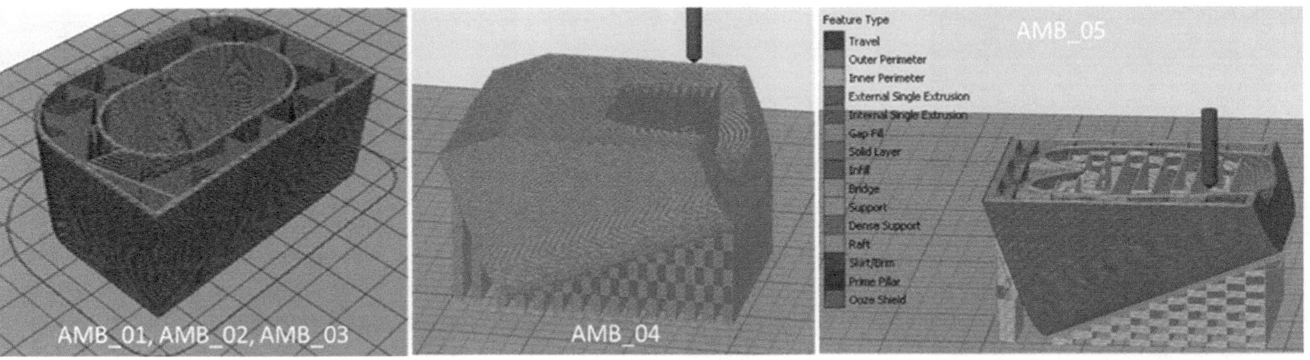

FIGURE 3.7 3D-printing of the first three 3D-printed samples and the second two specimens

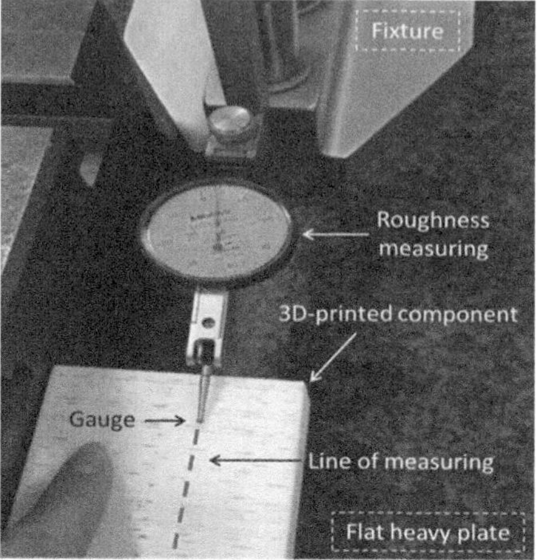

FIGURE 3.8   The device for the roughness measurement on the surface of 3D-printed samples

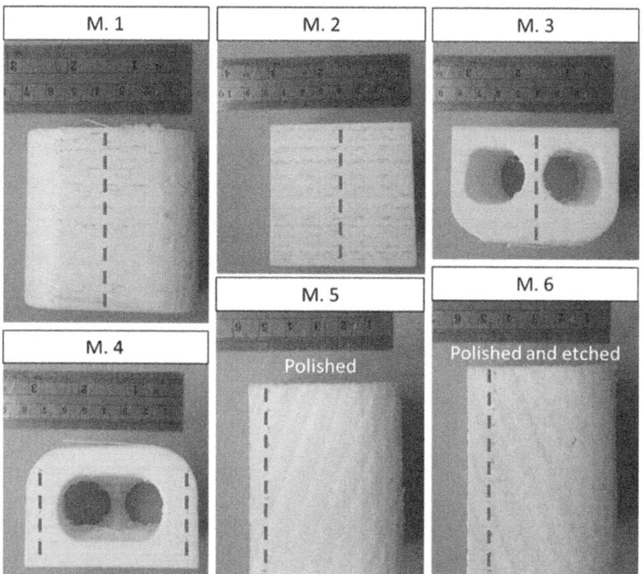

FIGURE 3.9   The roughness measurement areas on the AMB_03 sample

3.3   RESULTS AND DISCUSSION

As the first result, the roughness (R_a) of the AMB_03 sample is reported in Table 3.3. These measurements were done based on Figure 3.9 and in a case study, some scattered data (in Table 3.3) were removed for averaging. As an average data for the 3D-printed sample, the roughness was 0.37±0.30, without the scattered data. This value could be compared to 0.31±0.32 and 0.11±0.08, for polished and polished/

TABLE 3.3
The roughness of the AMB_03 sample based on Figure 3.9

| | M. 1 | | | | M. 2 | | | |
| | Right side of line | | Left side of line | | Right side of line | | Left side of line | |
Points of measuring	MIN	MAX	MIN	MAX	MIN	MAX	MIN	MAX
0.5	0.02	0.75	0.04	0.79	0.05	0.35	0.14	0.74
1.0	0.01	0.75	0.05	0.79	0.21	0.76	0.15	0.45
1.5	0.05	0.80	0.05	0.79	0.05	0.35	0.15	0.75
2.0	0.04	0.75	0.04	0.75	0.01	0.52	0.16	0.55
2.5	0.09	0.80	0.05	0.75	0.04	0.78	0.15	0.55
3.0	0.04	0.79	0.04	0.76	0.01	0.40	0.25	0.55
3.5	0.04	0.76	0.04	0.80	0.05	0.76	0.15	0.76
4.0	0.10	0.25	0.02	0.78	0.05	0.45	0.03	0.45
4.5	0.16	0.62	0.03	0.76	0.40	0.45	0.15	0.45
5.0	0.15	0.50	0.01	0.78	0.35	0.55	0.16	0.50
5.5	0.22	0.63	0.02	0.30	0.15	0.50	0.25	0.52
6.0	0.15	0.65	0.11	0.65	0.30	0.60	0.25	0.55
6.5	0.25	0.75	0.03	0.75	0.35	0.70	0.35	0.65
Average (All data)	0.52		0.45		0.56		0.55	
Average (Without scattered data)	0.47		0.42		0.56		0.55	

CONTINUED TABLE 3.3
The roughness of the AMB_03 sample based on Figure 3.9

| | M. 3 | | M. 4 | | | | M. 5 (Polished) | | M.6(Polished/ etched) | |
| | On the line | | Right side of line | | Left side of line | | On the line | | On the line | |
Points of measuring	MIN	MAX	MIN	MAX	MIN	MAX	MIN	MAX	MIN	MAX
0.5	0.09	0.35	0.05	0.71	0.03	0.25	0.04	0.77	0.04	0.50
1.0	0.25	0.50	0.08	0.35	0.05	0.70	0.05	0.77	0.05	0.70
1.5	0.02	0.75	0.05	0.40	0.02	0.75	0.25	0.15	0.15	0.25
2.0	0.05	0.20	0.05	0.65	0.05	0.75	0.11	0.16	0.00	0.15
2.5	0.01	0.10	0.15	0.75	0.04	0.70	0.01	0.78	0.03	0.06
3.0	0.02	0.12	0.05	0.45	0.03	0.75	0.05	0.75	0.06	0.30
3.5	0.05	0.45	0.15	0.75	0.03	0.75	0.05	0.15	0.03	0.09
4.0	0.05	0.80	0.05	0.70	0.15	0.78	0.05	0.30	0.03	0.07
4.5	-	-	-	-	-	-	0.05	0.78	0.01	0.11
5.0	-	-	-	-	-	-	0.05	0.78	0.06	0.11
5.5	-	-	-	-	-	-	0.05	0.20	0.05	0.15
6.0	-	-	-	-	-	-	0.02	0.77	0.09	0.23
6.5	-	-	-	-	-	-	0.02	0.77	0.14	0.22
Average (All data)	0.52		0.45		0.46		0.51		0.42	
Average (Without scattered data)	0.44		0.45		0.41		0.44		0.20	

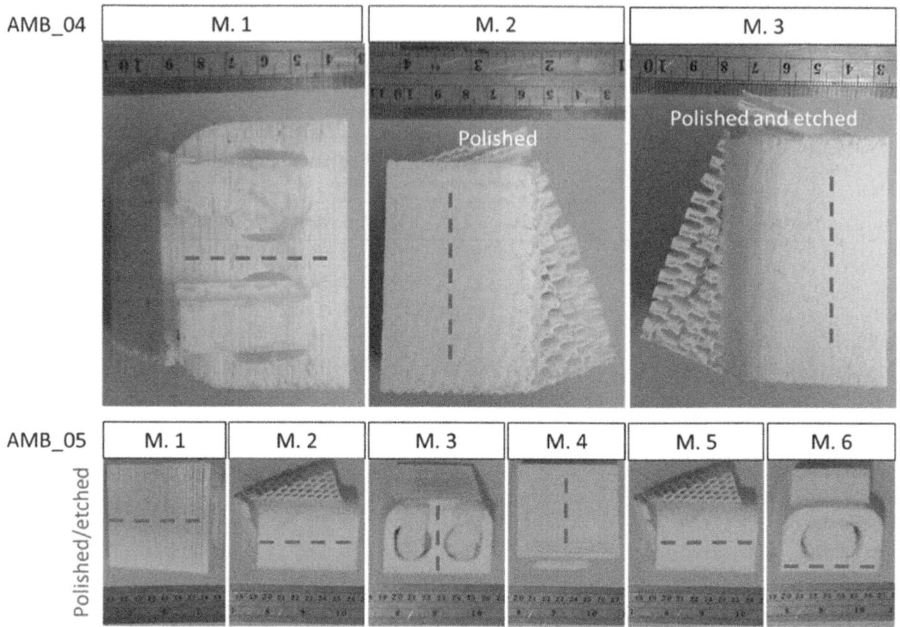

FIGURE 3.10 The roughness measurement areas on the samples of AMB_04 and AMB_05

etched surfaces of the AMB_3 sample, respectively. It means that polishing could decrease the roughness of the 3D-printed sample by 16%. However, this improvement in the roughness could be obtained by 71%, using polishing and smoothing.

In industrial applications, this improvement has no meaning only if the value is comparable with the roughness of a valid component. For this objective, a flow-box was considered, which was fabricated by machining, as a subtractive manufacturing approach. The roughness was measured as 0.13±0.07, lower than 3D-printed samples, but similar to polished/smoothed specimens. These values demonstrate that polishing and smoothing could resolve the layer-by-layer issue in the surface of 3D-printed components.

One other technique for reducing the roughness is to decrease the layer thickness by the nozzle diameter of 0.2 mm. This approach appeared in the samples of AMB_04 and AMB_05. Moreover, polishing and smoothing were also done on the surface of these specimens, with sandpapers (No. 400) and chloroform (98%), respectively. The images for these measurements on AMB_04 and AMB_05 are shown in Figure 3.10. Their related results could be found in Tables 3.4 and 3.5, respectively. The difference between AMB_04 and AMB_05 specimens was only due to have the polishing and smoothing processes on some surfaces (AMB_04) and all surfaces (AMB_05).

Based on the AMB_04 sample, the roughness was measured at 0.14±0.13, 0.19±0.05, and 0.17±0.03 for the 3D-printed, polished, and polished/etched surfaces, respectively. The change in the mean value of the roughness was not significant by

TABLE 3.4
The roughness (on the line) of the AMB_04 sample based on Figure 3.10

Points of measuring	M. 1 (3D-printed)		M. 2 (Polished)		M.3(Polished/ etched)	
	MIN	MAX	MIN	MAX	MIN	MAX
0.5	0.03	0.16	0.06	0.12	0.05	0.19
1.0	0.01	0.19	0.13	0.23	0.28	0.34
1.5	0.02	0.3	0.2	0.23	0.2	0.32
2.0	0.03	0.58	0.19	0.24	0.17	0.2
2.5	0.04	0.16	0.19	0.22	0.15	0.21
3.0	0.04	0.45	0.16	0.2	0.17	0.2
3.5	0.03	0.25	0.18	0.2	0.12	0.15
4.0	0.02	0.25	0.14	0.21	0.15	0.16
4.5	0.04	0.3	0.21	0.24	0.15	0.21
5.0	-	-	0.35	0.31	0.09	0.16
5.5	-	-	0.21	0.23	0.14	0.16
6.0	-	-	0.21	0.23	0.13	0.15
6.5	-	-	0.04	0.16	0.14	0.16
Average (All data)	0.31		0.33		0.31	
Average (Without scattered data)	0.24		0.22		0.19	

polishing or by polishing and smoothing. In other words, by only 3D-printing with a lower thickness layer (and without surface modifications), the demand value for the roughness could be obtained. However, the standard deviation of the roughness could be reduced by polishing or polishing plus smoothing, compared to 3D-printed surfaces.

The roughness of the polished/etched AMB_05 sample is reported in Table 3.5. The measured roughness was 0.19±0.13, compared to the value of 0.13±0.07 in the valid component. As the repeatability of fabrication, this value could be compared to the polished/etched surface of the AMB_04 sample, which was 0.17±0.03. This means that the 3D-printing process is a reliable and repeatable technique to fabricate the components.

As a concluding mark in this section, it could be claimed that the 3D-printing variables could improve the surface roughness of components, when a lower value of the layer thickness was considered by a lower nozzle diameter. Then also, the complementary processes could finalize the surface roughness of 3D-printed components to have polished (mirror-like) surfaces. In agreement with the literature, Tiwary et al. [12] used the acetone vapor for 2 min to reduce the roughness from 14.4 to 0.37 μm. Singh et al. [13] utilized the chemical vapor smoothening approach for rapid investment casting with 1.5 μm of the average value for the roughness.

TABLE 3.5
The roughness (on the line) of the polished/etched AMB_05 sample based on Figure 3.10

Points of measuring	M. 1		M. 2		M. 3		M.4		M. 5		M. 6	
	MIN	MAX	MIN	MAX	MIN	MAX	MIN	MAX	MIN	MAX	MIN	MAX
0.5	0.11	0.25	0.04	0.08	0.02	0.07	0.65	0.7	0.12	0.28	0.16	0.22
1.0	0.04	0.35	0.07	0.14	0.11	0.15	0.05	0.75	0.16	0.5	0.05	0.15
1.5	0	0.07	0.07	0.12	0.11	0.2	0.09	0.18	0.14	0.31	0.09	0.25
2.0	0.02	0.49	0.65	0.42	0.22	0.29	0.08	0.19	0.15	0.35	0.25	0.23
2.5	0.45	0.49	0.6	0.78	0.18	0.22	0.11	0.29	0.2	0.35	0.05	0.09
3.0	0.43	0.42	0.04	0.26	0	0.08	0.2	0.32	0.2	0.35	0.16	0.3
3.5	0.43	0.49	0.06	0.21	0	0.04	0.02	0.3	0.17	0.29	0.11	0.21
4.0	0.12	0.48	0.3	0.47	0.03	0.24	0.29	0.33	0.16	0.26	0.05	0.24
4.5	0	0.06	0.04	0.31	0.09	0.14	0.09	0.29	0.04	0.05	0.05	0.17
5.0	0.03	0.43	0.04	0.46	-	-	0.03	0.34	0.15	0.27	-	-
5.5	0.04	0.41	0.07	0.45	-	-	0.07	0.32	0.17	0.31	-	-
6.0	0.05	0.39	0.49	0.29	-	-	0.09	0.25	0.03	0.31	-	-
Average (All data)	0.47		0.71		0.25		0.70		0.33		0.27	
Average (Without scattered data)	0.30		0.27		0.20		0.22		0.26		0.23	

For advisement of selecting smoothing solutions, chloroform (98%) is usually used for PLA components, as also utilized in this study. Then, acetone and dimethyl-ketone are also preferred for ABS parts [11]. However, these chemical solutions could not be utilized for PLA 3D-printed components.

For a microscopic examination, the layer-by-layer configuration beside the rough surface of the PLA 3D-printed part is shown in Figure 3.11 by scanning electron microscopy (SEM). Moreover, in Figure 3.11, the gas pore or porosity plus the porosity between layers could be also seen. In this case study, the diameter of the nozzle was 0.4 mm and the thickness of the layer was 0.2 mm, which similar value could be observed through the SEM image. Moreover, the infill density was 50%, the 3D-printing speed was 50 mm/s, and finally, the nozzle temperature was 180°C [14].

The distortion can also have a significant impact on the surface roughness, as exhibited in Figure 3.12 for PLA 3D-printed samples. In this case, the nozzle diameter and temperature were 0.4 mm and 180°C, respectively. The layer thickness was 0.2 mm with a 3D-printing speed of 10 mm/s and different infill percentages and various patterns of cells [15]. When the printed part is distorted, it can result in uneven layers and an irregular surface, which in turn increases the roughness or the flatness. Therefore, it is important to consider distortion and its potential effects on the surface roughness during the design and the printing process, and take steps to mitigate distortion as much as possible.

The surface roughness of 3D-printed objects can be affected by several factors, including the extrusion process and the properties of the printed material. An inaccuracy in the nozzle position can lead to an imperfect configuration, which can affect the dimensional accuracy of the 3D-printed object. The calibration of the 3D-printing machine can help address this issue by ensuring that the nozzle is properly aligned and also, the extrusion process is accurately controlled. By optimizing these factors, it is possible to achieve more precise and accurate 3D-printed samples [16].

In addition, one of the reasons for a reduction in dimensions can be excessive material deposition at the endpoints of the raster paths, which may occur when the printer does not pause for a sufficient time at these locations. Another factor that can cause dimensional reduction is the transverse flow, which is influenced by the temperature and viscosity of the printing material. On the other hand, the 3D-printed part may also experience an increase in dimensions due to the shrinkage of the polymer material after printing. Therefore, the accurate control of the extrusion process variables and the consideration of material properties is necessary to ensure the final dimensions of the 3D-printed component match the designed values [17].

The raster angle, which refers to the orientation of the successive layers of the material, is identified as a key parameter that affects the surface roughness. Specifically, the orientation of the raster can affect the direction and the magnitude of surface roughness, with certain angles leading to smoother surfaces than others. Sandhu et al. [18] found that the optimal raster angle for achieving the lowest surface roughness may vary depending on the material being 3D-printed, the printing process, and other factors. However, adjusting the raster angle can be an effective method for controlling the surface roughness and improving the overall quality of 3D-printed components [18].

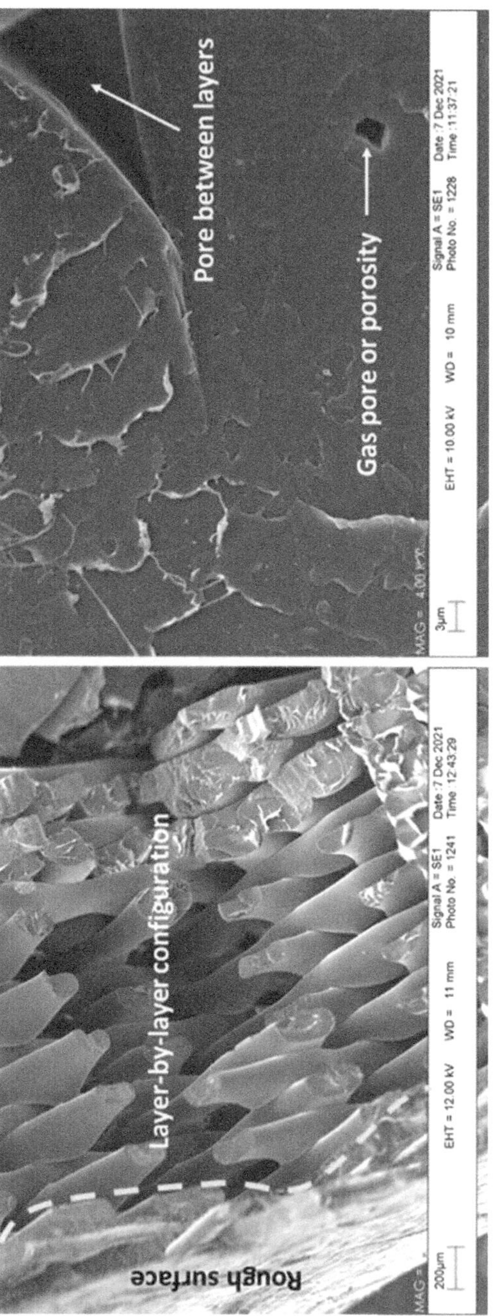

FIGURE 3.11 The SEM image of PLA 3D-printed components for layer-by-layer configuration, pores, and rough surface

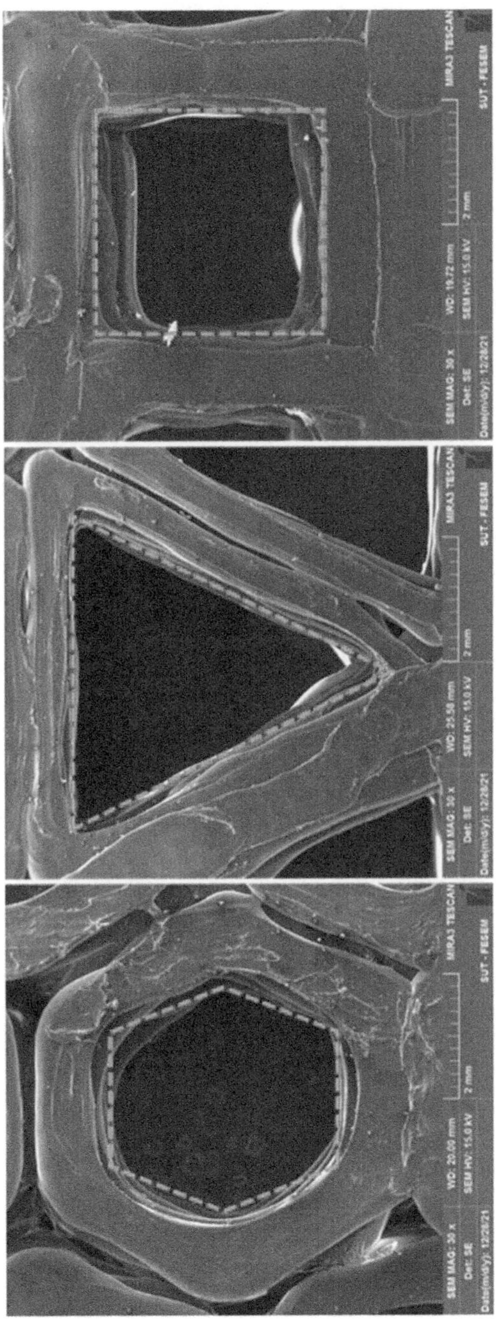

FIGURE 3.12 The SEM image of PLA 3D-printed components for distortion or warping in different infill patterns

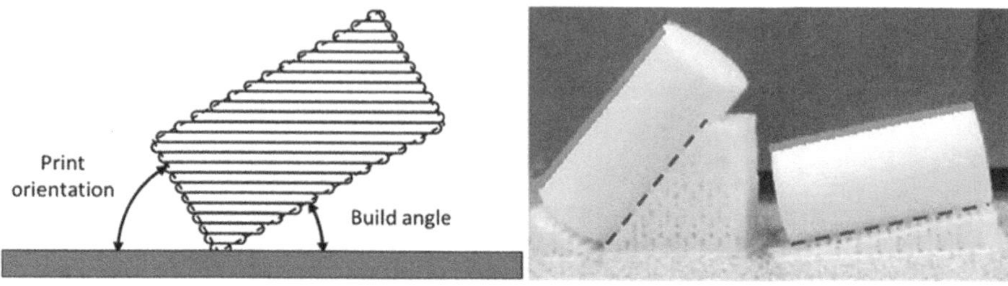

FIGURE 3.13 The 3D-printed part with the build angle and the angle of the print orientation

The bed temperature and the infill percentage can significantly affect the surface roughness [19]. As the temperature of the bed in the 3D-printing device rises, the surface roughness enhances effectively due to the thermal distortion. This issue occurs due to the thermal expansion and the contraction of material, leading to warping or distortion of the surface. It is essential to maintain the bed temperature within a specific range to achieve the desired surface quality. Similarly, the percentage of infilling has a significant influence on the surface roughness. Moreover, the infill refers to the amount of material that is 3D-printed inside the object to provide the structural support. If the infilling percentage enhances, the surface roughness also enhances significantly. This issue occurs since the infill creates a more complex internal structure, which can lead to distortion or warping of the surface. Therefore, it is important to choose the appropriate infill percentage for a specific 3D-printing job to achieve the desired surface quality.

The orientation angle of 3D-printing can have a highlighted impact on the final roughness of the surface. When the print orientation angle enhances, the roughness values also tend to enhance. This phenomenon can be explained by the stair-stepping effect, which occurs when the 3D-printer moves in a zigzag pattern to create the layers of the print (Figure 3.13), based on the claims in the literature [20].

As the angle between the layers and the print bed enhances, the stair-stepping effect becomes more pronounced, leading to an increase in the roughness. To minimize the influence of print orientation on the surface roughness, it is essential to consider the orientation of the print during the design phase. In some cases, it may be possible to adjust the orientation of 3D-printing to reduce the stair-stepping effect and to improve the overall surface finish. Additionally, post-processing techniques such as sanding or polishing may also be used to smooth out the rough surfaces. The impact between the print orientation and the surface roughness highlights the importance of careful design and parameter selection during 3D-printing [20].

Higher values of the build orientation, as another cause, have a less random deviation in the profile of the build edge. The reason is a higher build orientation that allows the layers to be 3D-printed closer to the vertical direction, resulting in less distortion and deviation at the edge of the build. A lower value of deviation in the profile of the build edge can lead to smaller average values of the surface roughness, as the surface becomes more uniform and smoother. Therefore, the build orientation is a critical factor affecting the edge profile and the roughness of the surface in 3D-printed components [21].

3.4 CONCLUSIONS

In this chapter book, the surface roughness was evaluated for the FDM additive-manufactured Polylactic acid (PLA) components. This surface property was affected by 3D-printing parameters and also vapor post processing treatments using chloroform. The case study was a flow-box in the automotive engine application. The following results could be briefly listed,

- The roughness was measured as 0.37±0.30 for the 3D-printed sample, compared to 0.31±0.32 and 0.11±0.08, for polished and polished/etched surfaces, respectively. In other words, vapor smoothing had a significant influence on the surface characteristics of 3D-printed components.
- Decreasing the layer thickness, using the nozzle diameter of 0.2 mm, also led to improving the roughness of the surface. In this case, the roughness was 0.14±0.13, 0.19±0.05, and 0.17±0.03 for the 3D-printed, polished, and polished/etched surfaces, respectively.

ACKNOWLEDGMENT

The authors should thank the Irankhodro Powertrain Company (IPCO), for providing the device for the roughness measurement and the CAD file for the components.

REFERENCES

[1] M. Azadi, A perspective on metamaterials for the biomechanics application: Multi-material metamaterial (4M) structures, *Materials Science Forum*, 1064 (2022) 151–156.

[2] S. Mohammadi Esfarjani, A. Dadashi, M. Azadi, Topology optimization of additive-manufactured metamaterial structures: A review focused on multi-material types, *Forces in Mechanics*, 7 (2022) 100100.

[3] M. Azadi, A. Dadashi, Experimental fatigue dataset for additive-manufactured 3D-printed polylactic acid biomaterials under fully-reversed rotating-bending bending loadings, *Data in Brief*, 41 (2022) 107846.

[4] A. Dadashi, M. Azadi, Experimental bending fatigue data of additive-manufactured PLA biomaterial fabricated by different 3D printing parameters, *Progress in Additive Manufacturing*, 8 (2023), 255–263. https://doi.org/10.1007/s40964-022-00327-1

[5] M. Azadi, A. Dadashi, M.S. Aghareb Parast, S. Dezianian, A. Bagheri, A. Kami, M. Kianifar, V. Asghari, A comparative study for high-cycle bending fatigue lifetime and fracture behavior of extruded and additive-manufactured 3D-printed acrylonitrile butadiene styrene polymers, *International Journal of Additive-Manufactured Structures*, 1 (2022) 1.

[6] M. Azadi, A. Dadashi, S. Dezianian, M. Kianifar, S. Torkaman, M. Chiyani, High-cycle bending fatigue properties of additive-manufactured ABS and PLA polymers fabricated by fused deposition modeling 3D-printing, *Forces in Mechanics*, 3 (2021) 100016.

[7] A. Bagheri, M.S. Aghareb Parast, A. Kami, M. Azadi, V. Asghari, Fatigue testing on rotary friction-welded joints between solid ABS and 3D-printed PLA and ABS, *European Journal of Mechanics – A/Solids*, 96 (2022) 104713.

[8] M.S. Aghareb Parast, A. Bagheri, A. Kami, M. Azadi, V. Asghari, Bending fatigue behavior of fused filament fabrication 3D-printed ABS and PLA joints with rotary friction welding, *Progress in Additive Manufacturing*, 7 (2022) 1345–1361.

[9] S.S. Alghamdi, S. John, N.R. Choudhury, N.K. Dutta, Additive manufacturing of polymer materials: Progress, promise, and challenges, *Polymers*, 13 (2021) 753.

[10] F. Tamburrino, S. Barone, A. Paoli, A. V. Razionale, Post-processing treatments to enhance additively manufactured polymeric parts: A review, *Virtual and Physical Prototyping*, 16(2) (2021) 221–254.

[11] D. Syrlybayev, A. Seisekulova, D. Talamona, A. Perveen, The post-processing of additive manufactured polymeric and metallic parts, *Journal of Manufacturing and Materials Processing*, 6 (2022) 116.

[12] V.K. Tiwary, P. Arunkumar, A.S. Deshpande, N. Rangaswamy, Surface enhancement of FDM patterns to be used in rapid investment casting for making medical implants, *Rapid Prototyping Journal*, 25 (2019) 904–914.

[13] D. Singh, R. Singh, K.S. Boparai, Investigations for surface roughness and dimensional accuracy of biomedical implants prepared by combining fused deposition modeling, vapor smoothing and investment casting, *Advances in Materials and Processing Technologies*, 8(1) (2020) 843–862.

[14] M.H. Rahaei, Study of layout sequence effect on creep behavior of polylactic acid polymer fabricated by additive manufactured fused deposition modeling technique, MSc Thesis, Semnan University, 2022.

[15] A. Dadashi, Investigating the effect of 3D printing parameters in additive manufacturing process on bending fatigue of PLA biomaterial, MSc Thesis, Semnan University, 2021.

[16] M. Frascio, M. Avalle, M. Monti, Fatigue strength of plastics components made in additive manufacturing: First experimental results, *Procedia Structural Integrity*, 12 (2018) 32–43.

[17] S. Singamneni, M.P. Behera, D. Truong, M.J. Le Guen, E. Macrae, K. Pickering, Direct extrusion 3D printing for a softer PLA-based bio-polymer composite in pellet form, *Journal of Materials Research and Technology*, 15 (2021) 936–949.

[18] G.S. Sandhu, K.S. Boparai, K.S. Sandhu, Effect of slicing parameters on surface roughness of fused deposition modeling prints, *Materials Today: Proceedings*, 48 (2022) 1339–1345.

[19] R. Raj Mohan, R. Venkatraman, S. Raghuraman, Experimental analysis on density, micro-hardness, surface roughness and processing time of Acrylonitrile Butadiene Styrene (ABS) through Fused Deposition Modeling (FDM) using Box Behnken Design (BBD), *Materials Today Communications*, 27 (2021) 102353.

[20] I. Buj-Corral, A. Dominguez-Fernandez, R. Duran-Llucia, Influence of print orientation on surface roughness in fused deposition modeling (FDM) processes, *Materials*, 12(23) (2019) 3834.

[21] M. Taufik, P.K. Jain, A study of build edge profile for prediction of surface roughness in fused deposition modeling, *Journal of Manufacturing Science and Engineering*, 138(6) (2016) 061002.

4 Post-Processing of Additive Manufacturing Functional Polymeric Parts

Influence on Surface, Dimensional Quality and Mechanical Performance

Giovanni Gómez-Gras and Marco A. Pérez

4.1 INTRODUCTION

The industry is undergoing a profound transformation fueled by new information and communication technology systems, which are driving digitization, decentralization, automation, personalized production, human-machine interactions, and value-added services [1]. Cutting-edge technologies such as augmented reality, cloud computing, the Internet of Things, big data, and advanced manufacturing have emerged as powerful solutions to meet the demands of this evolving landscape [2]. Within this context, additive manufacturing (AM) has garnered significant attention from the scientific community and various industrial sectors, positioning itself as a key enabler of the global transition to Industry 4.0 in the coming decade. The ability to fabricate functional parts with intricate geometries, surpassing the limitations of traditional manufacturing methods, has accelerated the integration of AM into both research and industrial applications [3]. Moreover, AM offers exceptional design flexibility, waste reduction, and shorter production times, which have played a vital role in driving its widespread adoption and popularity. As AM continues to evolve, it is poised to become a strategic pillar in reshaping the future of the industry and advancing the principles of Industry 4.0.

In this general context, layer-by-layer material deposition-based printing techniques are precisely the ones that have become more popular, thanks to their

DOI: 10.1201/9781032665351-5

technological simplicity and the low cost of most printers, making them accessible to many users. Currently, the main uses of 3D printing have been related to creating tools, prototypes, and molds for different industries. However, end-use parts are believed to represent the bulk of the additive manufacturing components market growth in the next decade, quadrupling by 2030, as stated in the article "Will 3D Printing Replace Conventional Manufacturing" [4]. Consequently, metal printing, engineering-grade polymers, and composites printing will increasingly take center stage, accompanied by systematic development and standardization of new materials.

Despite the promising developments in additive manufacturing (AM), certain limitations still impede its full realization of industry expectations. One significant drawback is the surface quality of the final parts, which lags behind that of conventional manufacturing processes. In numerous industrial applications, the surface characteristics of a component dictate its interactions with other parts, with surface roughness serving as a critical indicator of its mechanical performance [5]. For instance, in Fused Filament Fabrication (FFF), the layer-by-layer material deposition process (Figure 4.1a) inherently generates an uneven surface profile commonly referred to as the "staircase effect" [6, 7] (Figure 4.1c). This issue poses challenges in terms of achieving desirable surface quality and dimensional accuracy. As a result, it has been widely recognized as a primary concern when employing AM technologies for the production of final parts [8–15].

Another challenge inherent to these technologies, which also undermines surface quality, is the use of support structures during the manufacturing of complex parts that must be subsequently removed mechanically or chemically (Figure 4.1b). Both procedures directly impact the quality of the final parts, which can be damaged during the process.

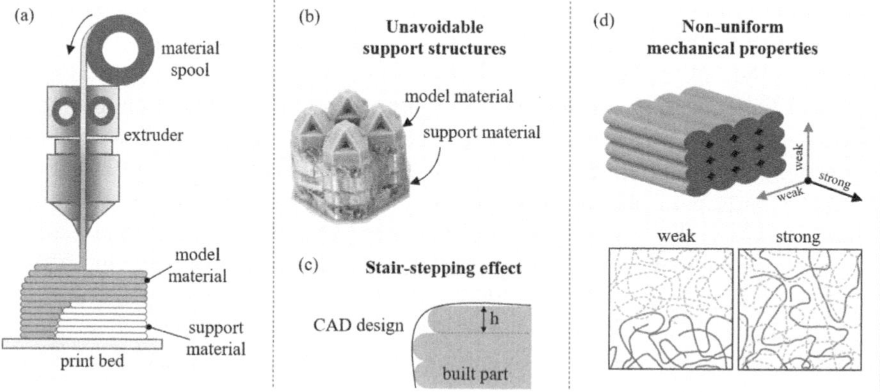

FIGURE 4.1 Fundamental challenges of the Fused Filament Fabrication: (a) effects of layer-by-layer deposition; (b) support material removal; (c) staircase effect; (d) mechanical anisotropy.

To these external defects, functional limitations derived from component performance must also be added, which are also highly dependent on printing conditions [16–18]. The most notable of these limitations is mechanical anisotropy due to poor filament bonding, which ultimately reduces the strength of the parts under certain stresses or invalidates them for specific uses in specific applications (Figure 4.1d) [7]. Because of these deficiencies declared by the AM community, much research has focused on analyzing the interactions between printing parameters as a determining requirement for achieving maximum surface quality and mechanical performance. However, more than these adjustments are needed to achieve the industry's optimal conditions needed.

From these considerations, it makes sense to post-process printed parts to minimize deficiencies not addressed by pre-printing parameter optimization. Including finishing processes in the final stage of obtaining functional parts, including chemical, thermal, or mechanical treatments (or a combination thereof), is a good alternative for improving the surface quality, dimensional accuracy, and mechanical performance of components [19]. This chapter addresses the impact of each of these post-processing areas, including useful recommendations for their application and scalability at an industrial level.

Ultem™ 9085, a member of the polyetherimide (PEI) family, has been selected as the reference material to explore the impact of post-processing techniques. This choice is attributed to Ultem™ 9085's exceptional properties as a high-performance polymer, encompassing a unique combination of thermal, chemical, and mechanical characteristics. These properties make it highly suitable for a diverse range of applications, including medical devices, heating elements, absorption membranes, electrical insulation, and more [20–25]. Notably, Ultem™ 9085's compatibility with the Fused Filament Fabrication (FFF) process has piqued interest in the transportation and aerospace industries, primarily due to its favorable certification in flame, smoke, and toxicity (FST) standards [26].

4.2 IMPROVEMENTS IN SURFACE

The surface quality of functional components plays a critical role in ensuring their compatibility and harmonious integration with other components. While it has been acknowledged that optimizing printing parameters can contribute to improved surface outcomes, the inherent nature of additive manufacturing (AM) makes it challenging to guarantee surfaces that meet specific technical requirements. This limitation becomes particularly evident when considering factors such as the assembly of parts or the adhesion of finishing coatings like paint and metallization [27]. Surface irregularities can create vulnerable areas where cracks or corrosion may initiate, while compromised dimensional accuracy can undermine the performance of various engineering applications. Consequently, there has been a growing emphasis on minimizing and compensating for surface roughness in AM, leading to extensive research in this field in recent years.

In this section, we will present the effects of the main chemical, thermal, and mechanical post-processing techniques, which have the purpose of improving the quality of the final parts after their obtaining process using FFF [19, 28, 29].

4.2.1 SCOPE OF CHEMICAL POST-PROCESSING

To analyze the scope of chemical post-processing of printed parts, two different approaches will be considered: one based on the use of chemical vapors that directly impact the surface, modifying it (Vapor Smoothing), and another based on the use of solvents for the extraction of support material, with an influence on the surface properties of the components (Support Removal Solvent).

4.2.1.1 Vapor Smoothing

Numerous studies have been presented to the scientific community, focusing on the effects of vapor smoothing as a relatively simple, cost-effective, and established method for enhancing surface quality. The optimization of this process involves parametric adjustments, particularly with acetone vapors, to evaluate the impact of duration and cycle repetition on surface finish, dimensional accuracy, and part stability (Chohan et al. [30–32]). These studies have demonstrated that shorter smoothing durations (e.g., 30 seconds) and repeated cycles can significantly reduce surface roughness. Mathematical models have also been developed to predict the average surface roughness of treated parts. Furthermore, the effectiveness of different acetone mixtures, such as ethyl acetate (Mu et al. [33]), has been investigated to improve surface roughness for samples with varying construction orientations. However, it should be noted that while surface improvement is achieved, exposure to acetone results in reductions in tensile strength proportional to the duration of exposure. Similarly, the use of tetrahydrofuran (THF) and dichloromethane (DCM) for smoothing PLA parts has been explored, leading to enhanced toughness but at the expense of the tensile mechanical properties of the components under study (Jin et al. [34] and Rajan et al. [35]). Other researchers have examined chemical treatment combined with drying and aluminum coating of ABS specimens, reporting improvements in surface quality and heat absorption through radiative heating (Nguyen et al. [36]).

Given its potential for smoothing printed parts by partially dissolving their outer layer through chemical vapor exposure, Ultem™ 9085 was treated with chloroform (99.9% purity). The selection of chloroform was based on its low compatibility with this material, as well as its low boiling point (approximately 61°C). Following exposure for 120, 180, and 270 minutes, the samples were allowed to dry under ambient laboratory conditions for at least 24 hours before further processing. Subsequent analysis revealed a significant increase in sample weight, proportional to the duration of exposure to chloroform vapors. The examination of different time intervals demonstrated that within the initial 120 minutes, a slight reduction in all dimensions occurred due to the initial surface smoothing. However, as time progressed, the samples expanded or swelled in all directions while retaining their general geometry. Beyond 270 minutes, it is likely that the samples would begin to melt completely, resulting in the loss of their original shape. These prolonged exposure times also affected surface roughness. While improvements in arithmetic mean roughness of approximately 90–95% were observed after 180 minutes of exposure, no further significant enhancements were achieved. These findings align with those of the aforementioned authors, who established a time limit beyond which no apparent improvements or even degradation of parts were observed (Figure 4.2).

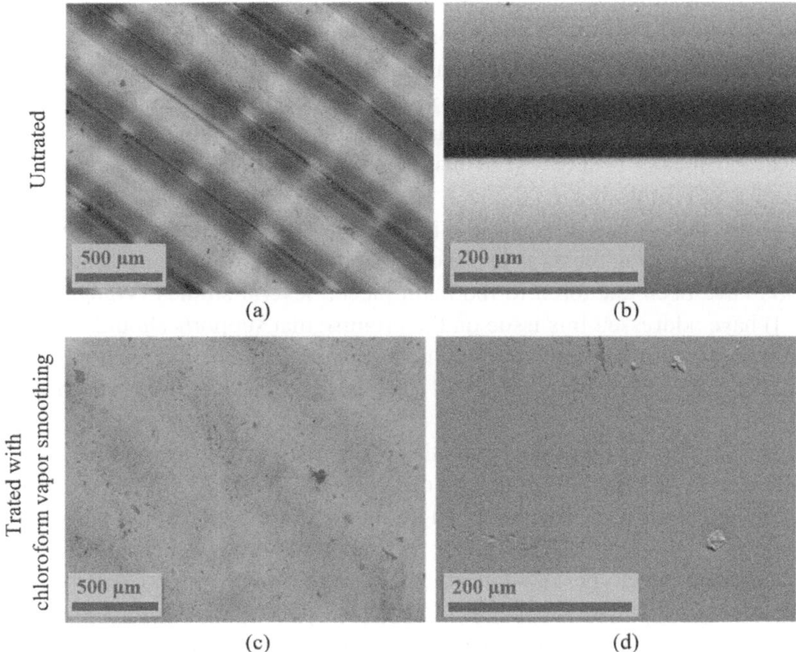

FIGURE 4.2 Comparison of the chloroform of vapor smoothing on Ultem™ 9095 samples: (a, c) Images from digital microscopy; (b, d) SEM images.

In conclusion, based on the consulted processing times, it can be confidently stated that Ultem™ 9085 exhibits superior chemical resistance to chloroform compared to ABS with acetone and PLA with THF. This finding establishes the feasibility of employing chemical post-processing to achieve a smoother surface finish, considering the respective time windows mentioned. Thus, for applications where the primary objective is to enhance surface smoothness, utilizing this type of post-processing technique with Ultem™ 9085 proves advantageous.

4.2.1.2 Support Removal Solvent

Removing support material is the most common post-processing step for FFF-printed parts. However, the need to create rigid auxiliary structures capable of enabling the construction of complex geometries, cantilevers, or intricate shapes during the fabrication process is inherent to the technology itself [37]. For this reason, even though it does not improve the surface or mechanical performance of the final parts, this additional material is often present and must be removed at the end of the treatment.

These structures can generally be classified into two types: those constructed from a single material, which is common when printers have a single extruder, and those created with a secondary material, which occurs when the printer is equipped with multiple printing nozzles. In the case of structures made from a single material, manual removal is typically required, often involving the use of pliers or similar tools.

However, this manual removal process can inadvertently cause damage to the surface, leading to a deterioration of the component's properties. On the other hand, structures made with a secondary material have the advantage of being able to extrude support materials that are typically weaker, soluble, and less resistant to high temperatures compared to the model material. This characteristic significantly facilitates their removal and makes printers with this capability the preferred choice in industrial settings for FFF printing.

However, both types of scaffold structures negatively impact construction times and harm surface roughness, as it is inevitable that marks will be present where these supports have been anchored to the main piece. Several studies (Wang, C. & Qian, X. [38]) have addressed this issue on the premise that supports should be minimized by optimizing their location based on the appropriate choice of construction orientation. Other researchers (Wang, Z. et al. [39]) have focused on the position of support points.

When it comes to thermoplastics, the preferred types of supports are those that can either melt or dissolve in an aqueous solution. Stratasys, in its "Best Practices FDM Support Removal" document, suggests the use of a recirculation or ultrasonic vibration system to aid in the extraction of support material [40]. Manufacturers commonly describe three types of supports that have proven to be more feasible: supports soluble in organic solvents (such as high-impact polystyrene, compatible with ABS and PLA), supports soluble in water with the use of an activator (typically an alkaline pH), and supports soluble in water, like polyvinyl alcohol (a petroleum-derived thermoplastic used as a support material for PLA and ABS parts) [41].

However, it is worth noting that not all materials and printers have readily available suitable solvents. For materials with high extrusion temperatures, finding a support material with similar physicochemical properties becomes challenging, as most solvents tend to react similarly and may negatively impact the mechanical performance of the model. In such cases, mechanical extraction remains the only viable alternative, but it has limitations due to the geometric complexity of the parts and the manual nature of the process. The research addressing this issue is still ongoing, and the search for effective support removal solutions for Ultem™ 9085, the chosen material in this chapter, is particularly challenging. A new filament called Aquasys®180 [42], which is a hydro-soluble copolymer filled with carbohydrates and polyamide capable of withstanding temperatures up to 300°C, has recently been introduced to the market. Although it shows promising compatibility for printing with Ultem™ 9085, it is still in the validation phase.

It is important to highlight that Ultem™ 9085 has a glass transition temperature of approximately 180°C, which means it requires a support material with similar thermal properties for processing within the hot chamber of the printer. The manufacturer suggests using polysulfone (PSF), a polymeric blend with a chemical structure very similar to PEI. Both PEI and PSF contain a diphenyl propane surrounded by an ether group on each side and a high proportion of aromatic rings, which explains their comparable thermal stability and chemical resistance to various organic solvents and diluted solvents [43].

Despite the challenges, extensive solubility studies have led to the development of a practical methodology using a toluene-based solvent capable of dissolving

FIGURE 4.3 Comparison of the effect of toluene-aniline solvent for support removal on Ultem™ 9095 samples: (a, b, d, e) Images from digital microscope; (c, f) SEM micrographs.

PSF. Through solubility tests and infrared spectroscopy, it has been determined that adding 20% v/v of aniline to the solvent is advantageous for preserving the integrity of Ultem™ 9085 and facilitating the dissolution of polysulfone. This leads to an improvement in the surface quality of the treated parts (Figure 4.3) [44].

Regarding the operational procedure, a temperature and ultrasound-controlled solution of the proposed solvent can dissolve approximately 1 cm³ of PSF per 10 mL of solvent within a maximum period of 2 hours. It has also been observed that a thin film of PSF between the Ultem™ 9085 filaments (Figure 4.3b) contributes to increased stiffness during mechanical stress testing. However, prolonged exposure to the solvent beyond the optimal time frame does not significantly improve surface roughness and may slightly compromise mechanical performance.

4.2.2 Contribution of Thermal Processes

According to numerous studies in additive manufacturing (AM), researchers widely acknowledge that the characteristics of parts produced by fused filament fabrication (FFF), including mechanical performance and surface finish, are significantly affected by the quality of bonds between individual polymer filaments [45, 46]. This intricate phenomenon encompasses various factors such as the part's shape, thermal boundary conditions, and gravitational forces [47]. The formation of these bonds

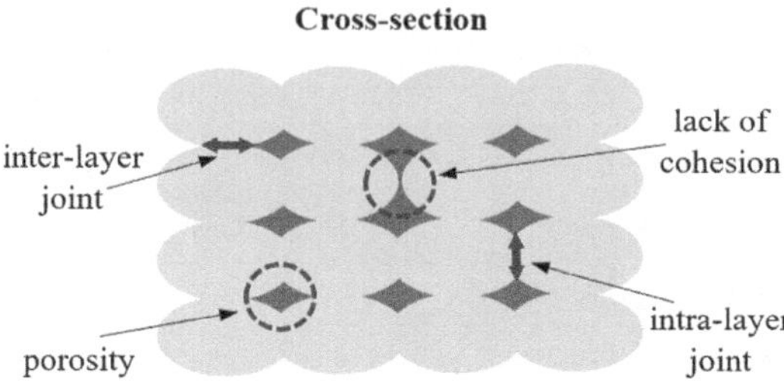

FIGURE 4.4 Schematic representation of internal defects identified in filament deposition.

relies on the growth of the neck region between adjacent filaments and molecular diffusion at this interface, leading to defects like lack of cohesion or inherent porosity within the filament geometry (Figure 4.4). These factors directly impact surface quality and mechanical performance [48]. For instance, Wang et al. [49] developed a predictive numerical model that establishes a correlation between surface roughness and molecular diffusion among molten filaments, a process that requires sufficient time at elevated temperatures. Through their model, they successfully demonstrated this direct connection and provided the potential for estimation purposes.

In line with this, Morales et al. [50] conducted a cohesion study, which highlighted that the bonding strength between layers in additive manufacturing is weaker compared to the bonding between coplanar filaments. This is due to the cooling time that occurs during layer deposition, which is significantly shorter than when filaments are added within the same layer. Consequently, it becomes crucial to consider how printing parameters influence the quality of these bonds in order to optimize the overall performance of the final printed objects. Establishing correlations between these printing conditions and the behavior of intralayer and interlayer bonds can provide valuable insights for process optimization.

Considering the nature of these bonds, a post-thermal processing approach seems a good alternative to improve cohesion and, consequently, reduce the mechanical anisotropy identified as a critical limitation of FFF. In addition, thermal annealing is a well-established technique for all materials with the potential to modify chemical and physical properties. This also includes cooling methods, which can contribute to the reduction of internal stresses generated during the manufacturing process itself. This section will address the effects of this type of post-processing on printed parts.

4.2.2.1 Temperature Effects

In FFF manufacturing, when a polymer mass is melted and then cooled, it does so non-homogeneously, causing volumetric shrinkage and generating residual stresses that often harm the functionality of the finished product [51]. Typically, two types of post-processing treatments are used to treat printed parts: local treatments (usually with lasers) or general treatments (annealing). Figure 4.5 below shows a scheme that typifies both alternatives and the resulting phenomena.

In amorphous polymer materials, it has been observed that isothermal heating below their glass transition temperature (Tg) leads to structural relaxation, resulting in

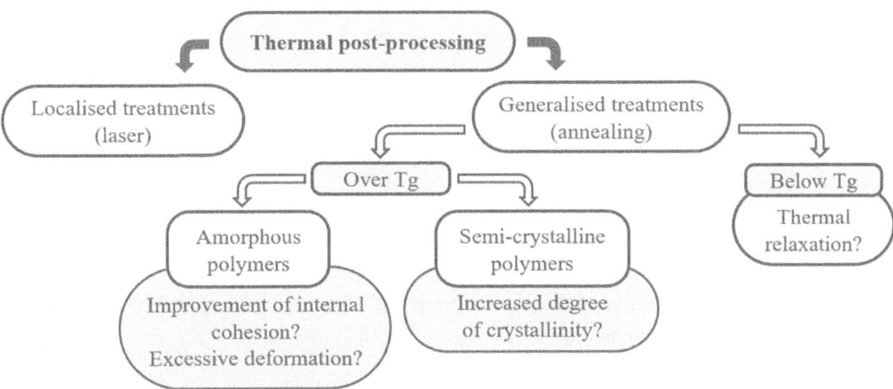

FIGURE 4.5 Application of thermal treatments to polymeric printed parts.

increased hardness and Young's modulus [52]. In the case of semi-crystalline polymers like PLA, an enhancement in crystallinity and improved mechanical behavior under stress have been demonstrated [53]. Furthermore, thermal post-processing above Tg, combined with uniaxial compression, has shown a significant increase in interlaminar fracture toughness, indicating the beneficial effects of temperature and pressure combination [54].

However, when it comes to exploring the densification of Ultem™ 9085 through thermal post-processing, there is limited research available. Padovano et al. [55] observed that sudden temperature variations had a noticeable impact on the mechanical performance of PEI, whereas stable high temperatures or gradual thermal ramps below Tg resulted in less pronounced effects. Zhang et al. [56] implemented a low-temperature annealing process lasting 24–96 hours, which led to expansion in the direction of printing layers, indicating stress relief. The analysis of mesostructure and fracture indicated increased ductility, improved coalescence, and minimal geometric distortions. These findings highlight the potential of thermal post-processing to mitigate the undesired effects of FFF and open up promising avenues for further research.

Another approach to treating Ultem™ 9085 is presented by Chueca et al. [57], who proposed high-temperature annealing to enhance bond strength between filaments, thereby reducing mechanical anisotropy through densification. This process also yielded improved surface quality of the components. The researchers conducted a dimensional analysis to evaluate the impact on the original measurements, as dimensional stability is a crucial concern when subjecting polymers to elevated temperatures.

The specimens used for annealing were sized for standardized static mechanical tests. The printing orientations chosen were ZX (printed vertically), the most negative mechanical behavior due to containing more inter-layer joints, and XY (specimens positioned horizontally on the printing tray). Both positions are disadvantaged by the printing process (Figure 4.6a). As expected, thermal relaxation on the faces perpendicular to the printing direction increases height while the rest of the dimensions adapt to compensate for this change (Figure 4.6b).

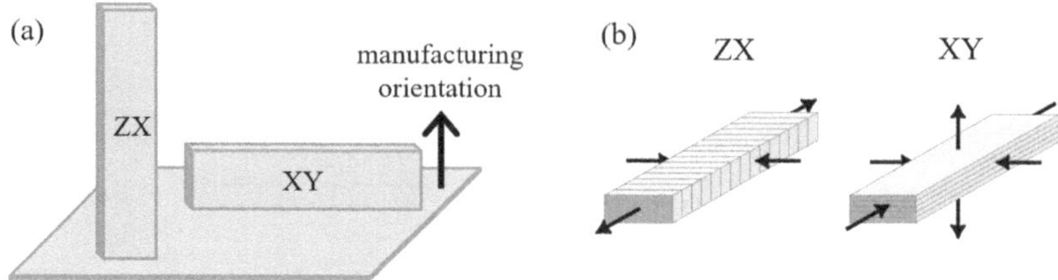

FIGURE 4.6   (a) Orientation of the test specimens used for thermal annealing, (b) Direction of change in the height dimension.

Considering the glass transition temperature (Tg) of Ultem™ 9085 at approximately 180°C and its exposure to a temperature of 195°C during fabrication within the printing chamber, annealing was conducted in a thermal range spanning from 175 to 240°C, with treatment durations ranging from 30 minutes to 3.5 hours.

A meticulous analysis of dimensional accuracy revealed that all samples exhibited expansion along the printing orientation, with an average increase of 0.5 mm in height. Conversely, contraction was observed in the remaining directions, resulting in reductions of 1 mm and 5 mm in width and length, respectively. These findings align with Zhang et al.'s theory regarding the relaxation of thermal stresses in the build direction, which remains valid even at temperatures exceeding the material's Tg, as long as exposure is within ranges that do not indicate degradation [56]. It is worth noting that more substantial dimensional changes may occur at higher temperatures, although these limits have not yet been investigated in the context of the expected post-processing outcomes.

Following the thermal treatment, the measured surface roughness exhibited significant variability, with 210°C identified as the threshold beyond which surface deterioration occurred. Faces not in contact with the thermal chamber showed an average arithmetic roughness improvement of approximately 15% compared to untreated parts, indicating a fusion process between adjacent filaments. Remarkably, the face of the samples in contact with the chamber tray experienced a substantial roughness improvement of 90%. This suggests that physical contact facilitated heat transfer through conduction, which proved more advantageous for surface enhancement than heat transfer through convection, as observed on faces exposed to air (Figure 4.7).

Microscopic examination of the images (Figure 4.7a, c, and e) revealed a noticeable decrease in the air gap between adjacent filaments on the top face, with the gap nearly disappearing on the bottom face. Ideally, the presence of a solid or liquid medium in direct contact with the entire part during thermal treatment would further contribute to achieving an even smoother surface, as observed during the inspection of the annealed part's bottom face.

As usage recommendations, it is suggested that dimensional variations suffered by XY specimens are generally greater than those suffered by ZX specimens. Since the area of the face perpendicular to the printing direction of XY specimens is larger, the thermal treatment has a more pronounced effect, impacting all the studied measures.

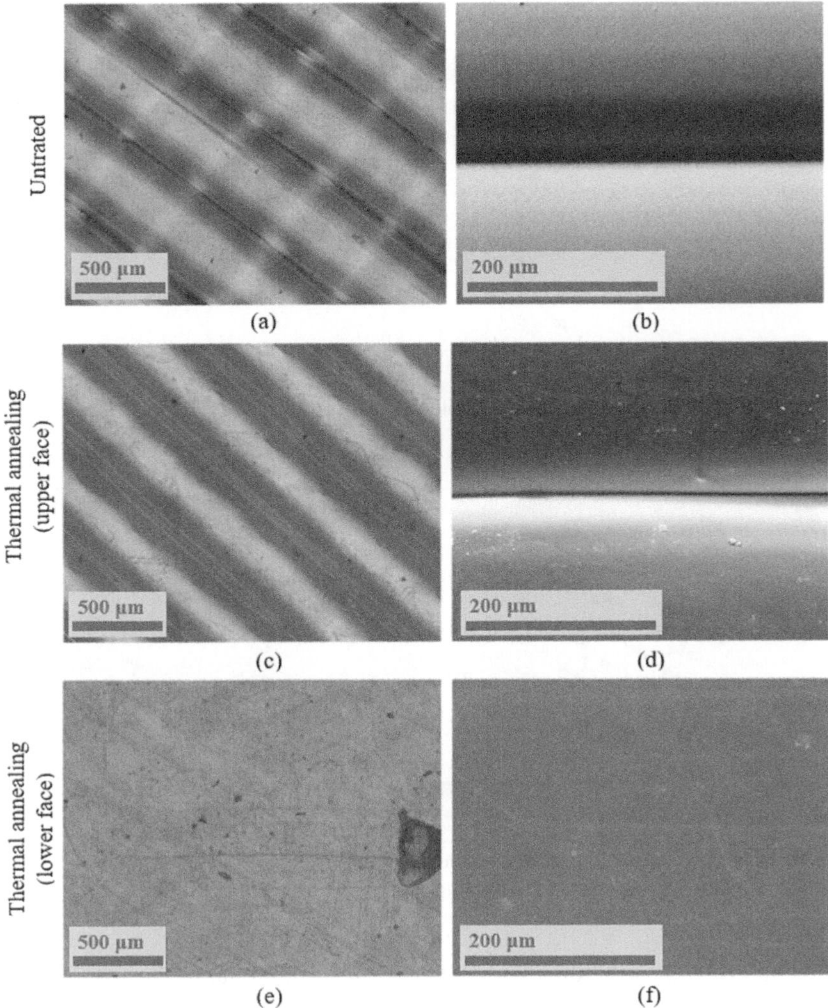

FIGURE 4.7 Comparison of the effect of thermal annealing on Ultem™ 9095 samples: (a, c, e) Images from digital microscopy; (b, d, f) SEM images.

These variations have been established, in an indicative way, at around +12% in height, -10% in width, and -4% in length. Therefore, this proportionality relationship should be considered when dimensioning specimens that will be thermally post-processed, as it could represent an important compromise regarding the precision of tolerances for certain applications.

4.2.2.2 Combination of Temperature and Vacuum

In the preceding sections, it has been discussed how temperature can play a significant role in enhancing the results of thermal post-processing. However, it is

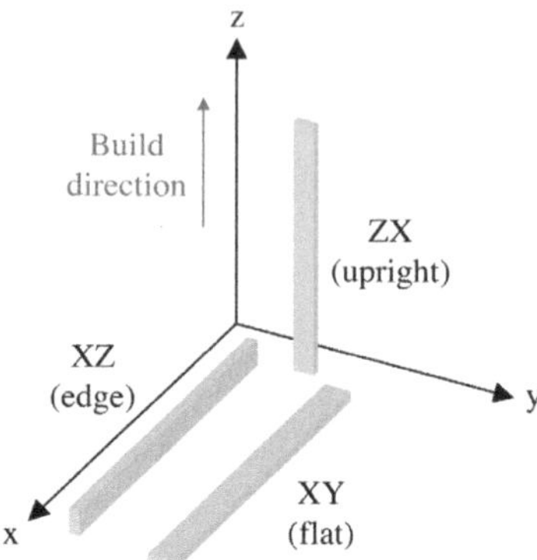

FIGURE 4.8 Schematic of printing sample orientations.

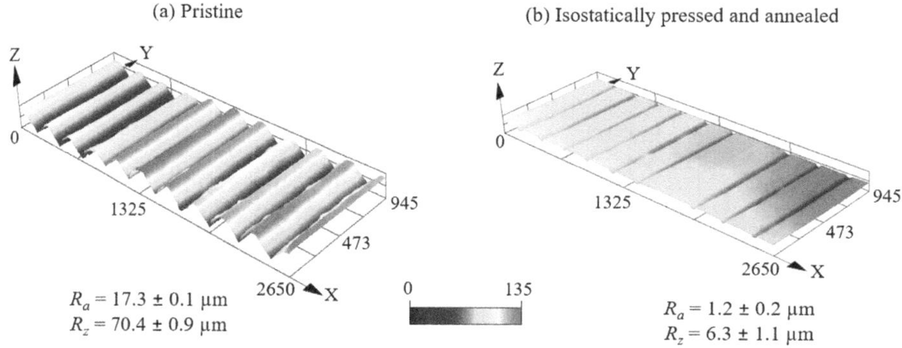

FIGURE 4.9 Surface roughness obtained by microscopy of (a) original (b) thermally annealed and pressurized Ultem™ 9085 samples.

worth exploring the additional improvements in surface quality that can be achieved through the combination of temperature and pressure. This section focuses on the effects observed when subjecting Ultem™ 9085 parts to a pressurized environment inside a thermal chamber enclosed in a vacuum bag made of polyamide. The vacuum bag is equipped with a vacuum valve and sealed with heat-resistant sealing tape, ensuring a pressure of 0.9 bar. To gain comprehensive insights into the impact of this combined process, all three primary printing orientations were investigated (Figure 4.8).

Figure 4.9, presented below, highlights the noticeable differences in roughness profiles between the untreated samples and those subjected to the described conditions. In the case of the untreated samples, the outer filaments exhibit a more rounded appearance and have a greater overall height compared to the annealed sample, which displays a significantly flatter surface. The thermal treatment of the samples in a pressurized environment led to a remarkable reduction in surface roughness, with R_a decreasing from 17.3 ± 0.1 µm to 1.2 ± 0.2 µm and R_z decreasing from 70.4 ± 0.9 µm to 6.3 ± 1.1 µm. Furthermore, the phenomenon of dimensional modification observed when treating the samples solely with temperature was minimized. The annealed and isostatically pressurized samples did not experience significant geometric distortions, and changes in width and length were negligible or fell within the inherent error margin of the printer's precision. The only measurable variation occurred in height, with the samples undergoing an average compaction of $3.4 \pm 0.7\%$ in this direction. This densification serves as a clear explanation for the improved surface and mechanical properties observed in the post-processed samples.

As the effects of post-processing were analyzed in the main construction directions, it was verified from a morphological analysis that the overall compaction was around 3% compared to untreated samples, resulting in a reduction of surface roughness of over 90% (both R_a and R_z) in the specimens printed in the ZX direction. In contrast, this average reduction was around 50% in those printed in the XY direction. In addition to these contributions to surface quality, the specimens were 21% stiffer and 75% more resistant as proved in subsequent mechanical tests. These changes can be attributed to the combined effect of temperature, which softens the material, and vacuum pressure, which repositions the softened material from peaks-to-valleys, strengthening the bond between filaments.

It is important to highlight that, to assess any potential changes in the chemical and crystalline structure of the material, the parts underwent chemical analysis using X-ray diffraction (XRD) and Raman spectroscopy. These analyses effectively ruled out any significant alterations, confirming that the improvements observed were solely attributed to the post-processing effects, while ensuring the integrity of Ultem™ 9085 remained intact.

Having validated the described treatment and recognizing the promising results achieved by combining thermal annealing with isostatic compaction, future endeavors should focus on exploring the effects of different pressure levels, scaling up the process to more complex geometries and higher filling percentages, as well as developing models to predict and anticipate potential deformations that components may experience in real-world application environments.

4.2.3 Influence of Mechanical Post-Processing, With and Without Chip Removal

Another alternative explored by the scientific community to improve the quality of parts is mechanical post-processing through different approaches based on the adaptation of conventional metal finishing technologies to be applied to thermoplastics. In general, they seek to press or cut the outer peaks of the profile, introducing reductions in surface roughness, and adjusting the dimensional inaccuracies arising from the FFF manufacturing process. This type of post-processing can pose a challenge when working with complex geometries or intricate surfaces, which is one of the differential advantages of 3D printing. However, recent studies have shown great advances in obtaining satisfactory results by making multiple adaptations to conventional processes. Machining, polishing, sanding, ball burnishing, shot peening, and barrel finishing (BF) are examples of mechanical finishing processes documented in the scientific literature as good candidates for post-processing FFF parts.

To illustrate the significance of integrating Fused Filament Fabrication (FFF) technology with various post-processing techniques, notable studies provide valuable insights. For instance, Boschetto et al. [7] developed a theoretical model to investigate the combination of FFF technology with Barrel Finishing (BF) for enhancing the surface quality of printed parts. Their experimental campaign confirmed that the deposition angle of the material directly influences the outcomes achievable with BF. Additionally, the research group proposed a Computer Numerical Control (CNC) post-processing method, wherein the depth of cut was adjusted based on the deposition angle. This approach resulted in reduced average roughness and improved surface uniformity [58].

Another noteworthy study by Nsengimana et al. [59] conducted a comparative analysis to examine the effects of various post-processing techniques on FFF parts, including tumbling, shot peening, hand finishing, spray painting, CNC machining, and chemical treatment. The focus was on dimensional accuracy, specifically for laser-sintered parts made of Nylon, Alumide®, and ABS. Positive outcomes were observed across all cases, albeit to varying degrees, highlighting the effectiveness of these post-processing methods in improving the dimensional accuracy of the printed parts.

These examples underscore the importance of exploring and implementing different post-processing approaches to enhance the quality, accuracy, and surface finish of FFF-printed components. By leveraging complementary techniques and optimizing process parameters, researchers can unlock further improvements and advancements in additive manufacturing.

Given the diversity of documented mechanical finishing processes for metallic materials, they are divided into two groups to facilitate comparisons. On the one hand, those based on superficial plastic deformation will be analyzed, and on the other hand, those focusing on chip removal machining.

4.2.3.1   Impact of Surface Plastic Deformation

The use of plastic deformation as a finishing process on metallic materials is widely documented and adopted due to the significant improvements introduced in the properties of the treated samples and the obtaining smoother and uniform surfaces [60–65]. Based on the application of controlled force on the material's surface, this technique causes microscopic deformation due to the displacement of dislocations, resulting in permanent changes in its shape, crystalline structure, and properties as hardness and strength, and ductility. However, only some publications refer to using these post-processing techniques on polymeric materials. This section will refer to the contributions of three techniques based on plastic deformation: Ball Burnishing (BB), Abrasive Shot Blasting (ASB) [66, 67], and Shot Peening (SP) [68, 69].

BB is a finishing operation that applies a constant external force from a highly polished ball of high hardness attached to a tool typically controlled by a CNC machining center. This force is the parameter with the most influence within the process, and it depends on the tool's rigidity, from which it is possible to determine an application range. In addition to force, the speed of burnishing and the number of passes are other relevant parameters of the process [70–72]. On the other hand, ASB involves projecting glass microspheres, white corundum, or other similar materials onto the surface of the parts, causing plastic deformation by impact. In this case, in addition to the abrasive material, the parameters with the greatest significance are the working pressure, exposure time, and impact angle. Finally, SP is very similar to ASB, but the impact material is metallic abrasive, typically stainless-steel shot fired onto the parts placed on a rotating unit. The controllable parameters are very similar to those of the previous process, and their choice is decisive in achieving the intended improvements.

The results of using the three mentioned technologies have also been validated on Ultem™ 9085, the same material that has been studied, using standardized specimens for mechanical tests. However, the surface roughness obtained after each of them differs significantly. As in previous cases, R_a and R_z have been used as response indicators. When the samples were subjected to ball burnishing, the decrease in R_a and R_z occurred around 70% and 50%, respectively, due to the forced plasticization by pressing the peaks' material into the valleys between filaments. When the parts were processed using ASB and glass microspheres as abrasive, the surface roughness improved by an average of 25% for both R_a and R_z. In contrast, when white corundum was used, this improvement was reduced to an average of 10% in both indicators. In the case of the samples treated with SP, the improvement reflected in R_a and R_z was approximately 20%.

The findings of the microscopy analysis of the samples post-processed with the described treatments are shown in Figure 4.10.

It can be appreciated how the typical rounded shape of the filament is flattened after the BB treatment (Figure 4.10 c and d), without causing apparent damage to the surface. These images corroborate the results quantified in the roughness measurements. In contrast, the images corresponding to the pieces treated with abrasives are radically modified (Figure 4.10 e-j). Although the average roughness of the surface or the dimensions of the post-processed parts do not show such variations, the surface of a filament was affected, with white corundum introducing the most noticeable changes, that is, producing the most eroded surface (Figure 4.10 e and f). This phenomenon has also affected the morphology, as the material presents cracking and numerous irregularities in the filaments. The utilization of glass microspheres in the shot blasting process can lead to the decomposition of these microspheres into smaller and protruding fragments, resulting in a surface that is less uniform in nature (Figure 4.10 g and h). Similarly, specimens subjected to shot peening (SP) exhibit noticeable modifications (Figure 4.10 i and j).

Comparing all the results shown, it seems reasonable to think that ball burnishing is the best candidate for obtaining surfaces improved by plastic deformation. However, it is essential to highlight that this technology has certain limitations when treating pieces with complex geometries or intricate shapes that make it impossible to access the burnishing tool. In these cases, some abrasive material could be the most advisable alternative, for which the use of glass microspheres as the shot is recommended, with the process parameters being exhaustively controlled.

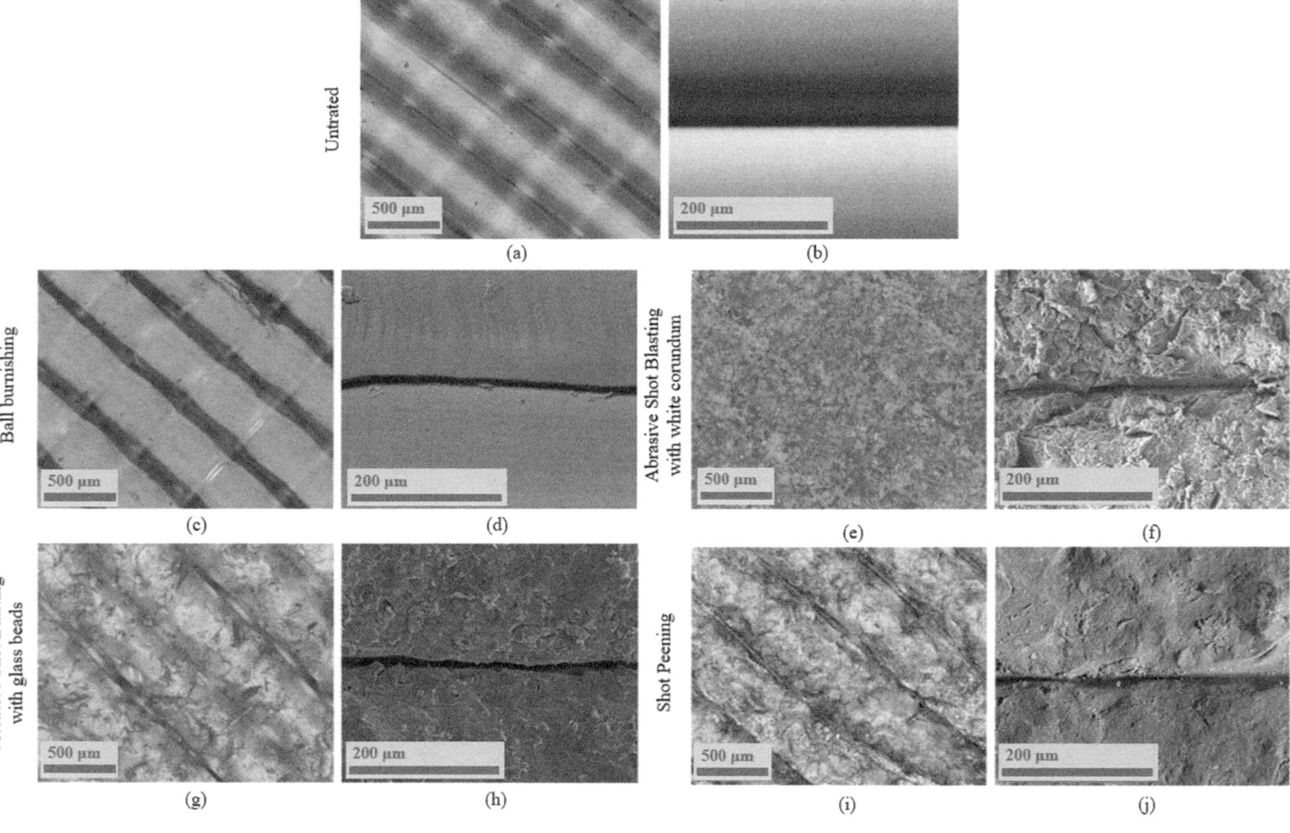

FIGURE 4.10 Comparison of the effect of surface plastic deformation induced by Ball Burnishing (BB) and Abrasive Shot Blasting (ASB) and Shot Peening (SP): (a, c, e, g, i) Images from microscopy; (b, d, f, h, j) SEM images.

4.2.3.2 Transformations Introduced by Machining

As previously mentioned, dimensional accuracy and surface quality of parts obtained by AM are two critical factors when manufacturing final components that need specific geometric and dimensional tolerances to be part of assembled sets with other components and even with other components materials. In cases like this, machining may be a good alternative to meet those requirements, as long as the final shape of the piece facilitates the implementation of these processes without causing essential modifications to the conventional machinery that may paradoxically increase the cost. Therefore, when dealing with a post-processing step using cutting tools coupled to milling machines, drills, or lathes, the operational limitations of these technologies must be considered, including the size limits of the components, the scale of the construction details, or the inherent characteristics of the cutting tools' geometry.

Another advantage of AM is the possibility of producing parts with a certain percentage of filling, if there is no compromise on their structural performance since it allows for substantial material and manufacturing time savings. However, studies like that of Núñez et al. [73] conclude that dimensional accuracy is impaired when the filling is not 100%. Furthermore, more pronounced errors in flatness and exterior texture are introduced. It should be noted that when the part is printed with a filling percentage lower than 100%, any machining process on its surfaces will entail the elimination of the external layers. Therefore, the structural stability of the piece would be affected, even rendering it unusable.

However, using machining as a complementary post-processing step to AM could be a good choice, especially when it results in shorter manufacturing times or is necessary to ensure a bounded dimensional accuracy. This could be the case for the manufacture of blinds or through holes within parts obtained by FFF, in which other elements, such as metal shafts or inserts, could be coupled. For example, in the paper of Gómez-Gras et al. [74], the feasibility of directly create a hole in Ultem™ 9085 samples or printing the solid piece and subsequently machining the hole is analyzed. These authors compared samples printed in three orientations, some obtained with an incorporated through-the-hole. In contrast, others were machined after printing with a standardized drill of the same diameter. Dimensional and cylindricity errors were verified using a pass/fail gauge. The differences in flatness resulting from printing and subsequent machining were also evaluated using a probe coupled to a displacement sensor.

The authors concluded that since the accuracy of the tolerance is so dependent on printing tolerances, pass/fail gauges work in both conditions, falsifying the result in some cases. They also detected that the cutting speed also increases the hole's diameter due to a local temperature rise caused by the drilling friction. This parameter is the most critical when quantifying the dimensional differences found. Increasing the drilling speed introduced variations in the quality and, notably, led to a lack of homogeneity in surface roughness. At higher speeds, the region of the piece where the drill was inserted exhibited a rougher texture compared to the area where the drill exited. This suggests that temperatures generated during machining can harm the surface quality of polymers. Depending on the processing time and consequent heat accumulation, the surface may soften, and the material may be dragged, resulting in a heterogeneous surface with other inaccuracies in dimensions (Figure 4.11). As for flatness errors, parts printed with the hole and subsequently machined showed similar deviations from the nominal tolerance for all three printing orientations. In all cases, the material retains nominal dimensions at the ends of the part while tending to bow towards the center of the faces.

Finally, considering the analyzed results, it is recommended to use machining as a post-process for components with simple geometries that do not pose an accessibility challenge for cutting tools. The best results will be achieved with 100% material fill, low cutting speeds, and low cutting times. Thermal variations introduced during machining can drastically damage the final parts or, counterproductively, fail to guarantee the dimensional accuracy and surface quality expected when choosing this type of post-processing.

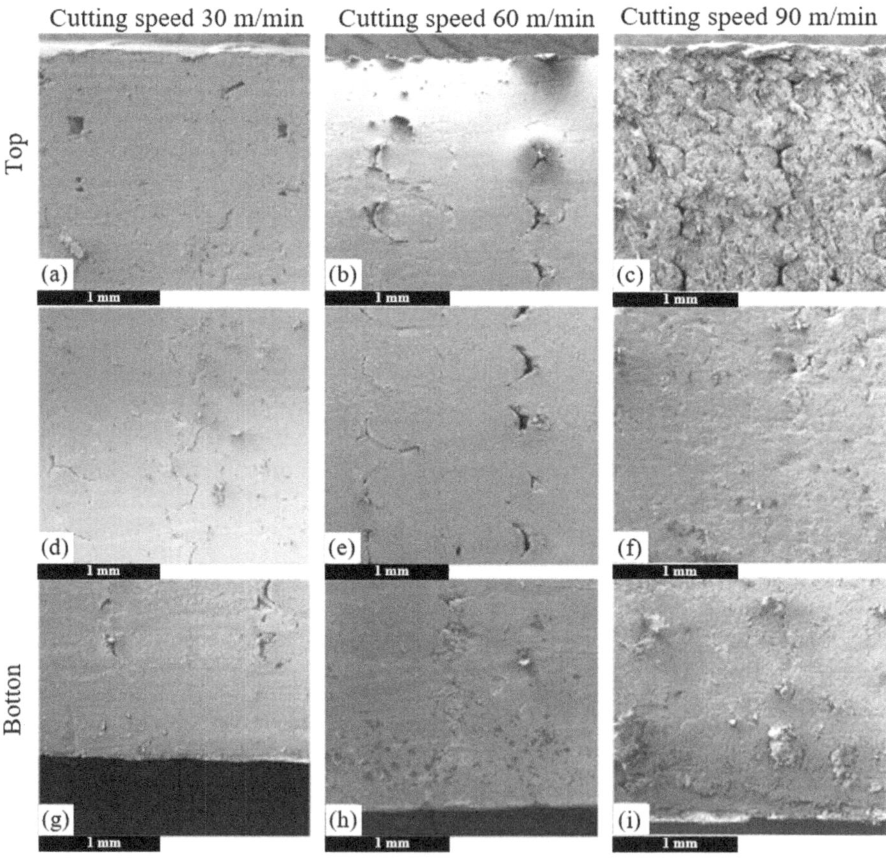

FIGURE 4.11 Comparison of the surface roughness changes due to increased cutting speed during drilling of Ultem™ 9085 samples.

4.3 IMPROVEMENTS IN MECHANICAL PERFORMANCE

It has already been anticipated that another of the great challenges of the AM community is to successfully address and correct the characteristic low mechanical performance of parts obtained by FFF, another inconvenience that currently slows down the generalization of their use for the industrial manufacturing of final products. Consensus among scientific contributions in this field acknowledges that the mechanical properties of 3D printed parts can exhibit notable variations influenced by several factors. These factors encompass the material type employed, the chosen combination of printing parameters, the build orientation, the precision of the printer, and the design and geometry of the sample. As a result, parts with a significant mechanical anisotropy are often obtained, which ultimately conditions their functionality in real contexts since, for example, their strength is reduced in certain directions, there are greater deformations under certain solicitations, or unexpected fractures are caused by poor adhesion between layers, often challenging to detect. This is one of the reasons why the success of AM is still mainly focused on the design and manufacturing of prototypes, molds with special geometries, or components that are not directly subjected to static or dynamic loads [75–80].

These statements again highlight the importance of post-processing printed parts using novel techniques or those already validated with other materials, with the ability to reduce the undesirable effects of these limitations. This section aims to explore the effects of chemical, thermal, and mechanical post-processing on the mechanical performance of Ultem™ 9085 parts produced through FFF. Through this analysis, valuable insights will be gained regarding the viability of these procedures and practical recommendations will be offered for their effective implementation.

4.3.1 INFLUENCE OF CHEMICAL POST-PROCESSING

One of the major concerns when choosing chemical solvents with which to treat printed parts is that despite being able to obtain improved surfaces (already described in Section 2.1), there is a risk that certain chemical products will end up producing degradation in the main material, which, paradoxically, would drastically compromise its mechanical performance. This section discusses recommended solvents for post-processing, such as high-impact polystyrene for ABS and PLA compatibility, as well as water-soluble options like polyvinyl alcohol (PVA) with various activating agents. However, the number of materials optimized for FFF printing is increasing, generating the inconvenience of developing some support material in parallel whose extraction does not add difficulty to the process. Precisely, because of this diversity, there are now very different filaments in terms of chemical and thermal properties, preventing universal support capable of being removed with a single solvent.

Ultem™ 9085, a polyetherimide with a high glass transition temperature (Tg), requires a support material with similar thermal properties for processing in the same hot print chamber. However, the conventional mechanical extraction of the support material, polysulfone (PSF), is slow, expensive, and can introduce defects in the parts, impacting their mechanical properties.

To address this challenge, the researchers explored the use of different solvents to chemically remove PSF from Ultem™ 9085. Figure 4.12 presents the stress-strain curves obtained from static tests conducted on Ultem™ 9085 after PSF removal using a toluene-based solvent mixed with 20% aniline. This solvent combination demonstrated the best solubility results and improved surface quality. While 1,4-dioxane and acetone also exhibited solubility potential and showed similar mechanical property trends, the detected mass loss and significant reduction in filament diameter observed in SEM micrographs were considered unacceptable.

Therefore, the toluene-based solvent with added aniline was deemed the most effective and acceptable option for PSF removal, as it preserved the mechanical properties of Ultem™ 9085 while achieving desired solubility and surface quality improvements.

Analysis of the graphs reveals that the mechanical properties within the elastic range of Ultem™ 9085 are minimally affected by solvent treatment. This can be attributed to the fact that in FFF, specimen failure typically occurs at high strains where filaments rupture or separate, whereas the material behaves uniformly in the initial stages of the test. Consequently, the degradation between filaments themselves predominantly impacts the properties calculated at higher strains.

However, SEM micrographs of the treated parts indicate that the chemical agent affects both the outer surface and layer junctions. While the chemical attack is more detrimental to coplanar filament junctions (intralayer) and adjacent layer junctions (interlayer), as these junctions are typically less mechanically robust, the overall impact on mechanical properties is moderate. Most changes fall within the experimental error of the tests, reinforcing the viability of the proposed support removal method over manual extraction.

Quantifying the analysis, it is observed that the tensile modulus and strength experience a smaller decrease (3% and 8%, respectively) compared to the flexural modulus and strength (8% and 11%, respectively). This phenomenon can be explained by the fact that, during the three-point flexure test, layer junctions bear the applied stress to prevent layer slippage caused by shear forces. As these junctions are more affected by the solvent, the changes are more pronounced in the flexural test. In contrast, the tensile test evenly distributes the applied stress between the filament and junctions during the initial stages, with the filament primarily supporting the stress at higher strains.

4.3.2 Effects of Heat Treatments

Section 2.2.1 extensively discussed the beneficial effects of densification caused by heat treatments on printed specimens to improve surface quality. However, this type of treatment has other positive consequences on the treated polymers. A review published by Chohan & Singh in 2017 [81] showed that this post-processing was promising for reducing the inherent mechanical anisotropy of additive manufacturing techniques.

These benefits are maximized when the effects of a pressurized environment are added to the heat treatment, which further impacts the densification of the parts, directly affecting the mechanical behavior. For the analysis, three-point bending tests

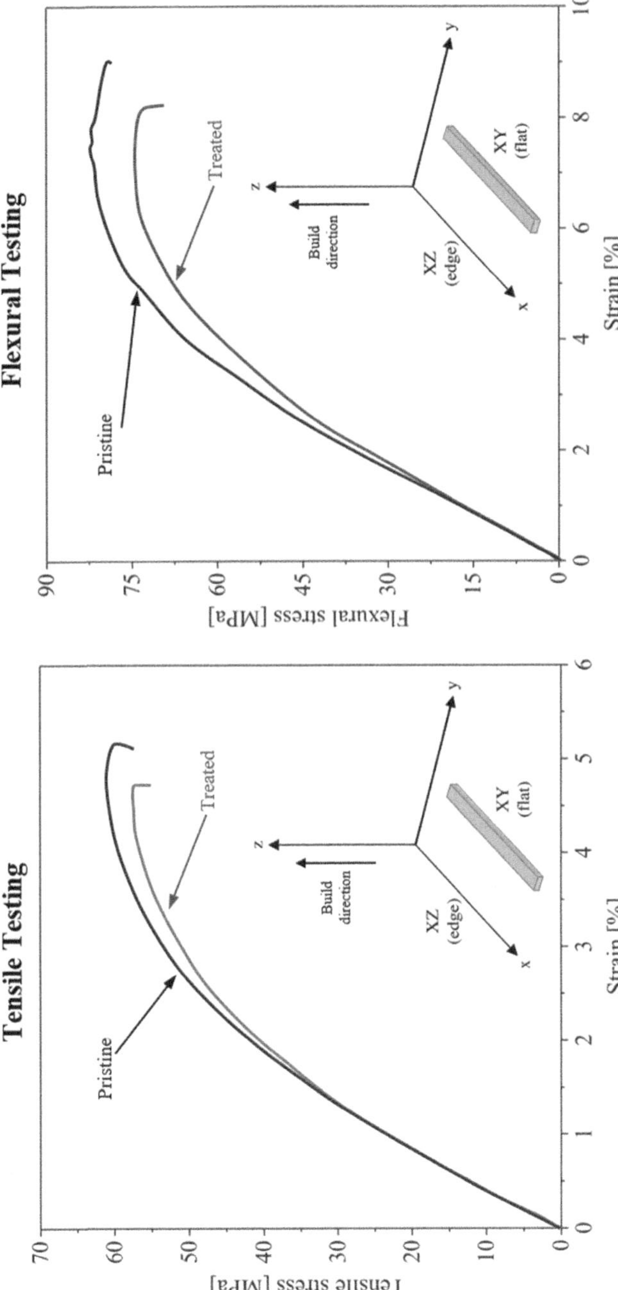

FIGURE 4.12 Effect of Chemical Treatment on the Mechanical Response of Ultem™ 9085: Tensile (right) and Flexural (left) Stress-Strain Curves.

were considered using, as usual, Ultem™ 9085 specimens. In this case, they were printed in the ZX direction (upright), as it is the configuration with the lowest mechanical performance found. Post-processed specimens showed increased strength, stiffness, and ductility compared to untreated specimens.

The improved resistance and deformation characteristics observed in the thermally annealed and pressurized ZX Ultem™ 9085 samples indicate a notable enhancement in the quality of layer bonding. Microscopy images of the fracture region further illustrate this improvement. In the case of a pristine specimen (Figure 4.13-a), a flat rupture area suggests failure due to layer separation. Conversely, the fracture region of a heat-treated sample (Figure 4.13-b) exhibits a more irregular rupture area, indicating that failure was not solely attributed to delamination. Moreover, the cohesion among the intermediate layer filaments has significantly increased, as evidenced by the more distinct deposition path of the filament in the treated sample compared to the pristine sample.

Once the effects on the specimens with the poorest mechanical performance were quantified, the remaining configurations were also examined using the optimal thermal annealing conditions identified for the ZX configuration. The mechanical properties of FFF specimens are well-known to depend largely on their orientation during printing, as this determines the percentage of stress borne by the filaments. Consequently, the ZX samples, where the filaments of the outer layer are parallel to the applied stress, exhibit the weakest performance.

Similarly, the XZ (edge) samples, with the outer layer weaves completely perpendicular to the applied stress direction, are less influenced by layer bonding, resulting in superior mechanical properties under bending. The XY (flat) specimens, with ±45° weaves in the outer layer, demonstrate intermediate performance. The average flexural strength for the pristine X-Edge samples is 110 ± 1 MPa, whereas it is 85 ± 3 MPa for the X-Flat samples and 64 ± 5 MPa for the ZX (upright) samples.

Figure 4.14 illustrates representative stress-strain curves for both the pristine and annealed samples produced in the aforementioned printing orientations. The comparative graph reveals a noticeable decrease in the anisotropy of the mechanical properties after thermal treatment. This reduction is evident in terms of maximum stress, flexural modulus, and deformation associated with flexural strength. For instance, the maximum difference in flexural strength between the samples with the best and worst performance decreases from 47 MPa (untreated) to 8 MPa (annealed), representing an 80% reduction in anisotropy.

The visual examination of cross-sectional images of the samples (Figure 4.15) aligns with the experimental analysis results of the mechanical properties. It is evident that the pristine samples, irrespective of the printing orientation, exhibit greater porosity compared to the annealed samples. This implies that the annealing process in a pressurized environment enhances the density of Ultem™ 9085 FFF samples and reinforces the interlayer bonds.

In general, the treated samples exhibited a 21% increase in flexural modulus and a 75% increase in flexural strength. This significant improvement can be attributed to the fusion, resolidification, and reinforcement of the interfilament bonds that occur during the post-processing heat treatment. These results clearly demonstrate the effectiveness of heat treatment in mitigating the mechanical anisotropy of FFF

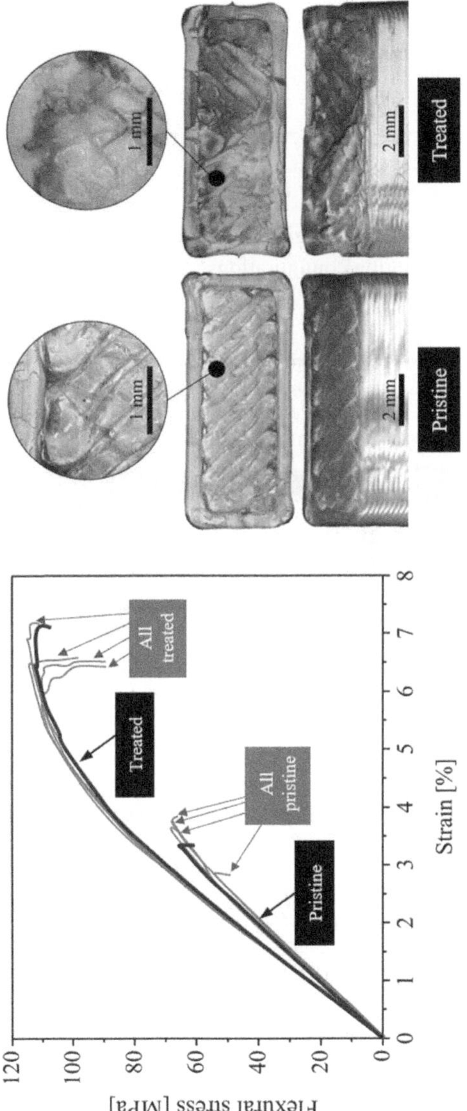

FIGURE 4.13 Bending stress-strain curves and microscopy images depicting the fracture region of (a) original (b) thermally annealed and pressurized ZX Ultem™ 9085 samples.

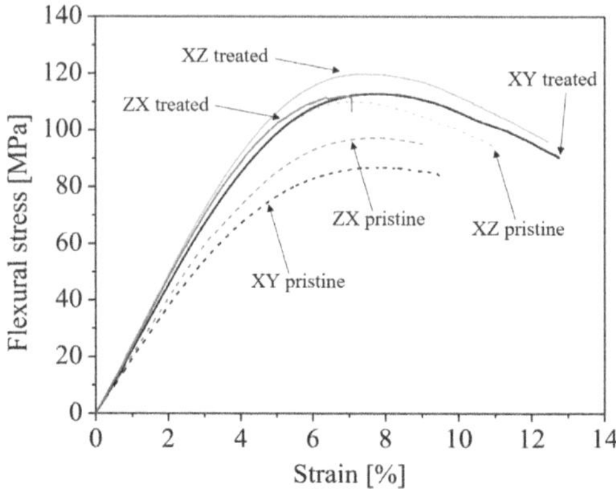

FIGURE 4.14 Stress-strain curves of original and temperature plus vacuum post-processed Ultem™ 9085 samples for all printing configurations.

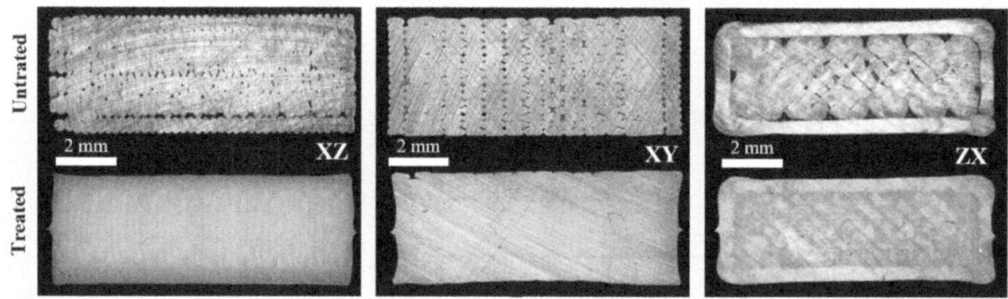

FIGURE 4.15 Cross-section of Ultem™ 9085 samples pristine and annealed in a pressurized environment, for different printing configurations.

parts by enhancing filament bonding. This addresses a key limitation of 3D printing, making it a promising approach for improving the overall mechanical performance of printed components.

4.3.3 IMPACT OF MECHANICAL POST-PROCESSING

As already analyzed in previous sections, mechanical plastic deformation processes have unequal effects on the surface and dimensional quality, depending on the nature of the applied technique, it is possible to obtain improved surfaces or, on the contrary, increased average roughness. At the same time, they pose significant challenges in some cases to successfully handle complex geometries or parts with low accessibility.

The application of ball burnishing, known for its notable enhancements in surface quality, yielded slightly higher tensile and flexural moduli while reducing the yield strength of the treated samples, as depicted in Figure 4.16.

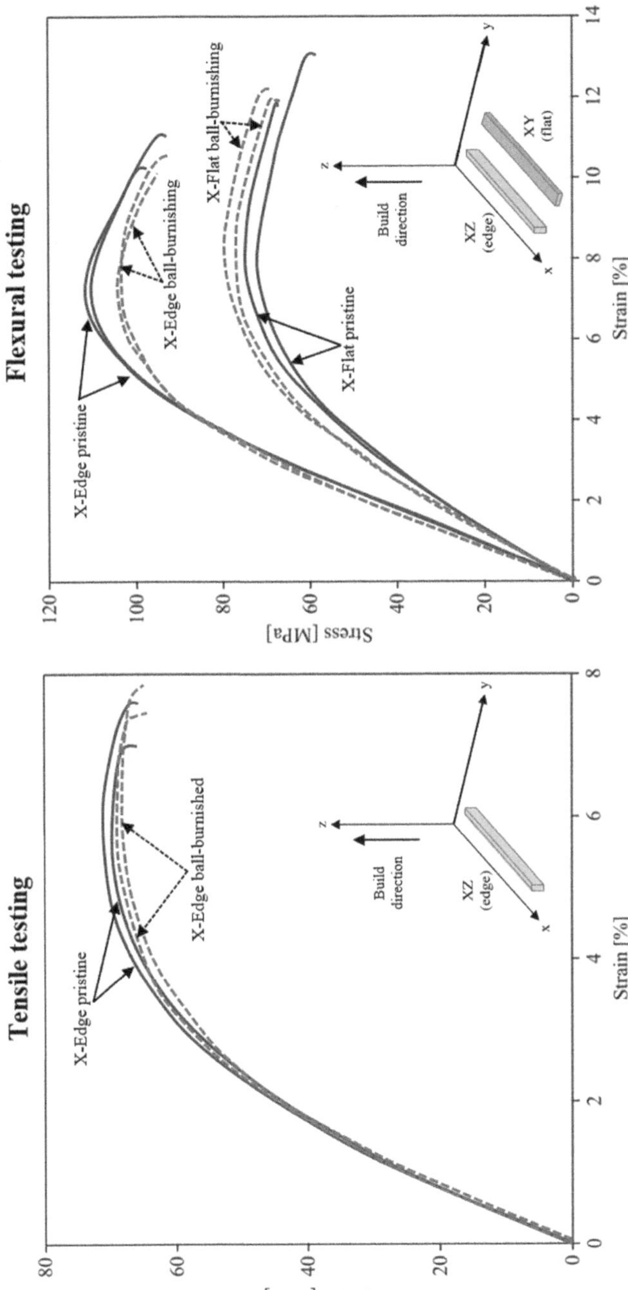

FIGURE 4.16 Stress-strain curves illustrating the impact of ball burnishing on the tensile and flexural behavior of Ultem™ 9085 samples.

This phenomenon is attributed to the early hardening and plasticization of the outer layer induced by the ball burnishing process. As described by Jatin et al. [82], compression stresses during burnishing cause polymer chains to come closer, resulting in an increased number of binary contacts with strong repulsive forces. This increased contact leads to a hardening response.

Additionally, an unexpected decrease in flexural strength is observed in the burnished X-Edge samples. This can be explained by the dual effect of ball burnishing on the top layer and underlying layers of the FFF specimens. Firstly, it decreases porosity and effectively increases the covered area of the material. Secondly, it compacts or thins the overall thickness of the burnished face. These effects are more favorable for X-Flat specimens, as the intralayer bond of the top layer supports stress due to the arrangement of the ±45° filaments, which is reinforced by the burnishing process. However, in X-Edge samples, where the filaments in the outer layer are perpendicular to the applied stress, the bond between layers is less influenced. Therefore, both effects impact both configurations, but the second effect is more prevalent in X-Edge samples. As the outer layer becomes thinner, the stress gradient supported by the internal layers and the bonds between them (arranged in a ±45° pattern) increases, resulting in lower tensile strength. The combination of these effects explains why the maximum stress is unaffected in X-Flat specimens but leads to reduced tensile and flexural strength in X-Edge samples.

Significant benefits are observed in the fatigue strength of ball-burnished FFF samples, akin to those seen in metallic materials. To evaluate the fatigue life, pristine and burnished samples underwent three-point bending fatigue tests at various loads. Following standard test recommendations, the results represent the number of cycles until the maximum deflection increased by more than 10%, indicating the onset of the elastic limit, as none of the samples failed before reaching this threshold.

The S-N R-1 curves obtained (Figure 4.17) clearly demonstrate that ball-burnished FFF samples printed in X-Edge and X-Flat configurations exhibit a fatigue life performance at least two times greater than untreated specimens. At lower load percentages, the improvement is even more remarkable, reaching up to seven times for X-Flat samples and five times for X-Edge samples. Furthermore, predictable enhancements in fatigue strength range from 20% to 100%, with the greatest improvements observed when prolonged fatigue life (10^5 cycles) is anticipated. This can be attributed to the concentration of residual compression stresses on the burnished surface, coupled with subsequent surface densification and hardening. These factors effectively retard crack initiation and propagation, thus accounting for the notable increase in fatigue life.

In addition to ball burnishing, the effects of two other plastic deformation post-processes, namely Abrasive Shot Blasting and Shot Peening, were investigated. The technological parameters described in Section 2.3.1 were utilized to explore the contributions of these processes to surface quality.

When comparing the tensile test results of the specimens treated by both techniques, it is possible to identify that the post-process has not introduced significant changes in either the elastic modulus, the maximum stress, or the stress at maximum elongation (Figure 4.18). These results indicate that neither of the two post-processes has had a positive or negative impact on the mechanical behavior of Ultem™ 9085.

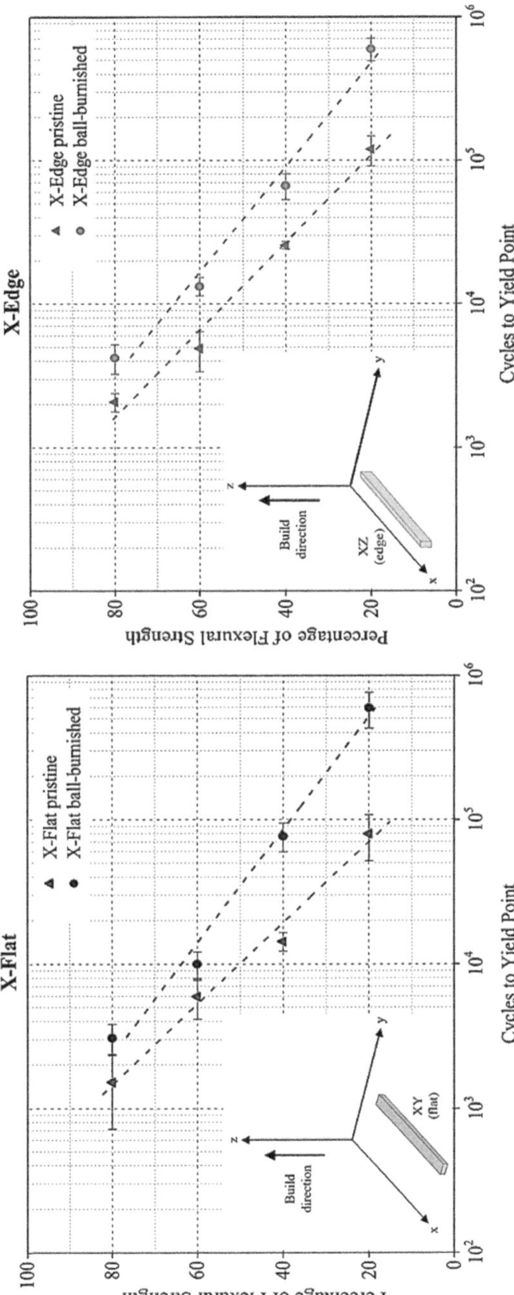

FIGURE 4.17 Fatigue S-N R-1 curves illustrating the results of three-point bending tests conducted on pristine and ball-burnished Ultem™ 9085 samples printed in the X-Flat and X-Edge orientations.

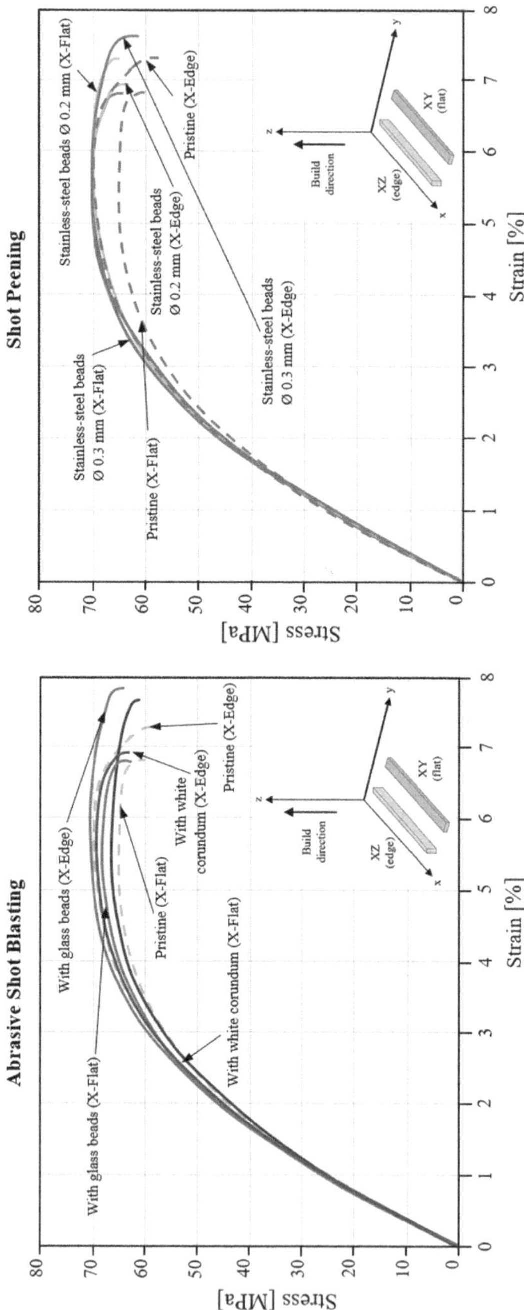

FIGURE 4.18 Stress–strain curves showing the effect of Abrasive Shot Blasting (ASB) and Shot Peening (SP) on Ultem™ 9085 samples' tensile behavior.

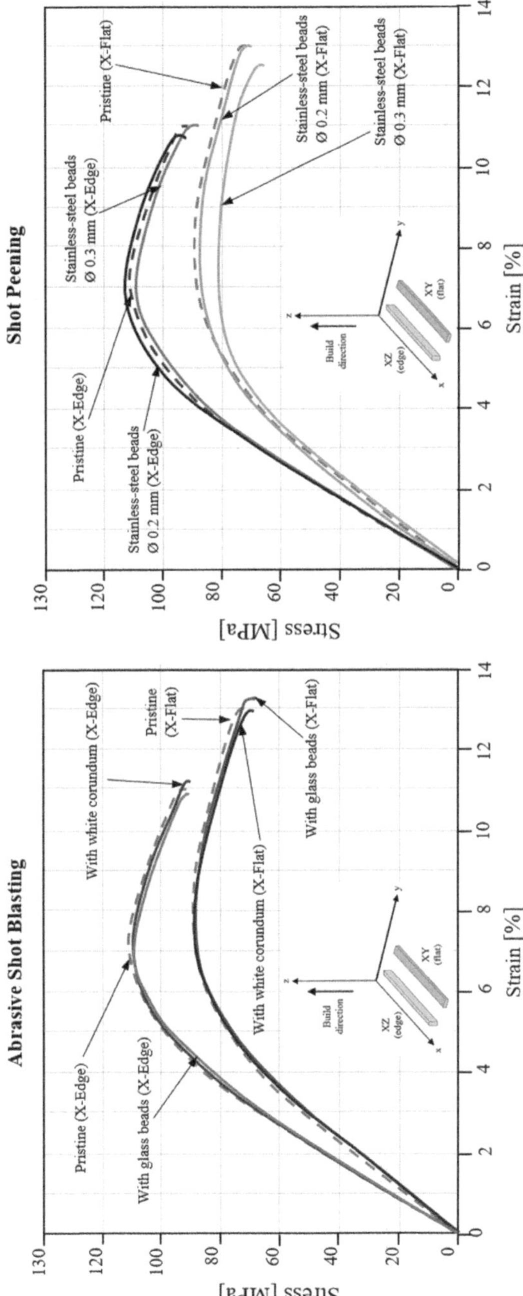

FIGURE 4.19 Stress-strain curves showing the effect of Abrasive Shot Blasting (ASB) and Shot Peening (SP) on Ultem™ 9085 samples' flexural behavior.

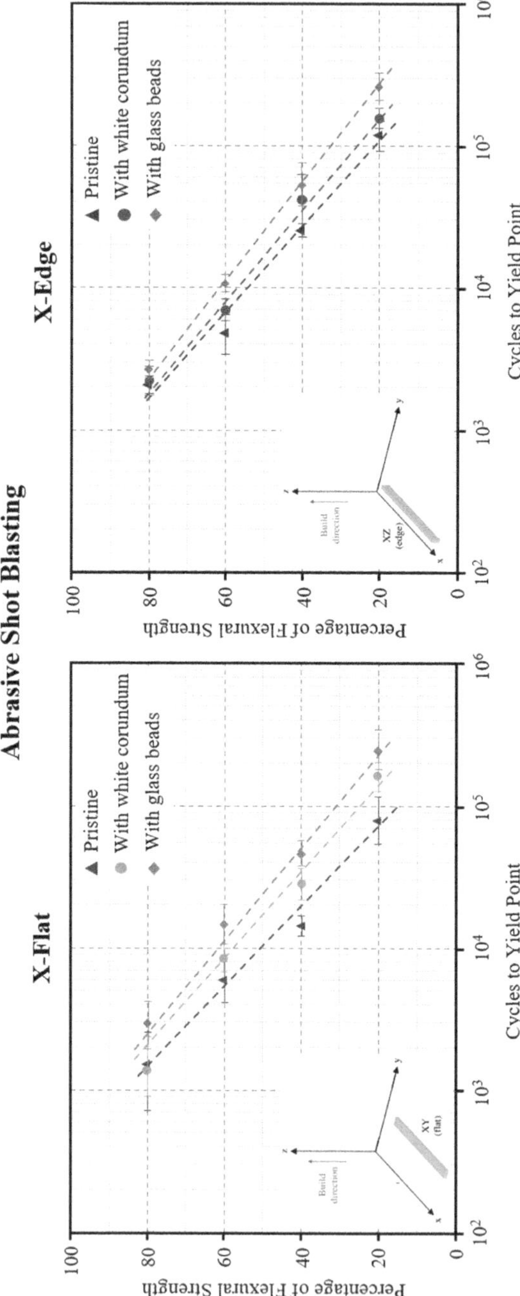

FIGURE 4.20 Fatigue S-N R-1 curves for three-point bending tests of pristine and abrasive shot-blasted Ultem™ 9085 samples in X-Flat and X-Edge orientations.

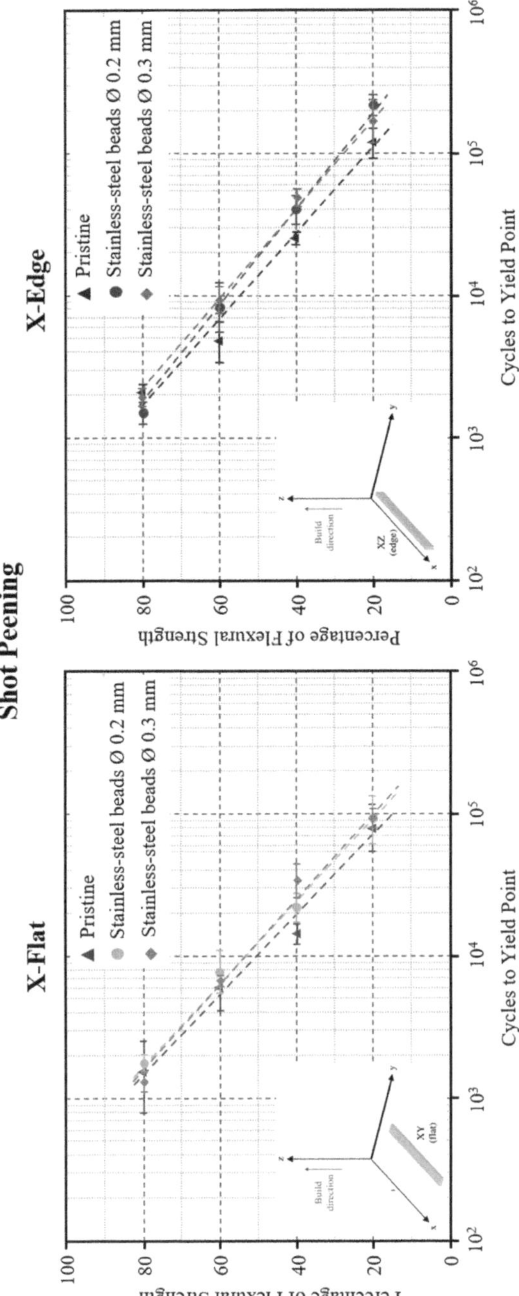

FIGURE 4.21 Fatigue S-N R-1 curves for three-point bending tests of pristine and shot-peened Ultem™ 9085 samples in X-Flat and X-Edge orientations.

The bending tests revealed consistent trends in the results. For the untreated specimens, there were no significant effects of the post-process on the bending properties, as indicated by the similar values and low standard deviations of the elastic modulus, maximum stress, and stress at maximum elongation. In fact, in the case of samples treated with Shot Peening using larger diameter steel balls (0.3 mm), there was a slight decrease in the maximum bending stress, particularly in the X-Flat orientation (Figure 4.19).

As expected, when analyzing the behavior of Ultem™ 9085 specimens in a three-point fatigue test, the treated specimens moderately increased their life cycle. Figures 4.20 and 4.21 show the S-N R-1 curves and the two construction configurations studied for the two treatments mentioned. As a result of the compressive stresses introduced by the plastic deformation suffered by the specimens, some conclusions can be drawn. The best results were obtained in pieces subjected to ABS, in which the fatigue strength doubled at a load equivalent to 20% of the maximum flexural stress, a behavior that is reasonably stable for higher load percentages. In this sense, the specimens treated with glass beads achieve the most notable results for both orientations.

This improvement is less evident when the pieces are subjected to SP; as can be seen, the margins of improvement are lower or nonexistent for higher load percentages. At the same time, the diameter of the stainless-steel beads does not influence the results.

Overall, from these results, none of the analyzed superficial plastic deformation treatments significantly affected the static behavior of Ultem™ 9085 parts. However, when analyzing the mechanical performance under fatigue, it can be observed that the resistance cycles increase in all cases, being especially notable when the samples are polished.

ACKNOWLEDGEMENTS

This research has received financial support from the grant RTI2018-099754-A-I00 funded by MCIN/AEI/10.13039/501100011033 and ERDF A way of making Europe. Additionally, support has been provided by the grant PID2021-123876OB-I00 funded by MCIN/AEI/10.13039/501100011033 and the European Union.

REFERENCES

[1]   Zhang, C., Chen, Y., Chen, H., & Chong, D. (2021). Industry 4.0 and its implementation: A review. *Information Systems Frontiers*, 1–24. doi:10.1007/S10796-021-10153-5, https://link.springer.com/article/10.1007/s10796-021-10153-5#citeas

[2]   Haleem, A., & Javaid, M. (2019). Additive manufacturing applications in Industry 4.0: A review. *Journal of Industrial Integration and Management*, 4, 1930001. doi:10.1142/S2424862219300011

[3]   Attaran, M. (2017). The rise of 3-D printing: The advantages of additive manufacturing over traditional manufacturing. *Business Horizons*, 60, 677–688. doi:10.1016/J.BUSHOR.2017.05.011

[4] Uslu, T., Holman, M., Zhong, X., & Schiavo, A. (2022). Lux Research | Will 3D Printing Replace Conventional Manufacturing. Retrieved from https://luxresearch inc.com/resources/chemicals/will-3d-printing-replace-conventional-manufacturing/

[5] Sanaei, N., & Fatemi, A. (2020). Analysis of the effect of surface roughness on fatigue performance of powder bed fusion additive manufactured metals. *Theoretical and Applied Fracture Mechanics*, 108, 102638.

[6] Kalami, H., & Urbanic, J. (2021). Exploration of surface roughness measurement solutions for additive manufactured components built by multi-axis tool paths. *Additive Manufacturing*, 38, 101822.

[7] Boschetto, A., Giordano, V., & Veniali, F. (2013). 3D roughness profile model in fused deposition modelling. *Rapid Prototyping Journal*, 19, 240–252.

[8] Barari, A., Kishawy, H. A., Kaji, F., & Elbestawi, M. A. (2017). On the surface quality of additive manufactured parts. *International Journal of Advanced Manufacturing Technology*, 89, 1969–1974.

[9] Gómez-Gras, G., Pérez, M. A., Fábregas-Moreno, J., & Reyes-Pozo, G. (2021). Experimental study on the accuracy and surface quality of printed versus machined holes in PEI Ultem 9085 FDM specimens. *Rapid Prototyping Journal*, 27, 1–12.

[10] Taufik, M., & Jain, P. K. (2017). Characterization, modeling and simulation of fused deposition modeling fabricated part surfaces. *Surface Topography: Metrology and Properties*, 5, 045003.

[11] Haque, M. E., Banerjee, D., Mishra, S. B., & Nanda, B. K. (2019). A Numerical Approach to Measure the Surface Roughness of FDM Build Part. *Materials Today: Proceedings*, 18, 5523–5529.

[12] Giri, J., Shahane, P., Jachak, S., Chadge, R., & Giri, P. (2021). Optimization of FDM process parameters for dual extruder 3d printer using Artificial Neural network. *Materials Today: Proceedings*, 43, 3242–3249.

[13] Wang, P., Zou, B., & Ding, S. (2019). Modeling of surface roughness based on heat transfer considering diffusion among deposition filaments for FDM 3D printing heat-resistant resin. *Applied Thermal Engineering*, 161, 114064.

[14] Raj Mohan, R., Venkatraman, R., & Raghuraman, S. (2021). Experimental analysis on density, micro-hardness, surface roughness and processing time of Acrylonitrile Butadiene Styrene (ABS) through Fused Deposition Modeling (FDM) using Box Behnken Design (BBD). *Materials Today Communications*, 27, 102353.

[15] Chohan, J. S., & Singh, R. (2017). Pre and post processing techniques to improve surface characteristics of FDM parts: A state of art review and future applications. *Rapid Prototyping Journal*, 23(3), 495–513.

[16] Eswaran, P., Subramaniyan, M., Appusamy, A., Annakalyani, N. S., & Pusapati, S. R. (2021). Investigations on acute angle parts fabricated fusion deposition modelling parts volumetric shrinkage and surface roughness. *Materials Today Proceedings*, 45(2), 930–935. https://doi.org/10.1016/j.matpr.2020.02.945.

[17] Ahn, D., Kim, H., & Lee, S. (2009). Surface roughness prediction using measured data and interpolation in layered manufacturing. *Journal of Materials Processing Technology*, 209(2), 664–671.

[18] Pramanik, D., Mandal, A., & Kuar, A. (2020). An experimental investigation on improvement of surface roughness of ABS on fused deposition modelling process. *Materials Today Proceedings*, 26(2), 860–863. https://doi.org/10.1016/j.matpr.2020.01.054.

[19] Chueca de Bruijn, A., Gómez-Gras, G., & Pérez, M. A. (2021). A Comparative analysis of chemical, thermal, and mechanical post-process of fused filament fabricated

polyetherimide parts for surface quality enhancement. *Materials*, 14(19), 5880. doi:10.3390/MA14195880

[20] Melton, G. H., Peters, E. N., & Arisman, R. K. (2011). *Applied Plastics Engineering Handbook*. William Andrew Publishing. doi:10.1016/B978-1-4377-3514-7.10002-9

[21] Bui, D. T., Vivekh, P., Islam, M. R., & Chua, K. J. (2022). Studying the characteristics and energy performance of a composite hollow membrane for air dehumidification. *Applied Energy*, 306, 118161. doi:10.1016/J.APENERGY.2021.118161

[22] Hashmi, M. P., & Koester, T. M. (2018). Applications of Synthetically Produced Materials in Clinical Medicine. *Reference Module in Materials Science and Materials Engineering*. doi:10.1016/B978-0-12-803581-8.10258-9, www.iso.org/obp/ui/#iso:std:iso-astm:52900:ed-2:v1:en

[23] International Organization for Standardization/American Society for Testing and Materials. (2021). *ISO/ASTM 52900:2021 – Additive manufacturing — General principles – Fundamentals and vocabulary*. Geneva, Switzerland

[24] Stratasys Inc. (2019). ULTEM 9085 Production-Grade Thermoplastic for Fortus 3D Printers tech. rep., 1–17. www.stratasys.com/materials/search/ultem9085

[25] Forés-Garriga, A., Pérez, M. A., Gómez-Gras, G., & Pozo, G. R. (2020). Role of infill parameters on the mechanical performance and weight reduction of PEI Ultem processed by FFF. *Materials Design*, 193, 108810.

[26] Stratasys. (2021). ULTEM 9085 Production-Grade Thermoplastic for Fortus 3D Printers. Retrieved May 15, 2021. www.stratasys.com/materials/search/ultem9085.

[27] Jiang, Y., Ding, J., Yin, X., Yan, J., Yang, L., Zhu, S., Zeng, R., Lu, C., Yin, A., & Xian, X. (2021). The influence of U-Nb substrate roughness on the service performance of Ti/TiN multilayer coatings. *Surface and Coatings Technology*, 405, 126571.

[28] Pérez, M., Medina-Sánchez, G., García-Collado, A., Gupta, M., & Carou, D. (2018). Surface quality enhancement of Fused Deposition Modeling (FDM) printed samples based on the selection of critical printing parameters. *Materials*, 11(8), 1382.

[29] Buj-Corral, I., Domínguez-Fernández, A., & Durán-Llucià, R. (2019). Influence of print orientation on surface roughness in Fused Deposition Modeling (FDM) processes. *Materials*, 12(22), 3834.

[30] Chohan, J. S., Singh, R., Boparai, K. S., Penna, R., & Fraternali, F. (2017). Dimensional accuracy analysis of coupled fused deposition modeling and vapour smoothing operations for biomedical applications. *Composites Part B: Engineering*, 117, 138–149.

[31] Chohan, J. S., Singh, R., & Boparai, K. S. (2016). Mathematical modelling of surface roughness for vapour processing of ABS parts fabricated with fused deposition modelling. *Journal of Manufacturing Processes*, 24, 161–169.

[32] Chohan, J. S., Singh, R., & Boparai, K. S. (2016). Parametric optimization of fused deposition modeling and vapour smoothing processes for surface finishing of biomedical implant replicas. *Measurement*, 94, 602–613.

[33] Mu, M., Ou, C.-Y., Wang, J., & Liu, Y. (2020). Surface modification of prototypes in fused filament fabrication using chemical vapour smoothing. *Additive Manufacturing*, 31, 100972.

[34] Jin, Y., Wan, Y., Zhang, B., & Liu, Z. (2017). Modeling of the chemical finishing process for polylactic acid parts in fused deposition modelling and investigation of its tensile properties. *Journal of Materials Processing Technology*, 240, 233–239.

[35] Rajan, A. J., Sugavaneswaran, M., Prashanthi, B., Deshmukh, S., & Jose, S. (2020). Influence of vapour smoothing process parameters on Fused Deposition Modelling parts surface roughness at different build orientation. *Materials Today: Proceedings*, 22, 2772–2778.

[36] Nguyen, T. K., & Lee, B.-K. (2018). Post-processing of FDM parts to improve surface and thermal properties. *Rapid Prototyping Journal*, 24(6), 1091–1100.

[37] Jiang, J., Stringer, J., Xu, X., & Zhong, R. Y. (2018). Investigation of printable threshold overhang angle in extrusion-based additive manufacturing for reducing support waste. *International Journal of Computer Integrated Manufacturing*, 31(10), 961–969.

[38] Wang, C., & Qian, X. (2020). Simultaneous optimization of build orientation and topology for additive manufacturing. *Additive Manufacturing*, 34, 101246. doi: 10.1016/j.addma.2020.101246

[39] Wang, Z., Zhang, Y., Tan, S., Ding, L., & Bernard, A. (2021). Support point determination for support structure design in additive manufacturing. *Additive Manufacturing*, 47, 102341. doi: 10.1016/j.addma.2021.102341

[40] Stratasys Inc. (2020). *Best Practices FDM Support Removal.* www.stratasys.com/

[41] Filament2Print. (2022). https://filament2print.com/gb/215-pla-premium

[42] AquaSys® 180 technical information (2022). https://infinitematerialsolutions.com/eu/en/support-materials/aquasys-180

[43] Chueca de Bruijn, A., Gómez-Gras, G., & Pérez, M. A. (2020). Mechanical study on the impact of an effective solvent support-removal methodology for FDM ultem 9085 parts. *Polymer Testing*, 85, 106433. doi: 10.1016/j.polymertesting.2020.106433

[44] Chueca de Bruijn, A., Gómez-Gras, G., & Pérez, M. A. (2022). Selective dissolution of polysulfone support material of fused filament fabricated Ultem 9085 parts. *Polymer Testing*, 108, 107495. doi: 10.1016/j.polymer.2022.107495

[45] Cailleaux, S., Sanchez-Ballester, N. M., Gueche, Y. A., Bataille, B., & Soulairol, I. (2021). Fused Deposition Modeling (FDM), the new asset for the production of tailored medicines. *Journal of Controlled Release*, 330, 821–841. doi: 10.1016/j.jconrel.2020.10.056

[46] Bellehumeur, C., Li, L., Sun, Q., & Gu, P. (2004). Modeling of bond formation between polymer filaments in the fused deposition modeling process. *Journal of Manufacturing Processes*, 6(2), 170–178. doi: 10.1016/s1526-6125(04)70071-7

[47] Lou, R., Li, H., Zhong, J., Zhang, C., & Fang, D. (2021). A transient updated Lagrangian finite element formulation for bond formation in fused deposition modeling process. *Journal of the Mechanics and Physics of Solids*, 152, 104450. doi: 10.1016/j.jmps.2021.104450

[48] Shelton, T. E., Willburn, Z. A., Hartsfield, C. R., Cobb, G. R., Cerri, J. T., & Kemnitz, R. A. (2020). Effects of thermal process parameters on mechanical interlayer strength for additively manufactured Ultem 9085. *Polymer Testing*, 81, 106255. doi: 10.1016/j.polymer.2019.106255

[49] Wang, P., Zou, B., & Ding, S. (2019). Modeling of surface roughness based on heat transfer considering diffusion among deposition filaments for FDM 3D printing heat resistant resin. *Applied Thermal Engineering*, 161, 114064. doi: 10.1016/j.applthermaleng.2019.114064

[50] Morales, N. G., Fleck, T. J., & Rhoads, J. F. (2018). The effect of interlayer cooling on the mechanical properties of components printed via fused deposition. *Additive Manufacturing*, 24, 243–248. doi: 10.1016/j.addma.2018.09.001

[51] Sreejith, P., Kannan, K., & Rajagopal, K. R. (2021). A thermodynamic framework for additive manufacturing, using amorphous polymers, capable of predicting residual stress, warpage and shrinkage. *International Journal of Engineering Science*, 159, 103412. doi: 10.1016/j.ijengsci.2020.103412

[52] Flügel, K., Hennig, R., & Thommes, M. (2020). Impact of structural relaxation on mechanical properties of amorphous polymers. *European Journal of Pharmaceutical Sciences*, 154, 214–221. http://dx.doi.org/10.1016/J.EJPB.2020.07.016

[53] Wach, R. A., Wolszczak, P., & Adamus-Wlodarczyk, A. (2018). Enhancement of mechanical properties of FDM-PLA parts via thermal annealing. *Macromolecular Materials and Engineering*, 303(9), 1800169. http://dx.doi.org/10.1002/MAME.201800169

[54] Hart, K. R., Dunn, R. M., Sietins, J. M., Hofmeister Mock, C. M., Mackay, M. E., & Wetzel, E. D. (2018). Increased fracture toughness of additively manufactured amorphous thermoplastics via thermal annealing. *Polymer*, 144, 192–204. http://dx.doi.org/10.1016/J.POLYMER.2018.04.024

[55] Padovano, E., Galfione, M., Concialdi, P., Lucco, G., & Badini, C. (2020). Mechanical and thermal behavior of Ultem® 9085 fabricated by Fused-Deposition Modeling. *Applied Sciences*, 10(9), 3170. http://dx.doi.org/10.3390/APP10093170

[56] Zhang, Y., Moon, S. K., Baino, F., Hussainova, I., & Riccio, A. (2021). The effect of annealing on additive manufactured ULTEM™ 9085 mechanical properties. *Materials*, 14(11), 2907. http://dx.doi.org/10.3390/MA14112907

[57] Chueca de Bruijn, A., Gómez-Gras, G., Fernández-Ruano, L., Farràs-Tasias, L., & Pérez, M. A. (2019). Optimization of a combined thermal annealing and isostatic pressing process for mechanical and surface enhancement of Ultem FDM parts using Doehlert experimental designs. *Journal of Manufacturing Processes*, 85(6) 1096–1115. https://doi.org/10.1016/j.jmapro.2019.06.006

[58] Boschetto, A., Bottini, L., & Veniali, F. (2016). Finishing of fused deposition modeling parts by CNC machining. *Robotics and Computer-Integrated Manufacturing*, 41, 92–101. https://doi.org/10.1016/j.rcim.2016.03.004

[59] Nsengimana, J., Van der Walt, J., Pei, E., & Miah, M. (2019). Effect of post-processing on the dimensional accuracy of small plastic additive manufactured parts. *Rapid Prototyping Journal*, 25(1), 1–12. https://doi.org/10.1108/RPJ-09-2016-0153

[60] Shiou, F.-J., & Chen, C.-H. (2003). Freeform surface finish of plastic injection mold by using ball-burnishing process. *Journal of Materials Processing Technology*, 140, 248–254.

[61] De Lacalle, L. N. L., Rodriguez, A., Lamikiz, A., Celaya, A., Alberdi, R., De Lacalle, L. N. L., & Mentxaka, A. L. (2011). Five-Axis Machining and Burnishing of Complex Parts for the Improvement of Surface Roughness. *Materials and Manufacturing Processes*, 26(8), 997–1003.

[62] Jerez-Mesa, R., Travieso-Rodriguez, J. A., Gomez-Gras, G., & Llumà-Fuentes, J. (2018). Development, Characterization and test of an ultrasonic vibration-assisted ball burnishing tool. *Journal of Materials Processing Technology*, 257, 203–212. doi:10.1016/j.jmatprotec.2018.02.036

[63] Gómez-Gras, G., Travieso-Rodríguez, J. A., González-Rojas, H. A., Nápoles-Alberro, A., Carrillo, F. J., & Dessein, G. (2015). Study of a ball-burnishing vibration-assisted process. *Proceedings of the Institution of Mechanical Engineers, Part B: Journal of Engineering Manufacture*, 229(1), 172–177. doi:10.1177/0954405414526383

[64] Jerez-Mesa, R., Gomez-Gras, G., & Travieso-Rodriguez, J. A. (2017). Surface roughness assessment after different strategy patterns of ultrasonic ball burnishing. *Procedia Manufacturing*, 13, 710–717. doi:10.1016/j.promfg.2017.09.116

[65] Travieso-Rodriguez, J. A., Gomez-Gras, G., Dessein, G., Carrillo, F., Alexis, J., Jorba-Peiro, J., & Aubazac, N. (2015). Effects of a ball-burnishing process assisted by vibrations in G10380 steel specimens. *International Journal of Advanced Manufacturing Technology*, 81(9–12), 1757–1765. doi:10.1007/s00170-015-7255-3

[66] Valerga Puerta, A. P., Lopez-Castro, J. D., Ojeda López, A., & Fernández Vidal, S. R. (2021). On improving the surface finish of 3D printing polylactic acid parts by

corundum blasting. *Rapid Prototyping Journal*, 27(7), 1398–1407. doi:10.1108/RPJ-05-2021-0105

[67] Mali, H. S., Prajwal, B., Gupta, D., & Kishan, J. (2018). Abrasive flow finishing of FDM printed parts using a sustainable media. *Rapid Prototyping Journal*, 24(4), 593–606. doi:10.1108/RPJ-10-2017-0199

[68] Huang, Z.-C., Zhou, Z.-J., & Jiang, Y.-Q. (2022). Effect of shot peening on static and fatigue properties of self-piercing riveting joints. *Journal of Materials Research and Technology*, 18, 1070–1080. doi:10.1016/j.jmrt.2022.03.031

[69] Hadidi, H., Boudaoud, H., Aguis, K., Bezzazi, B., & Tounsi, A. (2019). Low velocity impact of ABS after shot peening predefined layers during additive manufacturing. *Procedia Manufacturing*, 34, 594–602. doi: 10.1016/J.PROMFG.2019.06.169

[70] Swirad, S., & Wdowik, R. (2019). Determining the effect of ball burnishing parameters on surface roughness using the Taguchi method. *Procedia Manufacturing*, 34, 287–292. Elsevier B.V. https://doi.org/10.1016/j.promfg.2019.06.152

[71] Saldana-Robles, A., Diosdado De La Pena, J. A., Balvantín-García, A. D. J., Aguilera Gómez, E., Plascencia-Mora, H., & Saldana-Robles, N. (2017). Ball burnishing process: State of the art of a technology in development. *Dyna*, 92, 28–33. https://doi.org/10.6036/7916

[72] Gómez-Gras, G., Travieso-Rodríguez, J. A., Jerez-Mesa, R., Llumà-Fuentes, J., & Gomis de la Calle, B. (2016). Experimental study of lateral pass width in conventional and vibrations-assisted ball burnishing. *International Journal of Advanced Manufacturing Technology*, 87, 363–371. https://doi.org/10.1007/s00170-016-8490-y

[73] Núñez, P., Rivas, A., García-Plaza, E., Beamud, E., & Sanz-Lobera, A. (2015). Dimensional and surface texture characterization in fused deposition modelling (FDM) with ABS plus. *Procedia Engineering*, 132, 856–863. MESIC Manufacturing Engineering Society International Conference 2015. doi: 10.1016/j.proeng.2015.12.227

[74] Gómez-Gras, G., Pérez, M. A., Fábregas-Moreno, J., & Reyes-Pozo, G. (2021). Experimental study on the accuracy and surface quality of printed versus machined holes in PEI ULTEM 9085 FDM specimens. *Rapid Prototyping Journal*, 27(11), 1–12. https://doi.org/10.1108/RPJ-12-2019-0306

[75] Zaldívar, R. J., Muniz-Lerma, J. A., Rodríguez-Ortiz, J. A., & Rodríguez-Castro, R. (2017). Influence of processing and orientation print effects on the mechanical and thermal behavior of 3D-Printed ULTEM™ 9085 Material. *Additive Manufacturing*, 13, 71–80. doi: 10.1016/j.addma.2016.11.007

[76] Byberg, K. I., Gebisa, A. W., & Lemu, H. G. (2018). Mechanical properties of ULTEM 9085 material processed by fused deposition modeling. *Polymer Testing*, 72, 335–347. doi: 10.1016/j.polymertesting.2018.10.040

[77] Roberson, D. A., Lopez-Anido, R. A., Joo, Y. L., & Wicker, R. B. (2015). Comparison of stress concentrator fabrication for 3D printed polymeric izod impact test specimens. *Additive Manufacturing*, 7, 1–11. doi:10.1016/J.ADDMA.2015.05.002

[78] Dialami, N., Rivet, I., Cervera, M., & Chiumenti, M. (2022). Computational characterization of polymeric materials 3D-printed via fused filament fabrication. *Mechanics of Advanced Materials and Structures. Advance online publication*, 2023(30), 1357–1367. doi:10.1080/15376494.2022.2032496

[79] Forés-Garriga, A., Gómez-Gras, G., & Pérez, M. A. (2022). Mechanical performance of additively manufactured lightweight cellular solids: Influence of cell pattern and relative density on the printing time and compression behavior. *Materials & Design*, 215, 110474. doi:10.1016/J.MATDES.2022.110474

[80] Zhang, Y., Phil Choi, J., & Moon, S. K. (2022). A data-driven framework to predict fused filament fabrication part properties using surrogate models and multi-objective optimisation. *International Journal of Advanced Manufacturing Technology*, 120, 8275–8291. doi:10.1007/S00170-022-09291-0

[81] Chohan, J. S., & Singh, R. (2017). Pre and post processing techniques to improve surface characteristics of FDM parts: A state of art review and future applications. *Rapid Prototyping Journal*, 23(3), 495–513. doi:10.1108/RPJ-05-2015-0059

[82] Jatin, V., Sudarkodi, S., & Basu, S. (2014). Investigations into the origins of plastic flow and strain hardening in amorphous glassy polymers. *International Journal of Plasticity*, 56, 139–155. doi:10.1016/j.ijplas.2013.11.007

5 Coating Methods and Materials for 3D Printed (FDM) Parts

Sateeshkumar Kanakannavar, Kiran Shahapurkar, Gangadhar M. Kanaginahal, Rayappa Shrinivas Mahale and Prashant P. Kakkamari

5.1 INTRODUCTION

3D Printing is an additive manufacturing process which develops the required shapes with a built-up manner. It means that the product is developed as a thin layer without the preform [1]. In recent times, 3D printing technology has advanced, in first-rate element due to PC generation of 3D structures and management of the printing approaches [2]. In particular, increasing attention has been paid to fused deposition modeling (FDM), for polymer designing and prototyping the components not only inside the academic field but moreover within the manufacture of commercial components [3]. FDM was produced in the late 1980s and the first 3D printed product commercialization took place in the early 1990s by the USA based company called Stratasys Limited [4]. Compared to traditional polymer processes like, extrusion, injection molding, compression molding, etc., FDM has several benefits in the manufacture of customized products and parts complex geometries, which is also characterized by low waste emissions. The advancement of the FDM process has derived benefits from the conventional extrusion welding technique, evident from their resemblances such as the deposition and fusion between a highly viscous or nearly solid lower layer and the subsequent layer or welding filler strand, which undergoes complete melting during the joining procedure [3], [5], [6].

Recently, materials that are more complex and versatile have become more popular for 3D printing. These materials include nano sized materials, different functionality materials, biomaterials, Smart (Shape Memory / Piezo-electric) materials, and fast-drying concrete [7]. This means that the technology needs to be able to reliably produce parts that meet the needs of real applications, such as being sure the materials used are effective and having controls in place to reduce human error [8]–[10]. The three main factors that affect whether a product is considered good are how it performs its intended functions, what it looks like, and how it feels to use.

DOI: 10.1201/9781032665351-6

Post-processing refers to all the procedures that are conducted after the 3D printer has detached the components from the printing bed. Post-processing can be categorized into two types: primary and secondary. Primary post-processing comprises essential steps that are obligatory for any 3D-printed part to be functional in a specific application. On the other hand, secondary post-processing encompasses discretionary steps that have the potential to enhance the properties, performance, or appearance of the part. In general, carrier removal and cleaning are the primary post-processing activities; secondary activities include painting and steaming (vapor smoothing) based on the requirements of the user. Ex situ physical or chemical techniques, such as the addition of a new layer of material, can be used to achieve wetting, gloss, resistance to scratch, chemical resistance, thermal and electromagnetic shielding and so on [11].

The model and any supports are taken out during post-processing, and the model's surface is then finished using various techniques. Modern FDM materials enable the manufacture of actual parts that are durable and robust enough for installation, functional testing, and/or prototyping.

3D-printed products can be employed in a multitude of ways, exposing them to diverse environmental conditions. As a result, there is a necessity for surface treatment with coatings that provide safeguarding and the intended aesthetic appeal of 3D-printed objects, especially in demanding environments with high relative humidity and exposure to ultraviolet light irradiation [12]. A detailed understanding of surface qualities, in particular surface chemistry and surface wettability is necessary to create an appropriate surface system made up of polymeric surface and coating assets since these features directly affect the adhesion of coatings to substrates. Only with the right coatings is the surface treatment of polymers possible due to unique surface qualities. In fact, these coatings frequently comprise solvents and a variety of resins, including acrylic, alkyd, polyurethane, polyvinylacetate, melamine, and polyester [13]. In addition, the application of coatings on 3D-printed objects can be accomplished through various technologies, contingent on the characteristics and attributes of the coating, as well as the geometry and dimensions of the object [14]. A summary of some coating of 3D printed parts with metallic and polymer materials and techniques are shown in the Table 5.1.

5.2 METAL COATING ON 3D PRINTED POLYMER PRODUCTS

Metallizing objects augments the electrical conductivity of their surfaces, which has the potential to enhance their electron transfer properties.

There are several different types of metal coating that can be applied to 3D printed polymer products to give them a metallic finish or improve their durability and conductivity. Here are some of the common types:

1. Electroplating
2. Vacuum deposition
3. Spraying
4. Plating with conductive inks
5. Dip coating

TABLE 5.1
Metal Coating methodologies

Coating process	Coating Material	Materials used for printing	References
Electroless deposition	Gold	Acrylonitrile Butadiene Styrene (ABS)	[15]
Aerosol paints	nickel	Polylactic acid (PLA)	[16]
Aerosol paints	silver	Polylactic acid (PLA)	[16]
Sputtering	titanium	Polyetheretherketone (PEEK),	[17]
Chemical Coating	XTC 3D coating Layer	Acrylonitrile Butadiene Styrene (ABS).	[18]
Dip Coating	hydrophobic silica nanoparticles and methyl ethyl ketone (MEK)	polylactic acid (PLA)	[19]
Casting	graphene nanoflakes (GNF)	poly(vinyl alcohol) (PVA)	[20]
Manually applied	solvent borne alkyd coating	poly(lactic acid) and poly(lactic acid) with wood flour	[13]
Manually applied	water borne acrylic coating	poly(lactic acid) and poly(lactic acid) with wood flour	[13]
Manually applied	acrylonitrile-butadiene-styrene diluted in acetone	poly(lactic acid) and poly(lactic acid) with wood flour	[13]

5.2.1 ELECTROPLATING

This involves the use of an electrical current to deposit a layer of metal applied over the surface of the 3D printed polymer product. Electroplating is a technique employed for the metallization of objects wherein an electric current is administered at the junction of a designated substrate and an electrolytic solution housing metallic cations intended for deposition.

In the electrochemical cell, the object is connected as the cathode, distinct from the aforementioned methods. Electroplating possesses the unique capability of selectively depositing metals exclusively onto conductive surfaces. The thickness of the resulting metallic film can be controlled by manipulating various experimental parameters, including the concentration of dissolved metallic cations and additives, applied voltage or current density, and deposition time. By adjusting these variables, one can regulate the depth of the deposited metallic layer. Electroplating can be used to coat objects with a range of metals, including copper, nickel, gold, and silver [21].

5.2.1.1 Electroplating Process

Electroplating is often used to enhance the appearance and durability of metallic surfaces, and it can also be used to add a metallic coating to non-metallic surfaces like

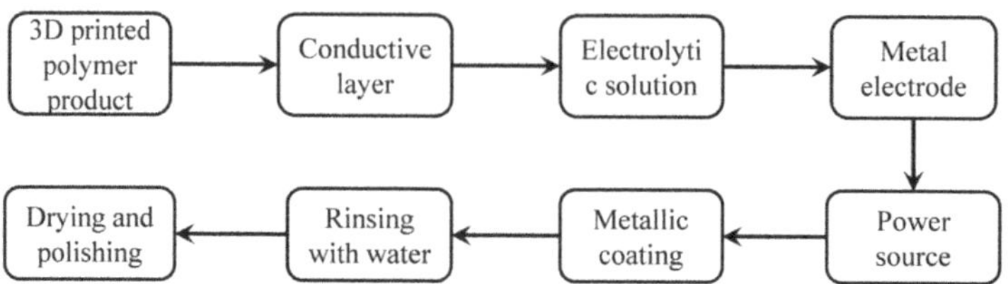

FIGURE 5.1 Flow diagram of electroplating process for 3D polymer products.

3D printed polymer products. Here is a step-by-step explanation of the electroplating process on 3D printed polymer products in Figure 5.1:

1. Prepare the 3D printed polymer product: The first step is to clean and prepare the surface of the 3D printed polymer product. This involves removing any dust, dirt, or oils that may be present on the surface. The product is then rinsed with water and dried.
2. Apply a conductive layer: A conductive layer is applied to the surface of the product. This layer can be made of materials like graphite, silver, or copper. The purpose of this layer is to make the surface conductive so that it can be electroplated.
3. Immersion in electrolytic solution: The product is then immersed in an electrolytic solution that contains ions of the metal that will be plated onto the surface. The electrolytic solution is typically a solution of the metal salt dissolved in water.
4. Electrolysis: Subsequently, the product is linked to the negative terminal of a power source while a metal electrode is coupled to the positive terminal. Upon activation of the power supply, an electric current passes through the solution, inducing the attraction of metal ions to the product's surface, where they precipitate and generate a layer of metallic coating.
5. Rinse and dry: Once the electroplating process reaches completion, the resultant product is extracted from the electrolytic solution and subjected to a thorough rinsing with water. It is then dried and polished to give it a smooth and shiny finish.

Kim et al. [22] explored the coating of copper by electrodeposition on the 3D printed polymer (Electrifi, Black Magic, and Proto-Pasta) products as shown in Figure 5.2. The process was performed using a two-electrode system, with a copper foil as the deploying electrode and the printed object as the working electrode. Before the electro-deposition, the printed object was cleaned and made more wettable by immersing it in ethanol and sonication for 60 seconds.

The copper electro-deposition was performed using an aqueous electrolyte containing 1.0 M Copper sulphate ($CuSO_4$), 0.5 M Sulphuric acid (H_2SO_4), and 1 mM Sodium Chloride (NaCl). To examine the impact of organic additions on copper electro-deposition, the electrolyte was mixed with 100 µM polyethylene glycol (PEG), 20 µM sodium 3-mercapto-1-propanesulfonate (MPSA), and 50 µM janus

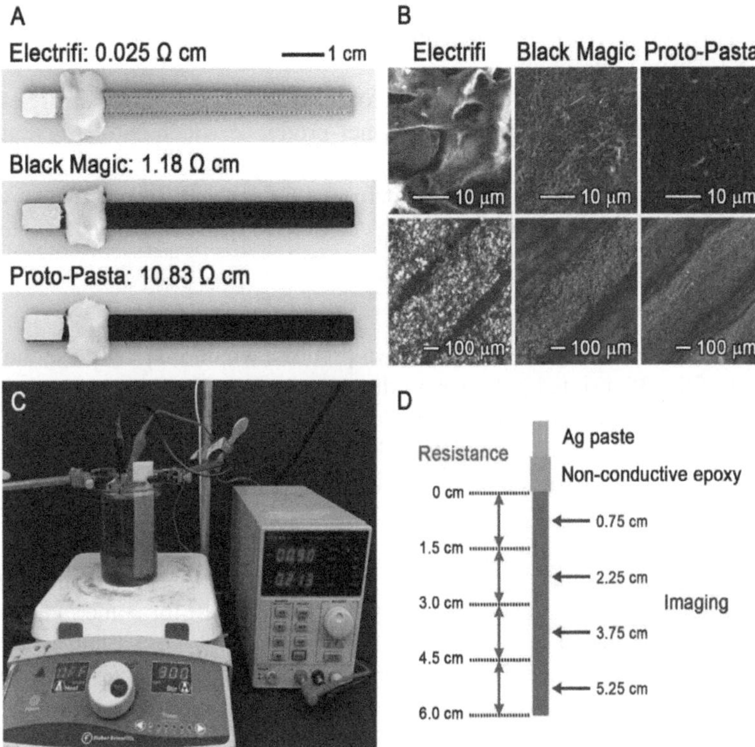

FIGURE 5.2 Experimental set-up with an input to measure voltage and output for display with a two-electrode system for copper electro-deposition [22].

green B (JGB). A DC power supply was used to apply a steady current during copper electrodeposition, and 50 mA/cm^2 current density was set for the printed traces.

To electrodeposit the printed coil, a step current deposition process was used, which involved applying 11.3 mA/cm^2 for 20 minutes and then 22.6 mA/cm^2 for 160 minutes. In order to carry out targeted copper electrodeposition within the horn antenna, a strategic approach was employed whereby the external surface of the horn fabricated by 3D printing was coated with an adhesive substance commonly known as super glue. Without the use of organic additives, copper electrodeposition was carried out for 120 minutes at 25 mA/cm^2. The electrolyte was stirred with a magnetic stirrer at a rotational speed of 300 rpm. The copper deposition rate was calculated to be 1.1 μm/min at a current density of 50 mA/cm^2, with a current efficiency of 100%.

Advantages:

- Enhanced Duability: Electroplating can provide a durable and long-lasting surface for 3D printed products, which may otherwise be prone to wear and tear over time.

- Improved Aesthetics: Electroplating can give a metallic finish to the 3D printed polymer products, thus enhancing their external appearance and add value to the 3D printed products.
- Better Electrical Conductivity: Electoplated metals can improve the electrical conductivity of the 3D printed polymer products, which can be beneficial for certain applications.
- Enhanced Corrosion resistance: Electroplating can add a protective layer to the 3D printed polymer products, which can help to resist corrosion and rust.

Disadvantages:

- Cost: Electroplating can be expensive process, which may increase the overall cost of the 3D printed polymer products.
- Limited Material Compatibility: Electroplating is typically limited to certain materials that can withstand the electrolyte bath and the electrical current, which may limit the choices of 3D printed polymer materials that can be electroplated.
- Complexity: Electroplating can be a complex process that requires specialized equipment and expertise, which may make it difficult for some manufacturers to implement and expertise, which may make it difficult for some manufacturers to implement in-house.
- Dimensional Changes: Electroplating can add a layer of metal that may cause dimensional changes to the 3D printed polymer products, which can affect their final shape and size.

Overall, electroplating can offer several benefits to 3D printed polymer products, but it may not be suitable for every application due to its limitations and potential drawbacks.

5.2.2 Electroless Coating

A uniform thick layer of coating on the surfaces of non-conductive substrates was achieved by disintegrating the metallic ions present in the electrolyte which was in the liquid form. This method is commonly referred to as electroless plating. The application of this technique has been demonstrated in various articles, wherein several metallic materials such as nickel, copper, platinum and so on were coated onto 3D-printed parts. The researchers have highlighted that these parts, with functionalized surfaces, hold potential for implementation in a range of fields including chemical and mechanical engineering, as well as electronics such as Micro Electronic Mechanical System [23], [24]. Through electroless deposition, [23] palladium was plated onto 3D-printed photopolymers. The authors created a cylinder with cubic lattices and printed cubes with logpile lattices. The tests made use of the PR48/PR57 and an Autodesk Ember 3D printer. The cubes had 200 m pores and measured 5 mm along its edges. The cubic frame was made up of 150 m-wide pores, and the cylinders had dimensions of 200 mm in length and 2 mm in diameter. One hour was spent plating with gentle stirring. The substrate was air dried after being cleaned with deionized water. A photopolymer cube is shown in Figure 5.3a both before and after plating (on the left). Figure 5.3b depicts the cross-sectional view of the plated cube printed by SLA method. Other thermoplastic materials printed using FDM showed similar strength during adhesion test, brightness of film, and rate of plating. Comparatively speaking to FDM-printed parts, particularly polycarbonate, photopolymers were plated internally which was easily accomplished [23]. The results of this study indicate that the combination of 3D printing and a simple electroless deposition technique can be employed to effectively fabricate 3D metal-polymer composite parts. Essentially, this involves coating on a polymer surface with a functional material. This could lead to the creation of useful 3D parts that make use of mechanical, electrical, thermal, and catalytic capabilities.

Advantages:

- Improved conductivity: Electroless plating can improve the electrical conductivity of 3D printed polymer products, making them suitable for use in electrical applications.
- Enhanced strength: Electroless plating can increase the strength and durability of 3D printed polymer products by adding a layer of metal to the surface.
- Better aesthetics: Electroless plating can give 3D printed polymer products a more aesthetically pleasing look by adding a metallic finish to the surface.
- Corrosion resistance: Electroless plating can improve the corrosion resistance of 3D printed polymer products by providing a protective barrier against environmental factors.

Disadvantages:

- High cost: Electroless plating can be an expensive process, especially when compared to other coating methods, which may not be feasible for some applications.
- Limited thickness: Electroless plating typically results in a thin layer of metal deposition, which may not be sufficient for some applications.
- Limited material compatibility: Electroless plating may not be compatible with all types of polymers, and certain pre-treatment steps may be required to ensure proper adhesion.
- Process complexity: Electroless plating is a complex process that requires careful control of the plating solution, temperature, and other variables to achieve consistent results.

Electroless plating can be a useful process for improving the properties of 3D printed polymer products. However, it is important to consider the potential advantages and disadvantages before deciding to use this method.

5.2.3 COLD SPRAYING

The CS technology is a method of thermal spray coating that uses micron-sized particles (measuring 10–100 μm in diameter) to coat a surface. It is called "cold" because it can maintain the temperature of the powders used during the deposition process below their melting point. As a result, when the particles hit the substrate at a high velocity, they deform and bond together to form the coating without melting. The coating then grows as more particles are deposited [25], [26]. Coaxial spraying (CS) technology is particularly advantageous for coating temperature-sensitive materials such as polymer matrix composites or plastics, as it prevents excessive thermal degradation, as supported by the existing literature. The process involves propelling metallic and/or non-metallic powders through a converging-diverging de-Laval nozzle using a gas stream, commonly air, helium, or argon, which is expanded to supersonic speeds. Typically, the impact velocities of the particles range from 300 to 1200 m s−1. Upon exiting the nozzle, the particles experience substantial plastic deformation and adhere to the surface of the target substrate [27]. The investigation of using cold spray technology to deposit metal particles on composites or polymeric surfaces is still in its early stages. Initial findings have demonstrated that the quality of metallization on polymer matrix composites (PMCs) is greatly influenced by the process parameters of CS, such as the type of gas used, gas pressure and temperature, standoff distance, and powder type. Furthermore, recent research on this topic has indicated that the substrate type and stratification are critical factors for achieving effective metallization of PMCs, which in turn depend on the manufacturing processes of the PMCs [28].

The cold spray system developed at the University of Science and Technology Beijing is presented in the Figure 5.4.

The cold spray process utilizes an assortment of equipment, including a high-pressure gas regulation unit, a gas warm up unit, a high-pressure fine particles depositor (UNIQUECOAT, model TSR-PF50), and a spray gun. The spray gun comprises a gas pre-chamber and a convergent-divergent hastening nozzle, featuring a throat diameter of 3.8 mm and an exit diameter of 7.8 mm. Axially, the powder is introduced into the nozzle upstream from the back of the gun, while the hastening gas is passed into the pre-chamber. Nitrogen serves the dual purpose of being both the process and fine particles carrier gas, and the nozzle inlet pressure can reach a maximum of 2.1 MPa. To ensure process stability, a mass flow meter is employed in the gas control and powder feeding units. The gas heating unit is equipped with two 24 kW high-power heaters arranged in series, utilizing a spiral-tube type heat exchange mode, and capable of attaining temperatures up to 600°C in the pre-chamber. Temperature and pressure sensors are incorporated in the spray gun to mitigate fluctuations in gas heating temperature and impact velocity resulting from variations in gas flow and pressure. These sensors ensure the stability of the gas and powder parameters escaping from the spray gun as shown in Figure 5.4 [29].

Advantages:

- Cold spraying provides excellent adhesion of coatings to 3D printed polymer products, resulting in a strong and durable coating.
- It is a low-temperature process, which means that the polymer substrate will not be subjected to high temperatures that could cause warping or melting.

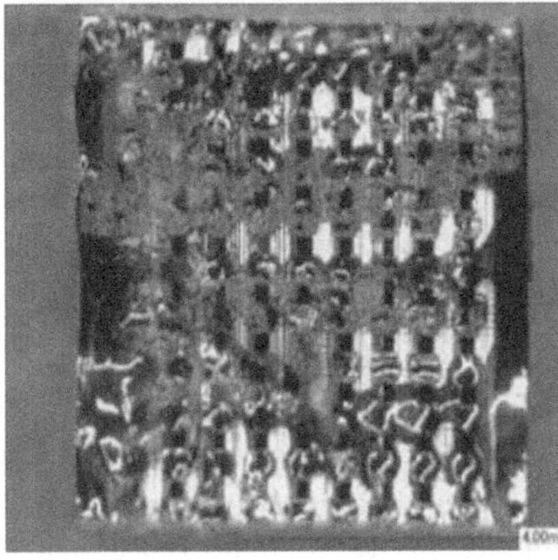

FIGURE 5.3 (a) SLA prints of plated (right) and unplated (left) cubes made from photopolymer resin. (b) An image of the cube's cross-section after plating [23].

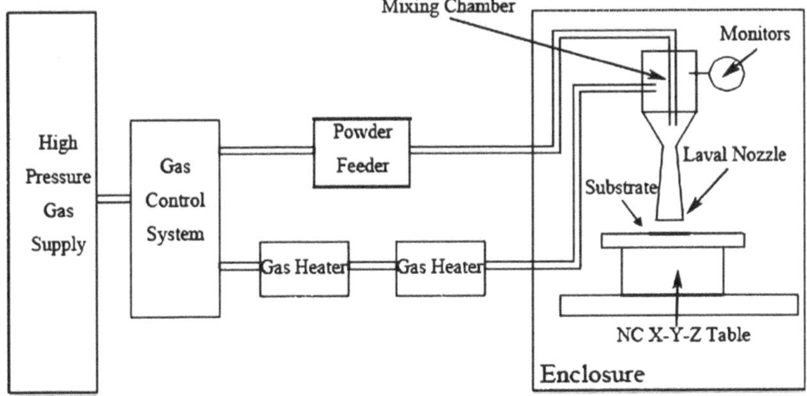

FIGURE 5.4 Schematic diagram of cold spray (CS) system inclusive of gas supply system combined with powder feeder transferred to a chamber where the coating occurs [29].

- Cold spraying is a relatively fast process, allowing for quick production of coated 3D printed polymer products.
- It is a versatile process, allowing for the deposition of various coating materials, including metals, ceramics, and polymers.
- Cold spraying offers good control over the thickness of the coating, allowing for precise coating thickness control.

Disadvantages:

- The equipment required for cold spraying is expensive, making it a costly process for small-scale production.
- The process can be noisy and requires high-pressure equipment, which may make it unsuitable for some environments.
- Coating quality basically depends on the substrate's surface roughness, so additional surface preparation may be required before cold spraying.
- Cold spraying is not suitable for coating large 3D printed polymer products due to limitations in the size of the equipment.
- The process may not be suitable for all types of polymers, and some polymers may not be compatible with the coating materials used in cold spraying.

5.2.4   Dip Coating

This involves dipping the 3D printed Polymer product into a bath of liquid metal or metal alloy. The metal then adheres to the surface of the object to form a metallic coating. This technique is often used for coating small or intricate parts. Dip-coating, a technique widely employed in diverse industrial domains, is a cost-effective and straightforward method for coating a variety of substrates. This procedure involves the application of coating in liquid form onto the substrate surface. The solution typically contains the target materials that are directly coated onto the substrate. After the wet coating is applied, it is allowed to sediment, and the solvent is then evaporated to produce a dry film. To guarantee complete infiltration of the substrate, this technique entails submerging the material into the solution and subsequently withdrawing it from the tank. It is important to acknowledge that despite its apparent simplicity, the cold spray process encompasses intricate multi-variable parameters. These parameters encompass factors such as immersion time, withdrawal speed, dip-coating cycles, density and viscosity of the solutions, surface tension, characteristics of the substrate surface, and evaporation conditions. Each of these variables plays a crucial role in determining the thickness and morphology of the deposited thin films [30].

The dip-coating process is popular due to its scalability in the fabrication process and high nanoparticle uptake per coating cycle due to non-covalent interactions between nanoparticles and polymers. However, it has limitations such as susceptibility to degradation of electrical properties due to weak bonding between nanoparticles and the substrate, leading to reduced sensing performance and limited shelf life, especially in harsh environments. In order to address these limitations, it is imperative to conduct additional research aimed at enhancing the integration of nanoparticles with the underlying flexible substrates during the coating process. This research would focus on developing improved techniques and methodologies to ensure a more efficient and effective bonding between the nanoparticles and the flexible substrates. By enhancing this integration, the overall quality and performance of the coated films can be significantly improved, opening up new possibilities for various applications [31].

Lee et al. [19] fabricated the super hydrophobic surface using 3D printing and dip coating processes as shown in Figure 9.5. The researchers employed polylactic acid (PLA) filament material as the substrate for fabricating intricate surface patterns. PLA, a biodegradable thermoplastic, served as the foundation for the desired structures. The sample preparation is schematically summarized in Figure 5.5.

Each of these methods has its own advantages and limitations, and the best method to use will depend on the specific application and desired properties of the finished product.

Here are some advantages and disadvantages of metallic dip coating on 3D printed polymer products:

Advantages:

- It is a simple and cost-effective method for adding metallic coatings to 3D printed polymer products.
- It offers high coverage and uniform deposition of the metal layer on the surface of the polymer product.
- It can improve the mechanical, thermal, and electrical properties of the polymer product by adding a metal layer.
- It can enhance the aesthetic appeal of the product by providing a metallic finish or color.
- It can protect the polymer product from environmental factors such as corrosion and abrasion.

Disadvantages:

- It may not be suitable for complex geometries and surfaces that are hard to coat due to limited accessibility.
- It may require post-coating processing such as heat treatment to improve adhesion and bonding between the metal layer and the polymer substrate.
- It may increase the weight and alter the original dimensions of the 3D printed product, which can affect its performance and functionality.
- It may not be a sustainable or environmentally-friendly process due to the use of hazardous chemicals in the coating solution and waste generation.

5.3 POLYMER BASED COATINGS

Surface coatings were used in the past to address material processing issues, such as protecting additively manufactured aluminium and titanium parts from corrosion and wear, smoothing surface roughness, changing or eliminating porosity, in

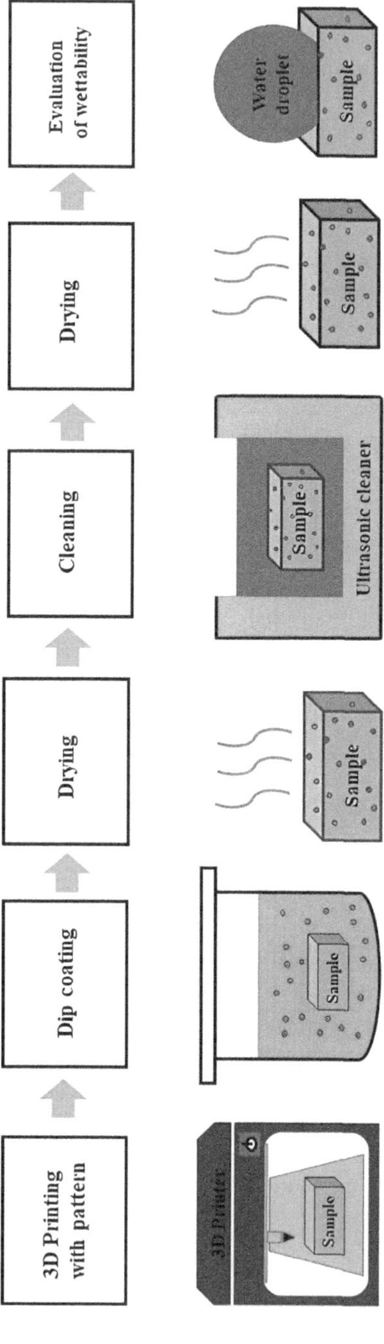

FIGURE 5.5 Schematic diagram of sample preparation where the coating of sample happens in a tub filled with coating material and testing for wettability [19].

handmade products such as bags and biomedical applications. Post-processing in 3D printing has been divided into primary and secondary categories. Coating can be manual or automatic, and coating polymers like polyurethane elastomer and liquid silicone can improve mechanical properties. Direct polymer coating can enhance adhesion of 3D printing materials onto textiles. Dip coating can serve as coatings based on food safety and sealants to prevent bacterial accumulation and particle migration.

5.3.1 Chemical Vapor Deposition (CVD)

Chemical vapor deposition (CVD) a post-processing method can modify the surface properties of 3D-printed polymers, enabling independent control over their bulk and surface properties. A study investigated the effects of filament and substrate temperature on the coating of 3D-printed parts using hydrophilic and hydrophobic polymers through initiated chemical vapor deposition (iCVD). The results showed that iCVD was able to produce functionalized 3D-printed parts with desirable hydrophilic or hydrophobic properties, which can have applications in microfluidics and tissue scaffolds. In addition, the iCVD technique can be used to modify different surfaces of 3D-printed components, as demonstrated in experiments involving ABS parts coated with P(HEMA-co-EGDA) and PPFDA [11].

5.3.2 UV Coating

BioMEMS refer to devices that integrate biological and mechanical components at the microscale level, with a diverse range of applications including drug delivery, biosensors, and tissue engineering. The 3D printing technology enables the fabrication of intricate and precise structures in BioMEMS, such as microfluidic channels that regulate fluid flow in devices. These channels can imitate the intricate network of blood vessels in the human body, facilitating the creation of biological system models. Moreover, 3D printing in BioMEMS can be employed to produce customized scaffolds for tissue engineering, tailored to match the size and shape of specific tissues, and can be integrated with various cell types and growth factors [32]. A study aimed at investigating the biocompatibility of various resins for 3D printing structures for cell culture applications. The researchers used different types of resin and printers and subjected the chips to various post-processing treatments, such as sonication rinse, UV post-print curing, thermal post-baking, and autoclaving as shown in Figure 5.6. They also coated the chips with various materials commonly used in BioMEMS and microfluidic devices and evaluated their effects on biocompatibility. The chips were pasted on one side of a wall of a sterile 48-well cell culture plate using biocompatible epoxy and coated with a solution of fibronectin and gelatin to facilitate the cell adhesion. The cells were added to the wells and cultured until they reached confluency. The cells were then counted, and an Adenosine Triphosphate (ATP) assay was performed to assess cell viability. The study found that the FormLabs Clear Resin was suitable material for 3D printing structures for cell culture applications when properly post-processed and coated. The combination of Isopropyl Alcohol (IPA) washing, UV curing, and autoclaving yielded the highest

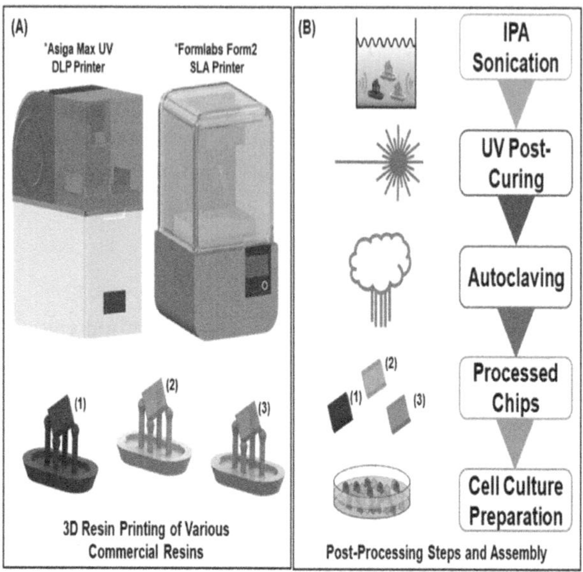

FIGURE 5.6 3D printing of polymers were dip coated followed by UV laser curing over specific areas for cell culture preparation [33].

cell culture viability of 99.31% ± 0.17%, and Polydimethylsiloxane (PDMS) was found to be the most effective coating for enhancing biocompatibility. The study also confirmed that the biocompatibility results were due to the resin chips themselves and not any interactions with the surrounding environment [33].

5.3.3 DIP COATING

Stereolithography, a 3D printing process that uses photopolymer resin and UV light to create an object layer by layer. A build platform will be submerged in a vat of liquid photopolymer, and a laser focuses UV light on specific areas of the photopolymer to cure the surface and create each layer. The Form 1+ desktop 3D printer from Formlabs was used to produce test samples. The parts were strategically placed near the hinge on the build platform to minimize detachment during printing. After printing, the parts underwent a 30-minute post-curing process in a Dymax Light Curing System to ensure they were completely dry and ready for coating. The parts were then coated with Plasti Dip, an air-dry, specialty rubber coating typically used for cosmetic purposes or forming a rubberized handle on tools and appliances, and left to dry for 30 minutes. The results obtained from the profilometer should consider the test part design and choice of dip-coating solvent, which can affect the accuracy of the results. The 75 degree face of the test part is the lowest point, causing a visible bulge and increased calculated waviness due to drainage, while the 15 and 30 degree faces may not be fully dipped, resulting in an effective scan length of roughly 10000 pm. To mitigate the effect of the drainage pattern causing significant waviness, a less viscous fluid as the dip-coating solvent may be helpful, but this could negatively affect surface roughness due to decreased adhesion. Withdrawal speed is an important parameter in the dip-coating process as it affects the amount of coating on the part. A faster withdrawal speed leads to more coating, but there is a tradeoff with feature resolution. The study found that a withdrawal speed of 5 mm/s showed the most consistent positive improvement in surface quality, while a slow withdrawal speed of 0.1 mm/s resulted in poor feature resolution and a rougher surface. The impact of print resolution on the dip-coated surface was also evaluated, with lower resolutions reducing preparation time but potentially negatively impacting surface quality. The study found that printing at 0.2 mm resolution had the shortest preparation time, while maintaining surface quality when dip-coated at a withdrawal speed of 1 mm/s [34].

5.3.4 Water-Based Coatings

Water-based coatings have been a type of coating material predominantly composed of water and other soluble constituents, including pigments, binders, and additives. These coatings have become a popular substitute for solvent-based coatings that contain volatile organic compounds (VOCs), presenting health and environmental hazards. Water-based coatings offer an array of advantages in various industries. They provide exceptional adhesion, durability, and resistance to chemical reactions, abrasion, and weathering. Furthermore, water-based coatings maintain excellent color retention, available in a broad spectrum of colors and finishes such as matte, gloss, and satin. They can be applied using different methods, including spraying, brushing, or rolling. Moreover, water-based coatings have a shorter drying time, minimizing production time and enhancing productivity [35]. A work on water-based polyurethane coatings (PU1 and PU2) were utilized in improving the smoothness and appearance of 3D printed parts made of acrylonitrile butadiene styrene (ABS). In this study, rectangular 3D printed ABS parts were utilized and subjected to coatings using dip coater which was handled by a computer. The fabrication process of the printed components involved the utilization of three distinct print tips and two orientations. In the case of parts printed with a side orientation, the dip coating was administered in a manner that aligned with the direction of the printed layers. Conversely, for parts printed with a vertical orientation, the dip coating was applied at a perpendicular angle to the printed layers. Through this investigation, the study revealed that the application methods and conditions of the coating could modify the surface morphology of 3D components, effectively smoothing out any surface corrugations present. Dilution of coating solution decreases viscosity while maintaining its shear-thinning behavior. Single-layer coated samples show similar degree of planarization (DOP) values, with a slight increase observed at higher coating speeds. The addition of a second coating layer further increases DOP, but no further improvements in surface smoothness are seen beyond two layers. The drying treatment significantly affects planarization, resulting in higher standard deviations and thickness variation in coated samples. By choosing a print tip that produces thicker layers and applying one or two coats of coating, one can obtain a similar surface quality to non-coated parts printed with a thinner print tip, thus decreasing printing time while potentially introducing functional features or color. Dip-coating parts printed vertically reduces the variability in measured Arithmetic Average Roughness Height (ARH) values among top, middle, and bottom positions, in contrast to parts printed in a side orientation. This suggests that there is less flow after deposition [36].

5.3.5 POLYDOPAMINE (PDA) COATINGS

Bone defects arising from conditions like osteoporosis, bone malignant tumors, and traumatic injuries are ubiquitous, but their clinical treatment has often been restricted by factors such as the defect's extent and location, the patient's age and health, and the unavailability of appropriate treatment options. However, bone tissue engineering method employs biomaterials like scaffolds, growth factors, and cells to promote new bone tissue growth and facilitate bone regeneration, harnessing the body's natural healing processes. Bone tissue engineering offers a more customized and precise approach to treating bone defects. Secondly, it minimizes the risk of complications and infections linked to invasive surgical procedures. Thirdly, it enhances the speed and efficacy of bone tissue regeneration, resulting in faster healing times and better patient outcomes [37]. In this investigation, a composite scaffold was synthesized by dissolving Poly(lactic-co-glycolic acid) (PLGA) in Dimethylformamide (DMF) and incorporating Tricalcium phosphate (β-TCP) particles and calcium phosphate powder to produce a paste as shown in Figure 5.7. The paste was then 3D printed to generate scaffolds with uniform pore sizes that were subsequently coated with PDA, leading to highly roughened surfaces and improved wettability. The scaffolds had a highly roughened surface with homogenous pores measuring approximately 500 µm in diameter, and micro-CT imaging revealed a honeycomb-like structure that facilitated the ingrowth of bone tissue. Scanning Electron Microscope (SEM) analysis demonstrated that PDA-coated scaffolds had a highly roughened structural layer, while uncoated scaffolds had pores dispersed throughout the internal walls. PDA coatings enhanced surface wettability, promoting the attachment and proliferation of stem cells. The scaffolds exhibited high porosity (~60%) and low density, emulating natural bone structure. The pore sizes ranged from 449–544 µm, and the PDA1 and PDA2 scaffolds contained interconnected open pores, similar to the uncoated scaffolds. The PDA-coated scaffolds retained an interconnected porous structure with an average pore size of ~500 µm, similar to that of the uncoated ones. PDA surface modification did not significantly affect the mechanical properties of the coated scaffolds. The PDA2 scaffold group demonstrated superior adhesive abilities and the highest cell seeding efficiency compared to the PDA0 and PDA1 groups. The stable interconnected pores of the scaffold withstood larger loads, endowing it with superior lightweight properties that were unaffected by coating thickness. The results of cell tests demonstrated that higher PDA coating concentrations improved cell adhesion, infiltration, and proliferation due to increased hydrophilicity. The PDA2 group exhibited significantly higher Alkaline Phosphatase (ALP) activity than the other groups, suggesting that high drug concentrations were beneficial for osteogenesis [38].

The CAD software converted 3D data into STL format, which was universal for 3D printing software. The STL data was imported into Flashprint, where printing

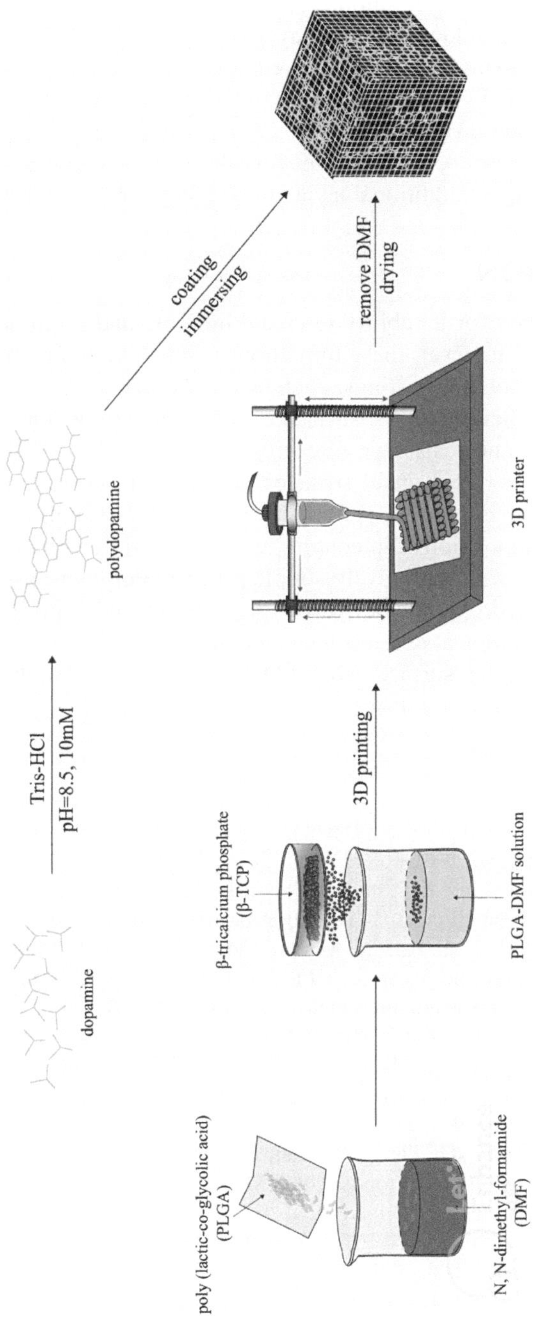

FIGURE 5.7 Scaffold preparation by dopamine and polydopamine i.e. coated on dry β-TCP scaffold [38].

parameters were set based on default machine settings. The printer had a build size of 230 x 150 x 140 mm and a precision of ±0.2 mm. Acrylonitrile Butadiene Styrene (ABS) was used to fabricate the samples, supplied as filament with a 1.75-mm diameter. Samples B and C underwent one layer of coating before a second layer was applied to Sample C. All samples were cured at room temperature. Sample A had an uneven surface, potentially due to the "stairstepping" effect. Sample B's surface quality was enhanced with one layer of XTC-3D coating, while Sample C performed better with two layers. The cross-sectional image revealed that XTC-3D coating reduced the "stairstepping" effect, but additional layers may affect part dimensions [39].

5.4　CONCLUSION

3D printing, renowned for its ability to create intricate and multifaceted structures, is not without flaws. However, these limitations can be overcome through the application of coatings. Metallic coatings, such as electroplating, utilize electric current to deposit organic materials on the surface of 3D printed parts, enhancing their durability, conductivity, and resistance to corrosion. Electroless coatings, on the other hand, can deposit functional metal surfaces on polymer substrates with high resolution. Cold spraying is another method that can achieve highly precise coatings with a high rate of deposition. Dip coating, while limited by certain parameters, can achieve thin coatings with a relatively simple process. Polymer-based coatings offer the ability to create hydrophobic or hydrophilic surfaces on 3D printed parts, and can be integrated with biological and mechanical components. Water-based coatings can help to achieve a smoother surface, while PDA coatings can aid in the removal of surface pores and improve wettability.

REFERENCES

[1]　K. Kun, "Reconstruction and Development of a 3D Printer Using FDM Technology" *Procedia Eng.*, vol. 149, pp. 203–211, 2016, doi: 10.1016/j.proeng.2016.06.657.

[2]　M. R. Hartings and Z. Ahmed, "Chemistry from 3D Printed Objects" *Nat. Rev. Chem.*, vol. 3, no. 5, pp. 305–314, 2019, doi: 10.1038/s41570-019-0097-z.

[3]　L. Lin, N. Ecke, M. Huang, X. Pei, and A. K. Schlarb, "Impact of Nanosilica on The Friction and Wear of a PEEK/CF Composite Coating Manufactured by Fused Deposition Modeling (FDM)" *Compos. Part B*, vol. 177, pp. 1–10, 2019, doi: 10.1016/j.compositesb.2019.107428.

[4]　B. Satyanarayana and K. Jaya, "Component Replication using 3D Printing Technology" *Procedia Mater. Sci.*, vol. 10, pp. 263–269, 2015, doi: 10.1016/j.mspro.2015.06.049.

[5]　X. Lu, C. Zhang, G. Zhao, Y. Guan, L. Chen, and A. Gao, "State-of-the-art of Extrusion Welding and Proposal of a Method to Evaluate Quantitatively Welding Quality During Three-Dimensional Extrusion Process" *Mater. Des.*, vol. 89, pp. 737–748, 2016, doi: 10.1016/j.matdes.2015.10.033.

[6]　P. Michel, "An Analysis of the Extrusion Welding Process" *Polym. Eng. Sci.*, vol. 29, no. 19, pp. 1376–1381, 1989, doi: https://doi.org/10.1002/pen.760291908.

[7]　J. Lee, J. An, and C. K. Chua, "Fundamentals and Applications of 3D Printing for Novel Materials" *Appl. Mater. Today*, vol. 7, pp. 120–133, 2017, doi: 10.1016/j.apmt.2017.02.004.

[8]   Q. Chen .,J. D. Mangadlao, J. Wallat, A. D. Leon, J. K. Pokorski and R. C. Advincula, "3D Printing Biocompatible Polyurethane / Poly (lactic acid)/ Graphene Oxide Nanocomposites: Anisotropic Properties" *ACS Appl. Mater. Interfaces*, vol. 9, no. 4, pp. 4015–4023, 2016, doi: 10.1021/acsami.6b11793.

[9]   N. Palaganas, J. Mangadlao, A. C. De Leon, K. Pangilinan, Y. J. Lee, and R. Advincula, "3D Printing of Photocurable Cellulose Nanocrystal Composite for Fabrication of Complex Architectures via Stereolithography" *ACS Appl. Mater. Interfaces*, vol. 9, no. 39, pp. 34314–34324, 2017, doi: 10.1021/acsami.7b09223.

[10] Q. Chen, P. Cao, and R. C. Advincula, "Mechanically Robust, Ultraelastic Hierarchical Foam with Tunable Properties via 3D Printing" *Adv. Funct. Mater.*, vol. 28, no. 21, pp. 1–18, 2018, doi: 10.1002/adfm.201800631.

[11] J. R. C. Dizon, C. C. L. Gache, H. M. S. Cascolan, L. T. Cancino, and R. C. Advincula, "Post-Processing of 3D-Printed Polymers" *Technologies*, vol. 9, no. 3, pp. 1–61, 2021, doi: 10.3390/technologies9030061.

[12] M. Reza, A. Zolfagharian, M. Jennings, and T. Reinicke, "Structural Performance of 3D-Printed Composites Under Various Loads and Environmental Conditions" *Polym. Test.*, vol. 91, no. July, pp. 1–11, 2020, doi: 10.1016/j.polymertest ing.2020.106770.

[13] J. Zigon, M. Kariz, and M. Pavlic, "Surface Finishing of 3D-Printed Polymers with Selected Coatings" *Polymers (Basel).*, vol. 12, no. 12, pp. 1–19, 2020, doi: 10.3390/ polym12122797.

[14] M. Lanzetta and E. Sachs, "Improved Surface Finish in 3D Printing Using Bimodal Powder Distribution" *Rapid Prototyp. J.*, vol. 9, no. 3, pp. 157–166, 2003, doi: 10.1108/13552540310477463.

[15] D. Shani, A. Inberg, D. Ashkenazi, Shacham-Diamand, and A. Stern, "Gold Plating ON AM-FDM ABS Components" *Weld. Equip. Technol.*, vol. 30, pp. 43–50, 2019, doi: 10.35219/awet.2019.06.

[16] M. E. Carkaci and M. Secmen, "Design and Prototype Manufacturing of a Feed System for Ku-band Satellite Communication by Using 3D FDM / PLA Printing and Conductive Paint Technolog" *Int. J. RF Microw. Comput. Eng.*, vol. 30, no. 4, pp. 1–15, 2019, doi: 10.1002/mmce.22062.

[17] H.-D. Jung, T.-S. Jang, J. E. Lee, S. J. Park, Y. Son, and S.-H. Park, "Enhanced Bioactivity of Titanium-Coated Polyetheretherketone Implants Created by a High-Temperature 3D Printing Process" *Biofabrication*, vol. 11, no. 4, pp. 1–14, 2019, doi: 10.1088/1758-5090/ab376b.

[18] A. H. M. Haidiezul, A. F. Aiman, and B. Bakar, "Surface Finish Effects Using Coating Method on 3D Printing (FDM) Parts" *IOP Conf. Ser. Mater. Sci. Eng.*, 2018, pp. 1–12, doi: 10.1088/1757-899X/318/1/012065.

[19] K. Lee, H. Park, J. Kim, and D. Chun, "Fabrication of a Superhydrophobic Surface Using a Fused Deposition Modeling (FDM) 3D Printer with Poly Lactic Acid (PLA) Filament and Dip Coating with Silica Nanoparticles" *Appl. Surf. Sci.*, vol. 467–468, no. October 2018, pp. 979–991, 2019, doi: 10.1016/j.apsusc.2018.10.205.

[20] X. Li ., M. Mahdi Honari, Y. Fu, A. Kumar, H. Saghlatoon, P. Mousavi, and H. Chung, "Self-Reinforcing Graphene Coatings on 3D Printed Elastomers for Flexible Radio Frequency Antennas and Strain Sensors" *Flex. Print. Electron.*, vol. 2, no. 3, pp. 1–15, 2017, doi: 10.1088/2058-8585/aa73c9.

[21] E. Vaněčková .,M. Bousa, R. Sokolova, P. Garcia, P. Broekmann, V. Shestivska, J. Rathousky, M. Gal, T. Sebechlebska, and V. Kolivoska, "Copper Electroplating of 3D Printed Composite Electrodes" *J. Electroanal. Chem.*, vol. 858, pp. 1–11, 2020.

[22] M. J. Kim ., M. A. Cruz, S. Ye, A. L. Gray, G. L. Smith, and N. Lazarus, "One-Step Electrodeposition of Copper on Conductive 3D Printed Objects" *Addit. Manuf.*, vol. 27, no. March, pp. 318–326, 2019, doi: 10.1016/j.addma.2019.03.016.

[23] C. G. Jones, B. E. Mills, R. K. Nishimoto, and D. B. Robinson, "Electroless Deposition of Palladium on Macroscopic 3D-Printed Polymers with Dense Microlattice Architectures for Development of Multifunctional Composite Materials" *J. Electrochem. Soc.*, vol. 164, no. 13, 2017, doi: 10.1149/2.1341713jes.

[24] R. Bernasconi, C. Credi, M. Tironi, M. Levi, and L. Magagnin, "Electroless Metallization of Stereolithographic Photocurable Resins for 3D Printing of Functional Microdevices" *Electrochem. Soc.*, vol. 164, no. 5, pp. 1–9, 2017, doi: 10.1149/ 2.0081705jes.

[25] H. Che, P. Vo, and S. Yue, "Metallization of Carbon Fibre Reinforced Polymers by Cold Spray" *Surf. Coat. Technol.*, vol. 313, p. pp.236–247, 2017, doi: 10.1016/ j.surfcoat.2017.01.083.

[26] H. Che, P. Vo, and S. Yue, "Investigation of Cold Spray on Polymers by Single Particle Impact Experiments" *J. Therm. Spray Technol.*, vol. 28, no. 1, pp. 135–143, 2019, doi: 10.1007/s11666-018-0801-4.

[27] H. Assadi, F. Ga, T. Stoltenhoff, and H. Kreye, "Bonding Mechanism in Cold Gas Spraying" *Acta Mater.*, vol. 51, no. 15, pp. 4379–4394, 2003, doi: 10.1016/ S1359-6454(03)00274-X.

[28] A. Viscusi .,R. D. Gatta, F. Delloro, I. Papa, A. S. Perna, and A. Astarita, "A Novel Manufacturing Route for Integrated 3D-Printed Composites and Cold-Sprayed Metallic Layer" *Mater. Manuf. Process.*, vol. 37, no. 5, pp. 568–581, 2021, doi: 10.1080/10426914.2021.1942908.

[29] X. L. Zhou, A. F. Chen, J. C. Liu, X. K. Wu, and J. S. Zhang, "Surface & Coatings Technology Preparation of Metallic Coatings on Polymer Matrix Composites by Cold Spray" *Surf. Coatings Technol.*, vol. 206, no. 1, pp. 132–136, 2011, doi: 10.1016/ j.surfcoat.2011.07.005.

[30] X. Tang and X. Yan, "Dip-coating for Fibrous Materials: Mechanism, Methods and Applications" *J. Sol-Gel Sci. Technol.*, vol. 81, pp. 378–404, 2017, doi: 10.1007/ s10971-016-4197-7.

[31] E. Davoodi., H. Montazerian, R. Haghniaz, S. Ahadian, A. Sheikhi, J. Chen, A. Khademhosseini, A. S. Milani, M. Hoorfar, and E. Toyserkani, "3D-Printed Ultra-Robust Surface-Doped Porous Silicone Sensors for Wearable Biomonitoring" *ACS Nano*, vol. 14, no. 2, pp. 1520–1532, 2020, doi: 10.1021/acsnano.9b06283.

[32] D. Pranzo, P. Larizza, D. Filippini, and G. Percoco, "Extrusion-Based 3D Printing of Microfluidic Devices for Chemical and Biomedical Applications: A Topical Review" *Micromachines*, vol. 9, no. 8, 2018, doi: 10.3390/mi9080374.

[33] C. Hart, C. M. Didier, F. Sommerhage, and S. Rajaraman, "Biocompatibility of Blank, Post-Processed and Coated 3D Printed Resin Structures with Electrogenic Cells" *Biosensors*, vol. 10, no. 11, pp. 1–14, 2020, doi: 10.3390/ BIOS10110152.

[34] S. Suet-Ning Lu, "Improving Surface Quality of SLA 3D Printed Parts Via Controlled Dip-Coating" 2018, Massachusetts Institute of Technology, USA, https://dspace.mit. edu/handle/1721.1/119948.

[35] J. Ho, B. Mudraboyina, C. Spence-Elder, R. Resendes, M. F. Cunningham, and P. G. Jessop, "Water-Borne Coatings That Share The Mechanism of Action of Oil-Based Coatings" *Green Chem.*, vol. 20, no. 8, pp. 1899–1905, 2018, doi: 10.1039/ c8gc00130h.

[36] J. Zhu, J. L. Chen, R. K. Lade, W. J. Suszynski, and L. F. Francis, "Water-Based Coatings for 3D Printed Parts" *J. Coatings Technol. Res.*, vol. 12, no. 5, pp. 889–897, 2015, doi: 10.1007/s11998-015-9710-3.

[37] A. R. Amini, C. T. Laurencin, and S. P. Nukavarapu, "Bone Tissue Engineering: Recent Advances and Challenges" *Crit. Rev. Biomed. Eng.*, vol. 40, no. 5, pp. 363–408, 2012, doi: 10.1615/CritRevBiomedEng.v40.i5.10.

[38] Z. Xu, N. Wang, P. Liu, Y. Sun, Y. Wang, F. Fei, S. Zhang, J. Zheng, and B. Han "Poly(dopamine) Coating on 3D-Printed Poly-Lactic-co-Glycolic Acid/β-Tricalcium Phosphate Scaffolds for Bone Tissue Engineering" *Molecules*, vol. 24, no. 23, pp. 4397–4412, 2019, doi: 10.3390/molecules24234397.

[39] A. H. M. Haidiezul, A. F. Aiman, and B. Bakar, "Surface Finish Effects Using Coating Method on 3D Printing (FDM) Part" *IOP Conf. Ser. Mater. Sci. Eng.*, vol. 318, no. 1, pp. 1–9, 2018, doi: 10.1088/1757-899X/318/1/012065.

Section II

Joining/Welding as Post-Processing Techniques for FDM 3D Printed Parts

6 Insights into Various Joining/Welding Techniques to Evade the Built Volume Constraint of FDM-3D Printers

Vivek Kumar Tiwary, Arunkumar Padmakumar and Vinayak R. Malik

6.1 INTRODUCTION

The manufacturing sector has been completely transformed by three-dimensional (3D) printing, which makes it simple to create complicated and unique things [1], [2]. As contrast to subtractive and formative manufacturing processes, the American Society for Testing of Materials (ASTM) describes 3D printing as a process of joining materials to build things from a 3D model data, generally layer upon layer. The main advantages of this technique over traditional manufacturing consist of its low buy to fly ratio, short lead time, design freedom, customized product manufacturing, less material depletion, and substantially lower energy requirements [3], [4]. Additionally, benefits like design sharing, manufacturing outsourcing, simplicity of design, and adaptability can also be foreseen, making 3D printing an unavoidable technology for the future [5], [6]. However, along with the aforementioned advantages and benefits, there are also a fair number of restrictions and downsides that the technology faces. An attempt to depict the shortcomings of the widely used 3D printing technique has been made in the Figure 6.1 below.

3D printing can be a time-consuming process, especially for complex and detailed models. It often takes several hours or even days to complete a print, depending on the size, complexity, and resolution of the object. Faster printing techniques are being developed, but currently, it remains a limitation. 3D printers, although being able to work with an extensive variety of materials, encompassing food, ceramics, polymers and metals, there are still limitations on the types of materials that can be used. Further, while the cost of 3D printers has decreased over time, high-quality 3D

DOI: 10.1201/9781032665351-8

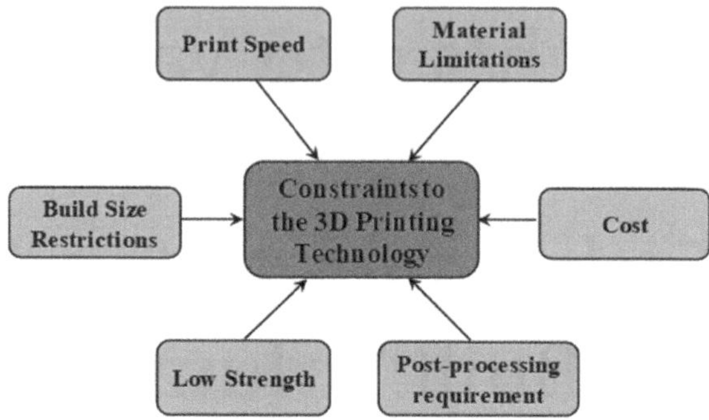

FIGURE 6.1   Challenges in 3D printing

printers with advanced capabilities can still be expensive. To attain the appropriate finish, many 3D-printed products need to be post-processed. This can include additional steps like sanding, spraying, polishing or applying coatings. Post-processing can be time-consuming and may require manual effort to achieve the desired result. In addition, numerous researchers have noted that 3D printed components have lower mechanical strength than injection or compression molded ones, which prevents them from meeting practical demands [7], [8].

Among the many shortcomings mentioned above, one significant restriction that is sometimes overlooked is its restricted build capacity, which prevents it from promptly printing a part larger than the bed size [9], [10]. The instability, distortions, tolerances, and price of the 3D printed component are all observed to increase as the size of the object to be printed grows [11]. Researchers claim that an easy solution to this problem would be to print the component in several pieces and later join them to obtain the desired part. Out of the problems mentioned above, joining techniques for 3D printed parts have sadly lagged behind other concerns in this technology that urgently require attention [12], [13].

A survey of the literature conducted in this regard found that although joining of plastics has been extensively investigated and documented, there is a noticeable lack of attention paid to the actual practices for joining or welding 3D printed thermoplastics. Considering this fact, in the current chapter, a review of various techniques that can be employed to join/weld 3D printed parts, surmounting the bed size limitation is presented. The study is also crucial because printing smaller pieces and then connecting them together may also help to solve additional complications, such as anisotropy-related inferior mechanical strength as well as warpage-related dimensional inaccuracies.

6.2   PROPOSED TECHNIQUES FOR JOINING/WELDING FDM-3D PRINTED PARTS

There are three generic categories that can be used to categorize joining/welding procedures for FDM-3D printed parts: primitive joining techniques, friction based welding techniques, and the most recent newer joining techniques. Adhesive bonding,

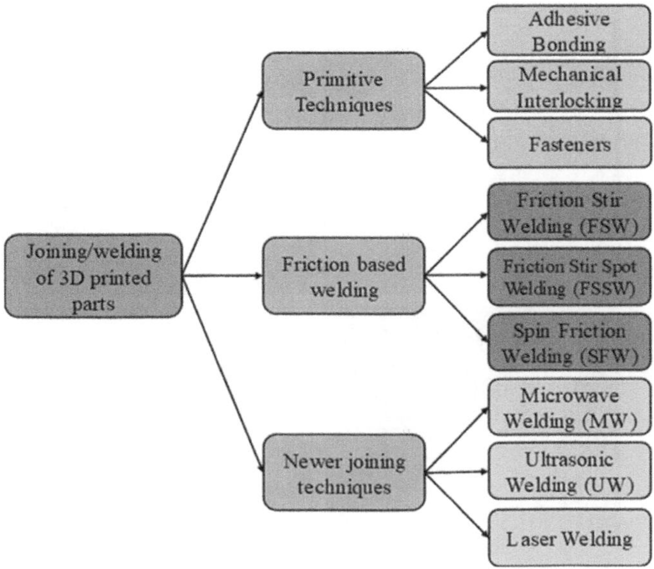

FIGURE 6.2 Techniques for joining/welding 3D printed components

mechanical interlocking, and the usage of fasteners are examples of primitive techniques. Friction welding techniques encompasses Friction Stir Welding (FSW), Friction Stir Spot Welding (FSSW) and Spin Friction Welding (SFW) while the newer joining techniques includes techniques like Microwave, laser and Ultrasonic welding. Figure 6.2 below depicts the general way of classifying the joining/welding techniques for 3D printed components. In the following segments, these practices, their application procedures, advantages and limitations are discussed briefly.

6.3 PRIMITIVE TECHNIQUES

6.3.1 Bonding 3D Printed Parts by Adhesives

Adhesive bonding is a regularly used technique for joining 3D printed parts together. It involves using an adhesive material to create a bond between the surfaces of the parts, resulting in a strong and durable connection [14], [15]. The design freedom presented by the 3D printing technique combined with adhesive bonding seems to be extremely encouraging to surmount the volume limitation of a 3D printer.

This is the most prevalent method employed by the 3D printing community to achieve a larger volume component. In comparison to other conventional techniques, this practice provides a constant spreading of load along the bond line. Moreover, the bonded parts have no regions where the stress is concentrated and can be employed to join even dissimilar parts also.

The method's benefits include improved electrical and thermal insulation at the adhesive layer, excellent damping characteristics, low weight, tightness, and corrosion resistance [16], [17]. While, necessity of preparing the surface before

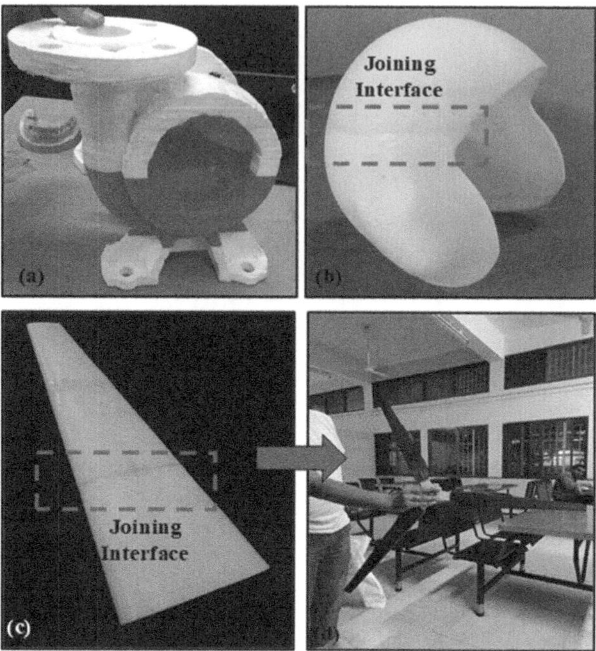

FIGURE 6.3 A few case studies of adhesively bonded FDM components (a) A pump (b) A helmet (c) An wing turbine wing (d) After assembly

applying adhesives, elastic mismatch when dissimilar polymers are joined, adhesives blocking the holes and difficulty in testing adhesively bonded components without destructing them are the limitations of this technique [18], [19], [13].

Adhesively bonded components also contribute to environmental pollution, hence they are avoided in medical applications. Further, climatic conditions like moisture, humidity, and temperature can affect the joint, even leading to its degradation. Figure 6.3 depicts a few examples which are 3D printed and joined by adhesives.

6.3.2 Usage of Fasteners

Research in joining 3D printed parts by means of fasteners is fueled by two basic reasons: firstly, because of the complications in joining dissimilar 3D printed polymers and secondly, due to the limits of primitive techniques in joining efficiently complex geometries [12]. This method doesn't require any surface preparation, unlike the adhesive bonding method. It is simple to inspect the joints as well as easy to disassemble [20]. Mechanical fastening of 3D printed parts, in comparison to the welding techniques, doesn't display any thermal degradation that generally happens from too much heating [21].

The main issue with using mechanical fasteners, however, is that it adds weight to the component. Around the fastener holes, stresses begin to form, lowering the strength and possibly leading to corrosion-related issues [22], [23]. The joining process

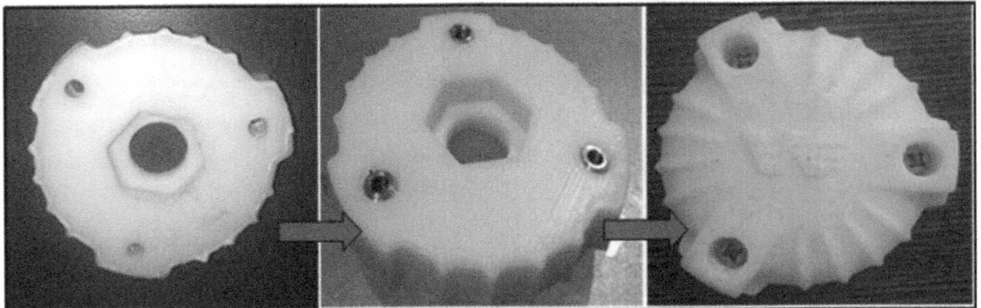

FIGURE 6.4 Image showing the usage of fasteners in 3D printing

appears to be laborious at times, involving extra fixtures and steps like heating and screwing the fasteners into the holes, which weaken and damage the structure [16], [19]. A fact that around 20–40% of the total aircraft cost in a 3D printed composite aircraft structure is for mechanical fasteners, increases the requirement of effective optimization of this joining technique reducing the overall manufacturing costs [24]. Figure 6.4 below shows an example of using fasteners to assemble a final 3D printed component.

6.3.3 MECHANICAL INTERLOCKING

Joints created using mechanical fasteners are not recommended every time for 3D printed components as it results in a huge amount of stress concentration, decreasing the performance [4]. Another approach to overcome this matter and also evade the bed size restriction of FDM-3D printer could be to employ interlocking of the 3D printed parts. Figure 6.5 depicts the principle of mechanical interlocking.

In contrast to mechanical fastening, this method achieves connections by their inherent geometry in the absence of any male/female connectors. This method's benefits include the simplicity of the connections, ease of assembling and dismantling the printed pieces, and cost-effective maintenance for storage and transportation [25], [26]. The 3D interlocked parts have demonstrated to be robust (without any hole drilling and protrusion), reinforced by the inter-parts blockage with their geometry enabling to produce large 3D components with a clean and superior surface finish. However, development of a systematic framework prior to execution is the matter that complicates the issue. To overcome this, researchers have attempted to develop several common frameworks making this technique a lot easier to deploy. Literature review revealed that a number of frameworks like "Chopper", "Packmerger", "Burr Puzzles" has been developed and put to use by numerous researchers [9], [27], [28].

6.4 FRICTION BASED WELDING TECHNIQUES

6.4.1 FRICTION STIR WELDING (FSW)

FSW is a solid-state joining method that may join a variety of materials, including metals. Traditionally, FSW has been employed for welding conventional, wrought

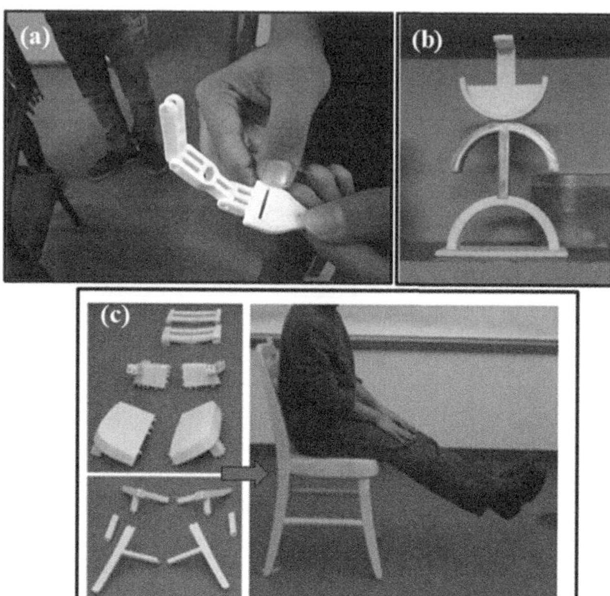

FIGURE 6.5 Mechanically interlocked 3D printed parts (a) A robotic arm (b) Speaker holder (c) A chair

materials. However, advancements in 3D printing have led to the exploration of FSW for joining 3D printed parts as well [29], [30].

Due to the variations in material structure, thermal conductivity, and melting temperatures, FSW of polymers is considered to be different from that of metals and alloys. Owing to the low thermal conductivity of thermoplastics, in the case of metallic FSW the tool is to be cooled externally, while in the case of polymer FSW the tool has to be heated externally [31]. Ease of automation, capability to weld major thermoplastics, no consumables required and lesser assembling costs are the advantages of this process [32], [33]. From the ecological viewpoint, there is no radiation, material wastages or dangerous gas emissions [34], [35].

The method is limited by the thermoplastic's low thermal conductivity (<0.5 W/m-k), which prevents melting and subsequent joining from occurring correctly. For the reason that the thermo-physical properties are different, joining dissimilar thermoplastics has proven to be an even more difficult operation. Further, FSW can be conducted only for butt weld/linear joint configurations. Figure 6.6 represents the FSW procedure being carried out for 3D printed thermoplastics.

6.4.2 FRICTION STIR SPOT WELDING (FSSW)

FSSW an important variation of FSW, was industrialized by Kawasaki in the year 2003 for assembling the back doors and hood of RX-8, MX-5 sports cars. Promising to be a prospective replacement for Resistance Spot Welding (RSW) and Self-Piercing

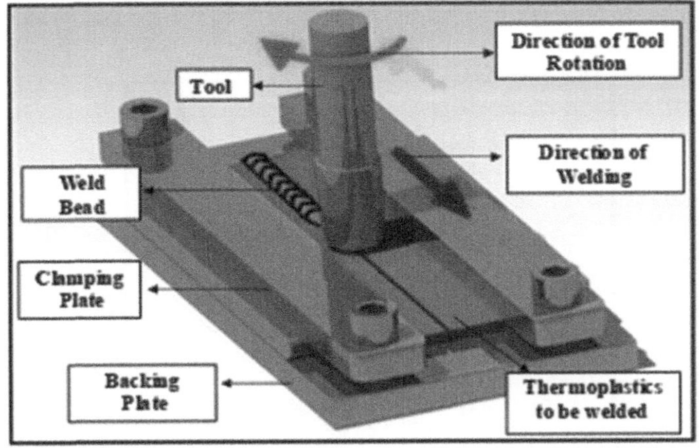

FIGURE 6.6   The FSW process for 3D printed parts

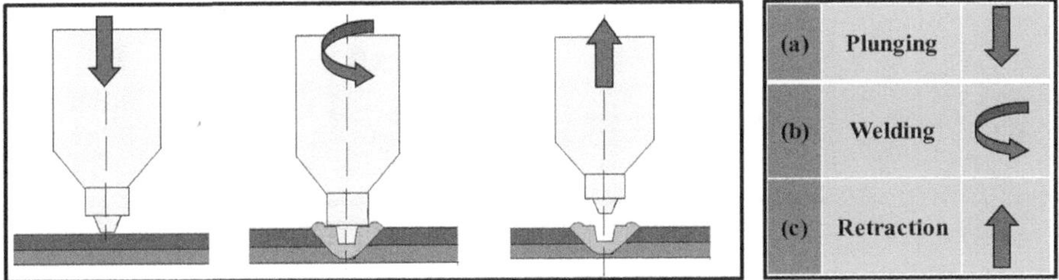

FIGURE 6.7   FSSW process for 3D printed parts

Rivets (SPR), the technology is gaining interest from automobile, airplane, off-road equipment manufacturers and even consumer electronics [36], [37].

The FSSW method is carried out in four stages namely rotation, plunging, stirring and retraction resulting in a solid-state weld. FSSW overcomes some of the intrinsic complications observed in other methods, including higher fasteners cost; need for subsequent operations and longer downtime. Some more advantages like the simplicity of joining materials with small distortion, outstanding mechanical properties and less costs makes this technique one of the most competent ones [38], [39]. The technique's limits, on the other hand, appear to be the creation of a permanent exit hole in the joined pieces as well as its capacity to be used only for lap joint configurations. Figure 6.7 shows the technique of FSSW being conducted on overlapped 3D printed plates.

The majority of the open source publications discovered are concerned with the FSSW of conventionally manufactured thermoplastics aiming to increase the joint strength by optimizing the process parameter, tool geometry or by developing a new tool. However, the same results cannot be extrapolated for a 3D printed part, due to its layer-by-layer addition approach which usually results in a less dense, hollow

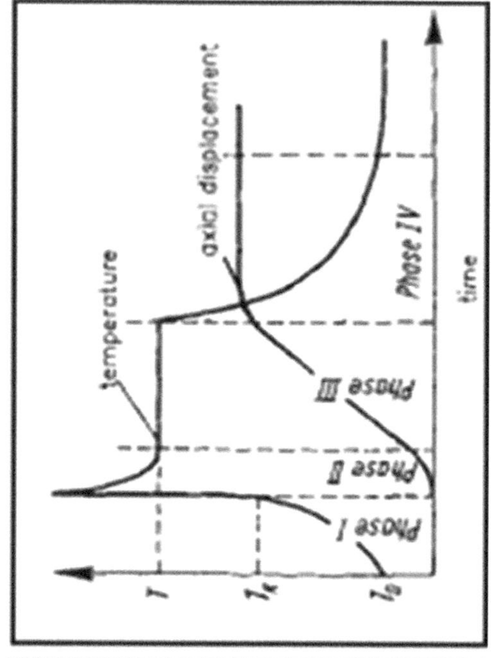

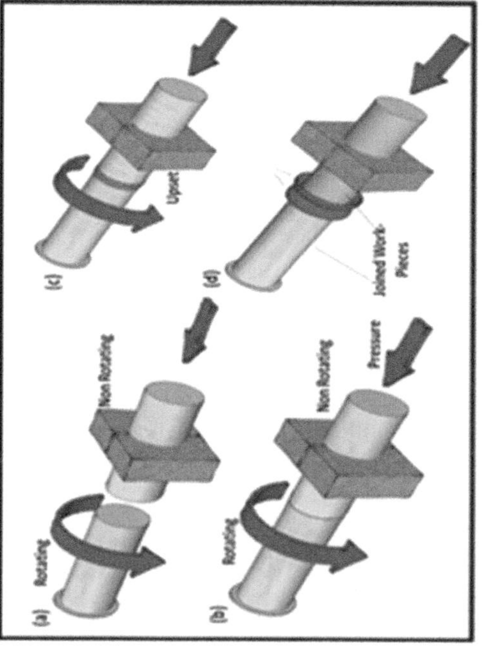

FIGURE 6.8 (a) SFW process for 3D printed thermoplastics [45] (b) Heat development during the phases [46]

and lighter parts, with a bad surface quality. Further, FSSW joining technique for 3D printed parts with the aim of surmounting the issue of volume limitation is also unclear as well as unexplored.

6.4.3 Spin Friction Welding (SFW)

Spin friction welding (SFW) is yet another modified technique of FSW, a solid-state joining process that can be put into use to join 3D printed parts. It is a friction welding technique that involves rotating one-part relative to another while applying axial pressure. The frictional heat produced between the parts softens the polymers and allows for the formation of a solid-state bond. The method is designed basically for joining cylindrical parts, comprising of forcing a stationary, non-rotating tubular component against a rotating cylindrical part.

Advantages of this joining process are easy equipment design, quick cycle times, low costs, the ability to weld different polymers, the lack of consumables, and environment friendliness [20], [40]. However, one major drawback of the method is that it cannot be used to weld components that cannot be rotated [41], [42].

The SFW process consists of four stages. During the first phase the material is clamped in a chuck and rotated at a predetermined speed while the second material is clamped on the tailstock and is stationary. The work pieces are held axis-symmetrically to each other. No pressure is applied during the first stage. However, heat is generated by friction. During the second phase, the stationary part is moved against the rotating part to a predetermined depth (Penetration depth). During this phase, a linear and rotary motion happens. During the third phase, the rotation is stopped. The linear motion is continued till the complete fixed penetration is achieved. Finally, during the fourth phase, the joint parts are allowed to solidify completing the weld. Figure 6.8a depicts the four phases of the SFW process while Figure 6.8b depicts the subsequent heat that develops during these stages [43], [44].

6.5 NEW JOINING TECHNIQUES

6.5.1 Microwave Welding

Microwave radiation (MW) is electromagnetic radiation with wavelengths ranging from 1 mm to 1 m and a frequency range spanning from 300 MHz to 300 GHz. Though it first emerged for communication, the method is now commonly used for preparing food, drying materials, processing raw materials, vulcanizing rubber, and plastic welding [47], [48]. Figure 6.9 illustrates the process of Microwave Welding.

MW welding is a volumetric based heating process, which results in consistent temperature increase all the way through at the same time, displaying vast potential for welding intricate joint geometries with superior qualities. Further, the affordable availability of implant materials, short process time and no surface preparation are additional advantages [49], [50]. The only insignificant difficulty of this method is the requirement for a suitable absorbent at the joint interface which many times heats up swiftly and deteriorates if the process factors are improperly set.

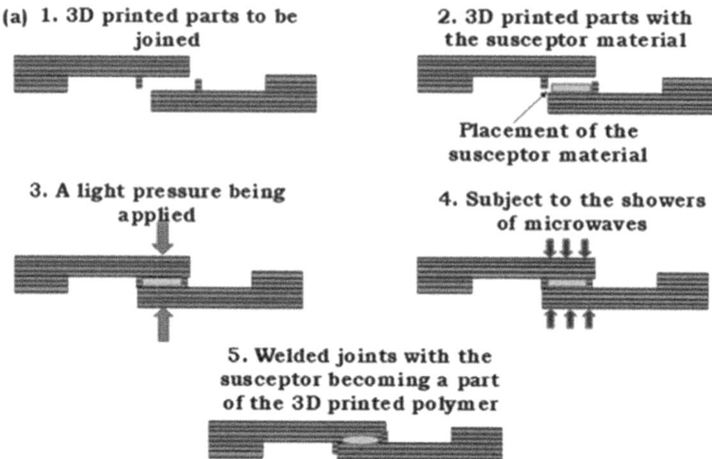

FIGURE 6.9   Schematic representation of microwave welding process

With precise control of process parameters, a very functional weld is possible without any changes in the bulk material. Although microwave finds its applications in many domains, the application of the technology for processing of materials and welding 3D printed components to achieve a bigger volume object still waits to be explored.

Based on the literature survey, it is realized that this welding technology has enormous potential for development as a flexible assembly technique suited for use in large-scale manufacturing industries such as automobile and aviation. Nonetheless, the need for Inherent Conducting Polymer (ICP) is a substantial impediment to technology's ability to complete the task.

6.5.2   Ultrasonic Welding

UW is regarded as the greatest of its type and might be employed to combine 3D printed polymers because of its "ease of automation" [51]. As shown in Figure 6.10, this method applies vertical ultrasonic vibration to the weld interface while the materials to be joined are stacked one on top of the other. A welding tip simultaneously applies stress in the axial direction to the overlap zone, which causes the zone to partially melt forming a joint.

UW is a reasonably quick (it takes around 1 sec each weld) and clean method that doesn't need flux, adhesives, solvents, solders, or other binders. Because of its efficient use of materials and energy, as well as its low generation of waste, pollutants, and extra gases, UW is frequently referred to as a "sustainable" manufacturing method.

However, one of the biggest drawbacks of UW is the customization of the tooling. Further, to focus the ultrasonic waves and further help bonding, unique structural elements like raised ridges are to be created and made into the pieces [52]. Only overlapping and shear joints, not exceeding 3 mm in thickness, can be used through this

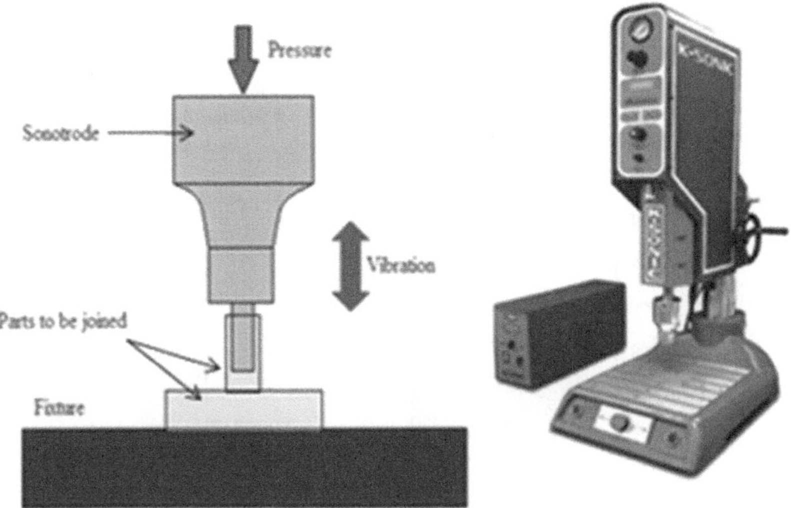

FIGURE 6.10   Ultrasonic welding being used to join 3D printed objects

approach. The noise that is created, further increases the procedure's drawbacks [53]. The technique is also limited to straight lines with thin sheets and short joint lengths.

6.6 BIG AREA ADDITIVE MANUFACTURING (BAAM)

A straightforward solution to evade the built size limitations of an FDM-3D printer would be to procure a bigger bed FDM-3D printer. However, this above consideration is discouraged when a cost and energy consumption comparison is made between a bigger (500 mm^3) and a commercially available (240 mm^3) FDM-3D printer, depicted in Figure 6.11. A bigger FDM-3D printer's cost is around four times compared to a smaller printer while the energy consumed by a bigger printer is more than double that of a smaller one. Further, the success rate in obtaining successful prints from a bigger 3D printer is very low compared to small bed size printer due to de-lamination, warpages, distortions, and stability issues.

The Manufacturing Demonstration Facility at Oak Ridge National Laboratory (ORNL) and Cincinnati, Inc. created BAAM, a large-scale 3D printing method, in 2013. The BAAM printers can quickly and cost-effectively manufacture unique molds and tools. This technology uses pellets as input raw material reducing the material processing cost. The technology is capable of attaining a speed of 210 in/sec peak velocity. Additionally, it has been found that parts manufactured by BAAM with carbon fibre reinforcements have good thermal conductivity and strength, eliminating the need for a separate heat chamber and lowering energy intensity.

One of BAAM 3D printers' main drawbacks, meanwhile, is their inability to predict the quality and performance of the print in advance due to the difficulty in figuring out how the temperature is distributed on the big bed platforms. Additionally, BAAM 3D printers must address anisotropy, instability, distortions, and costs, which

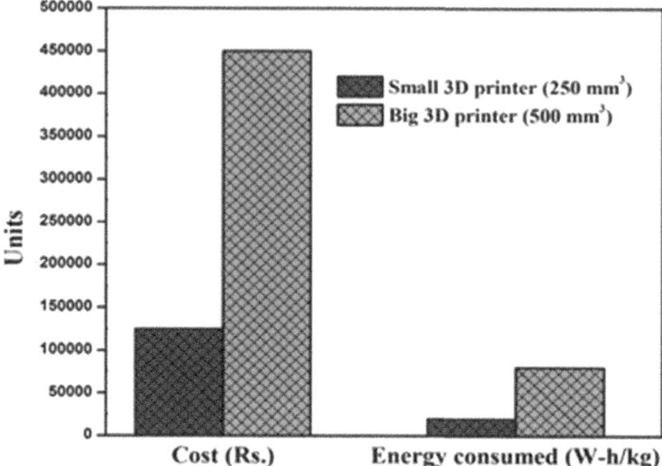

FIGURE 6.11   Cost and energy consumption comparison between a small and big bed 3D printer

appear to be linearly related to the volume of manufactured components [22]. It will also be necessary to assess the maintenance and reparability of parts produced from BAAM. Figure 6.12 shows some of the issues observed in BAAM.

6.7   CRITICAL CHALLENGES TO OVERCOME IN THE JOINING TECHNIQUES

Along with the benefits and downsides, the key difficulties in each joining/welding process are listed below:

* Requirement of a good surface prior to the use of the adhesives seems to be an important challenge of joining 3D printed polymers by adhesives.
* Employing fasteners to join 3D printed parts results in the increment of the final weight of the components, stress development and corrosion related issues.
* The restriction of mechanical interlocking 3D printed parts is that thin boundary and hollow objects cannot be interlocked without avoiding the visibility of cuts. This seems to be a restriction of employing interlocking of 3D printed parts.
* Change in the material properties because of the heat generation, requirement of a specific equipment, difficulty in joining dissimilar plastics are the important challenges friction related welding techniques have to overcome.
* Requirement of a conducting polymer which can absorb the microwaves as well as the customization of tooling are the major drawbacks of microwave welding and ultrasonic welding techniques respectively.
* Though BAAM seems to be capable of evading the bed size limitation, issues related to its heat management, warping, de-laminations, anisotropy as well as its recused percentage of successful prints are its key challenges.

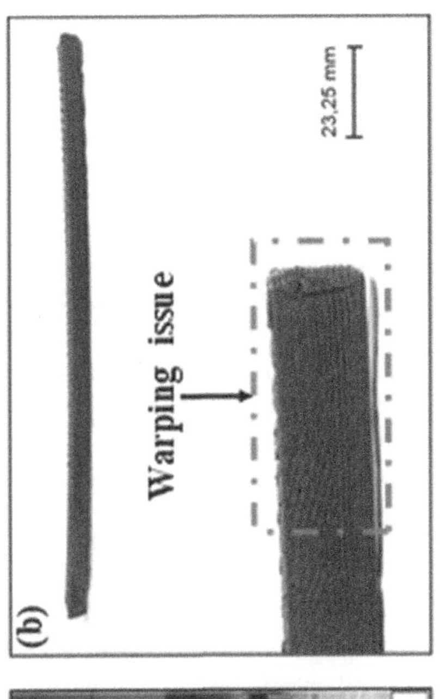

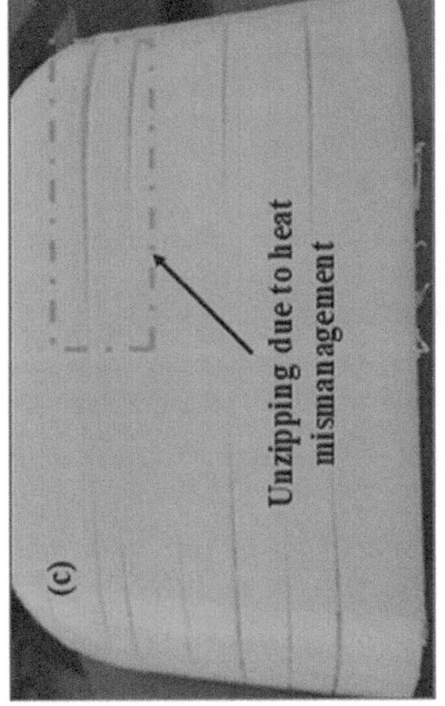

FIGURE 6.12 Issue related to BAAM (a) Delamination (b) Warpage (c) Unzipping due to the heat mismanagement

Efforts are continually being made by different research groups to address these challenges through advancements in printing technologies, materials, software, and design methodologies. As the field of 3D printing evolves, it is expected that many of these challenges will be overcome or mitigated.

6.8 CONCLUSIONS

3D printing is on the threshold of getting transformed from a niche technology to a mainstream technology. However, there are a few limitations that are holding back its popularity. Restricted volume of the print brought about by the limited bed size of 3D printers is one such drawback. A bigger 3D printer cost consumes more energy. Sectioning the CAD model, printing them in parts and then joining/welding attaining a bigger volume, seems to be a rational approach to circumvent this issue. This chapter provides an overview of the various joining/welding procedures that can be used on FDM-3D printed parts attaining big sized, less weight components. The pros, cons as well as the difficulties in each of the joining/welding techniques are also presented to the readers. The study's primary objective was to create a knowledge base that would benefit researchers and the 3D printing community, increasing the likelihood that manufacturing industry leaders would accept low-cost FDM printers.

ACKNOWLEDGMENTS

The authors would like to acknowledge Centre of Excellence on Industrial Microwave Heating Applications-Enerzi Microwave Systems Pvt. Ltd., Additive Manufacturing and Reverse Engineering Lab, for the research facilities and KLS Gogte Institute of Technology, Belagavi for the research encouragement.

REFERENCES

[1] U. M. Dilberoglu, B. Gharehpapagh, U. Yaman, and M. Dolen, "The role of additive manufacturing in the era of Industry 4.0," *Procedia Manuf.*, vol. 11, no. June, pp. 545–554, 2017, doi: 10.1016/j.promfg.2017.07.148.

[2] C. Wee, J. Lim, K. Q. Le, Q. Lu, and C. H. Wong, "An overview of 3-D printing in manufacturing and automotive industries," *IEEE Potentials*, vol. 35, no. 4, pp. 18–22, July-August 2016 [Online]. doi: 10.1109/MPOT.2016.2540098

[3] N. Bogojevic, S. Ciric-kostic, and D. Croccolo, "Advantages and drawbacks of additive manufacturing," no. January, 2017, doi: 10.5937/IMK1702057V.

[4] Y. L. Yap, W. Toh, R. Koneru, R. Lin, K. I. Chan, H. Guang, W. Yew, B. Chan, S. S. Teong, G. Zheng, and T. Y. Ng , "Evaluation of structural epoxy and cyanoacrylate adhesives on jointed 3D printed polymeric materials," *Int. J. Adhes. Adhes.*, vol. 100, no. March, pp. 1-6, 2020, doi: 10.1016/j.ijadhadh.2020.102602.

[5] V. K. Tiwary, P. Arunkumar, A. S. Deshpande, and N. Rangaswamy, "Surface enhancement of FDM patterns to be used in rapid investment casting for making medical implants," *Rapid Prototyp. J.*, vol. 25, no. 5, pp. 904–914, 2019, doi: 10.1108/RPJ-07-2018-0176.

[6] V. Tiwary, P. Arunkumar, A. S. Deshpande, and V. Khorate, "Studying the effect of chemical treatment and fused deposition modelling process parameters on

surface roughness to make acrylonitrile butadiene styrene patterns for investment casting process," *Int. J. Rapid Manuf.*, vol. 5, no. 3/4, p. 276, 2015, doi: 10.1504/ijrapidm.2015.074807.

[7] Z. Weng, J. Wang, T. Senthil, and L. Wu, "Mechanical and thermal properties of ABS/montmorillonite nanocomposites for fused deposition modeling 3D printing," *Mater. Des.*, vol. 102, pp. 276–283, 2016, doi: 10.1016/j.matdes.2016.04.045.

[8] S. Dul, L. Fambri, and A. Pegoretti, "Fused deposition modelling with ABS-graphene nanocomposites," *Compos. Part A Appl. Sci. Manuf.*, vol. 85, pp. 181–191, 2016, doi: 10.1016/j.compositesa.2016.03.013.

[9] L. Luo, I. Baran, S. Rusinkiewicz, and W. Matusik, "Chopper: Partitioning models into 3D-printable parts," *ACM Trans. Graph.*, vol. 31, no. 6, pp. 1–9, 2012, doi: 10.1145/2366145.2366148.

[10] P. Song, Z. Fu, L. Liu, and C. W. Fu, "Printing 3D objects with interlocking parts," *Comput. Aided Geom. Des.*, vol. 35–36, pp. 137–148, May 2015, doi: 10.1016/j.cagd.2015.03.020.

[11] Michael Brooks, "RepRage – How much power does a 3D printer use?," *Reprage. Com*, 2017. https://reprage.com/post/39698552378/how-much-power-does-a-3d-printer-use (accessed June 11, 2021).

[12] R. B. Azhiri, R. Mehdizad Tekiyeh, E. Zeynali, M. Ahmadnia, and F. Javidpour, "Measurement and evaluation of joint properties in friction stir welding of ABS sheets reinforced by nanosilica addition," *Meas. J. Int. Meas. Confed.*, vol. 127, pp. 198–204, 2018, doi: 10.1016/j.measurement.2018.05.005.

[13] M. Ş. Adin and M. Okumuş, "Investigation of microstructural and mechanical properties of dissimilar metal weld between AISI 420 and AISI 1018 STEELS," *Arab. J. Sci. Eng.*, vol. 47, no. 2021, pp. 8341–8350, Oct. 2021, doi: 10.1007/s13369-021-06243-w.

[14] P. Kah, R. Suoranta, J. Martikainen, and C. Magnus, "Techniques for joining dissimilar materials: Metals and polymers," *Rev. Adv. Mater. Sci.*, vol. 36, no. 2, pp. 152–164, 2014.

[15] M. Ş. Adin, "A parametric study on the mechanical properties of MIG and TIG welded dissimilar steel joints," *J. Adhes. Sci. Technol.*, vol. 38, no. 1, pp. 115–138, 2023, doi: 10.1080/01694243.2023.2221391.

[16] E. Dugbenoo, M. F. Arif, B. L. Wardle, and S. Kumar, "Enhanced bonding via additive manufacturing-enabled surface tailoring of 3D printed continuous-fiber composites," *Adv. Eng. Mater.*, vol. 20, no. 12, pp. 1–9, 2018, doi: 10.1002/adem.201800691.

[17] J. M. Arenas, C. Alía, F. Blaya, and A. Sanz, "Multi-criteria selection of structural adhesives to bond ABS parts obtained by rapid prototyping," *Int. J. Adhes. Adhes.*, vol. 33, pp. 67–74, Mar. 2012, doi: 10.1016/j.ijadhadh.2011.11.005.

[18] V. K. Tiwary, A. Padmakumar, and V. Malik, "Adhesive bonding of similar/dissimilar three-dimensional printed parts (ABS/PLA) considering joint design, surface treatments, and adhesive types," *Proc. Inst. Mech. Eng. Part C J. Mech. Eng. Sci.*, vol. 236, no. 16, pp. 8991–9002, Aug. 2022, doi: 10.1177/09544062221089849.

[19] A. P. da Costa, E. C. Botelho, M. L. Costa, N. E. Narita, and J. R. Tarpani, "A review of welding technologies for thermoplastic composites in aerospace applications," *J. Aerosp. Technol. Manag.*, vol. 4, no. 3, pp. 255–265, 2012, doi: 10.5028/jatm.2012.04033912.

[20] U. M. Dilberoglu, B. Gharehpapagh, U. Yaman, M. Dolen, E. Çanti, M. Aydin, and F. Yildirim, "Manufacturing and automotive industries," *Mater. Des.*, vol. 35, no. 4, pp. 1–8, 2018, doi: 10.1016/j.matpr.2021.05.140.

[21] S. Inaniwa, Y. Kurabe, Y. Miyashita, and H. Hori, *Application of friction stir welding for several plastic materials*, vol. 2. Woodhead Publishing Limited, 2013, doi: 10.1533/978-1-78242-164-1.137.

[22] A. Spaggiari and F. Denti, "Mechanical strength of adhesively bonded joints using polymeric additive manufacturing," *Proc. Inst. Mech. Eng. Part C J. Mech. Eng. Sci.*, vol. 235, no. 10, pp. 1851–1859, May 2021, doi: 10.1177/0954406219850221.

[23] R. Garcia and P. Prabhakar, "Bond interface design for single lap joints using polymeric additive manufacturing," *Compos. Struct.*, vol. 176, pp. 547–555, 2017, doi: 10.1016/j.compstruct.2017.05.060.

[24] V. K. Tiwary, A. P, and V. R. Malik, "An overview on joining/welding as post-processing technique to circumvent the build volume limitation of an FDM-3D printer," *Rapid Prototyping Journal*, vol. 27, no. 4. Emerald Group Holdings Ltd., pp. 808–821, 2021, doi: 10.1108/RPJ-10-2020-0265.

[25] Y. Zhou, S. Sueda, W. Matusik, and A. Shamir, "Boxelization: Folding 3D objects into boxes," in *ACM Transactions on Graphics*, vol. 33, no. 4, pp. 1–8, 2014, doi: 10.1145/2601097.2601173.

[26] P. A. Bajakke, S. C. Jambagi, V. R. Malik, and A. S. Deshpande, *Friction Stir Processing: An Emerging Surface Engineering Technique*, in: Gupta, K. (ed), *Surface Engineering of Modern Materials. Engineering Materials*, Springer, Cham. 2020, doi: 10.1007/978-3-030-43232-4_1.

[27] J. Vanek, J. A. Garcia Galicia, B. Benes, R. Měch, N. Carr, O. Stava, G. S. Miller, "PackMerger: A 3D print volume optimizer," *Comput. Graph. Forum*, vol. 33, no. 6, pp. 322–332, 2014, doi: 10.1111/cgf.12353.

[28] S. Xin, C. F. Lai, C. W. Fu, T. T. Wong, Y. He, and D. Cohen-Or, "Making burr puzzles from 3D models," *ACM Trans. Graph.*, vol. 30, no. 4, pp. 1–8, 2011, doi: 10.1145/1964921.1964992.

[29] V. K. Tiwary, N. J. Ravi, P. Arunkumar, S. Shivakumar, A. S. Deshpande, and V. R. Malik, "Investigations on friction stir joining of 3D printed parts to overcome bed size limitation and enhance joint quality for unmanned aircraft systems," *Proc. Inst. Mech. Eng. Part C J. Mech. Eng. Sci.*, vol. 234, no. 24, pp. 4857–4871, Dec. 2020, doi: 10.1177/0954406220930049.

[30] V. K. Tiwary, A. Padmakumar, and V. R. Malik, "Investigations on FSW of nylon micro-particle enhanced 3D printed parts applied to a Clark-Y UAV wing," *Weld. Int.*, vol. 36, no. 8, pp. 474–488, 2022, doi: 10.1080/09507116.2022.2104141.

[31] B. Vijendra and A. Sharma, "Induction heated tool assisted friction-stir welding (i-FSW): A novel hybrid process for joining of thermoplastics," *J. Manuf. Process.*, vol. 20, pp. 234–244, 2015, doi: 10.1016/j.jmapro.2015.07.005.

[32] K. J. Quintana and J. L. L. Silveira, "Mechanistic models and experimental analysis for the torque in FSW considering the tool geometry and the process velocities," *J. Manuf. Process.*, vol. 30, pp. 406–417, 2017, doi: 10.1016/j.jmapro.2017.09.031.

[33] R. Kumar, R. Singh, I. P. S. Ahuja, A. Amendola, and R. Penna, "Friction welding for the manufacturing of PA6 and ABS structures reinforced with Fe particles," *Compos. Part B Eng.*, vol. 132, pp. 244–257, 2018, doi: 10.1016/j.compositesb.2017.08.018.

[34] S. Eslami, T. Ramos, P. J. Tavares, and P. M. G. P. Moreira, "Shoulder design developments for FSW lap joints of dissimilar polymers," *J. Manuf. Process.*, vol. 20, pp. 15–23, 2015, doi: 10.1016/j.jmapro.2015.09.013.

[35] V. Malik, N. K. Sanjeev, H. S. Hebbar, and S. V. Kailas, "Finite element simulation of exit hole filling for friction stir spot welding – A modified technique to

apply practically," *Procedia Eng.*, vol. 97, pp. 1265–1273, 2014, doi: 10.1016/j.proeng.2014.12.405.

[36]   J. Sprovieri, "Friction Stir Spot Welding," 2016. www.assemblymag.com/articles/93337-friction-stir-spot-welding

[37]   V. Malik, N. . Sanjeev, H. S. Hebbar, and S. V. Kailas, "Time Efficient Simulations of Plunge and Dwell Phase of FSW and its Significance in FSSW," *Procedia Mater. Sci.*, vol. 5, pp. 630–639, 2014, doi: 10.1016/j.mspro.2014.07.309.

[38]   M. K. Bilici and A. I. Yukler, "Effects of welding parameters on friction stir spot welding of high density polyethylene sheets," *Mater. Des.*, vol. 33, no. 1, pp. 545–550, 2012, doi: 10.1016/j.matdes.2011.04.062.

[39]   X. W. Yang, T. Fu, and W. Y. Li, "Friction stir spot welding: A review on joint macro- and microstructure, property, and process modelling," *Adv. Mater. Sci. Eng.*, vol. 2014, pp. 1–11, 2014, doi: 10.1155/2014/697170.

[40]   P. Michał, W. Krzysztof, G. Jan, and B. Maciej, "Analysis of the friction (spin) welding – preliminary study," vol. 02036, 2019, doi: 10.1051/matecconf/201925402036

[41]   K. Faes, A. Dhooge, O. Jaspart, L. D'Alvise, P. Afschrift, and P. De Baets, "New friction welding process for pipeline girth welds – Parameter optimization," *Proc. Inst. Mech. Eng. Part B J. Eng. Manuf.*, vol. 221, no. 5, pp. 897–907, 2007, doi: 10.1243/09544054JEM758.

[42]   V. Malik and S. V. Kailas, "Understanding the effect of tool geometrical aspects on intensity of mixing and void formation in friction stir process," *Proc. Inst. Mech. Eng. C: J. Mech. Eng. Sci.*, vol. 235, no. 4, pp. 744–757, 2020, doi: 10.1177/0954406220938410.

[43]   J. J. S. Dilip, G. D. J. Ram, and B. E. Stucker, "Additive manufacturing with friction welding and friction deposition processes," *Int. J. Rapid Manuf.*, vol. 3, no. 1, p. 56, 2012, doi: 10.1504/ijrapidm.2012.046574.

[44]   V. K. Stokes, "Analysis of the friction (spin)-welding process for thermoplastics," *J. Mater. Sci.*, vol. 23, no. 8, pp. 2772–2785, 1988, doi: 10.1007/BF00547450.

[45]   M. G. Kazen and P. Asadi, *Advances in friction stir welding and processing related titles.* Woodhead Publishers, 1981.

[46]   P. Tappe and H. Potente, "New results on the spin welding of plastics," *Polym. Eng. Sci.*, vol. 29, no. 23, pp. 1655–1660, 1989, doi: 10.1002/pen.760292306.

[47]   S. Poyraz, L. Zhang, A. Schroder, and X. Zhang, "Ultrafast microwave welding/reinforcing approach at the interface of thermoplastic materials," *ACS Appl. Mater. Interfaces*, vol. 7, no. 40, pp. 22469–22477, 2015, doi: 10.1021/acsami.5b06484.

[48]   M. S. Adin, B. Iscan, and S. Baday, "Optimization of welding parameters of AISI 431 and AISI 1020 joints joined by friction welding using Taguchi method," *BSEU J. Sci.*, vol. 9, no. 1, pp. 453–470, 2022.

[49]   V. K. Tiwary, P. Arunkumar, and V. R. Malik, "Investigations on microwave-assisted welding of MEX additive manufactured parts to overcome the bed size limitation," *J. Adv. Join. Process.*, vol. 7, no. February, p. 100141, 2023, doi: 10.1016/j.jajp.2023.100141.

[50]   P. A. Bajakke, V. R. Malik, P. Mugali, and A. S. Deshpande, "Microwave processing of engineering materials," in: Kumar, K., Babu, B. S., Davim, J. P. (eds), *Coatings. Materials Forming, Machining and Tribology*, Springer, Cham, pp. 31–55, 2021, doi: 10.1007/978-3-030-62163-6_2.

[51]   R. Singh, H. Singh, I. Farina, F. Colangelo, and F. Fraternali, "On the additive manufacturing of an energy storage device from recycled material," *Compos. Part B*, vol. 156, no. June 2018, pp. 259–265, 2019, doi: 10.1016/j.compositesb.2018.08.080.

[52] J. Favero, S. Belhabib, S. Guessasma, and H. Nouri, "Optimal assembling of 3D printed features by modulation of interfacial behaviour," *Rapid Prototyp. J.*, vol. 25, no. 1, pp. 13–21, Jan. 2019, doi: 10.1108/RPJ-11-2016-0178.

[53] S. K. Bhudolia, G. Gohel, K. F. Leong, and A. Islam, "Advances in Ultrasonic Welding of Thermoplastic Composites: A Review," *Materials*, vol. 13, no. 6, p. 1284, 2020.

7 Investigation of the Effect of Bonding Parameters on the Adhesive Bonding Strength of Parts Produced by FDM-3D Method

Nergizhan Anaç and Oğuz Koçar

7.1 INTRODUCTION

In the last decade, 3D printing technologies, software and used printing materials in these technologies have become widespread. Additive manufacturing methods have become an innovative manufacturing method that is preferred day by day due to their advantages such as ease of production of complex parts, reduction in the number of parts, being time saving and less costly. Although there are various additive manufacturing technologies, the method has become popular due to 3D printers in which the FDM (Fused Deposition Modeling) technique is used. Molten thermoplastic material can be added layer by layer on a platform to create products using 3D printers. Although there are many positive properties of parts produced by a 3D printer, they have the disadvantage of limited part size due to the build platform. This situation is eliminated by assembling the parts produced one by one with each other. Welding, soldering, bonding and mechanical fastening, which have been accepted as traditional joining methods for years, have been used for joining similar or dissimilar materials.

The adhesive bonding method differs from other methods since it is a very old technique that has been used since ancient times. Throughout the history of humanity, it has been possible to paste and reuse broken objects made from various materials in our daily lives. Archaeological evidence shows that natural herbal and animal adhesives were known by Stone Age people, Ancient Egyptians, Greeks, and Romans [1, 2]. The adhesive bonding technique is the process of joining the components using an adhesive material. In order for this joining to be functional and robust, it is necessary to pay attention to some issues. In addition, adaptation of the process parameters of traditional methods to current application conditions should be ensured for the joining of new materials or high-tech products developed today.

It is difficult to find information on the adhesive bonding of advanced materials such as composites or polymers produced under laboratory conditions other than what adhesive manufacturers offer, unlike the traditional materials that are often used. The information presented is also given only for some limited material groups. Therefore, detailed research and comprehensive information are needed to realize practical applications on the bonding of innovative polymer-based materials. Filaments used to print polymer parts in 3D printers are produced from various thermoplastic materials like ABS, PLA, PET (Polyethylene Terephthalate), PC (Polycarbonate), PETG (Polyethylene Terephthalate Glycol), TPE/TPU/TPC (Thermoplastic Elastomer / Polyurethane) / Copolyester), PA (Polyamide or Nylon [3, 4]. Both the printing parameters and bonding processes of these materials in the 3D printer naturally show differences. Although joining plastic materials brings advantages such as lightness, tightness, reduced product cost and in some cases longer working life, there are some difficult aspects of bonding [5]. One of the most important problems in bonding polymer-based materials is low surface energy, which causes poor adhesion. Adhesive and materials to be bonded, necessary surface preparations and joining working conditions (exposure loads) are important for the bonding process to be carried out in accordance with its purpose [6]. When selecting the adhesive used in the bonding process, it should be checked whether it is compatible with the part to be bonded and whether it is resistant to working conditions. The front surface preparation of the part to be bonded is one of the factors affecting the load resistance in bonding joints [7]. So, what kind of surface preparation should be done when bonding plastic materials?

Physical and chemical surface preparation processes can be done for plastics [8]. All the methods provide chemical changes to the surface. Corona and plasma treatment are used to increase the surface energy of plastic for improved adhesion of adhesives. Corona treatment is based on the principle of spraying a high voltage discharge from atmospheric air onto a surface. Plasma is an ionized gas with essentially an equal density of negative and positive charges. Operationally, plasma differs from corona treatment in that plasma treaters are operated at less than atmospheric pressure. It is not preferred because of its high cost [8]. There is evidence that adhesive material properties should be considered for plasma pretreatments [9]. Corona treatment is used for materials such as polyethylene, polypropylene and polyethylene terephthalate (PET), while plasma treatment is used for materials such as polyethylene, polycarbonate, polystyrene, polypropylene (PP), PET, nylon 6 and ABS. The mechanical pretreatments such as grinding, blasting (media, abrasive), or sanding are procedures often used to remove surface defects and modify surface roughness to improve adhesion. Increasing the roughness of the surface improves the adhesive's surface contact area, as long as it does not increase excessively, because it improves mechanical interlocking. Therefore, the bond strength increases [10–13]. However, the optimal roughness value depends on the type of adhesive and bonded material, joint geometry, and amount of load [14]. In the preparation of plastic surfaces by chemical processes, methods such as acid etching or priming are used. The solution and primer may change according to the properties of the material to be prepared on the surface [8, 15]. The general trend for the adhesives and surface preparation methods used for joining parts printed on the 3D printer is given in Table 7.1.

TABLE 7.1

The adhesives generally used for joining FDM-3D printed parts.

Researchers	Materials	Adhesive	Surface Treament	Joint Design/Test
Özenç et al. [10]	PLA/ABS	Loctite HY 4090	Loctite SF 7063	Shear impact joint, Izod impact test
Khosravani et al. [11]	PETG	Loctite 3090	Loctite SF 770	Single lap joint (SLJ), Tensile test
Khosravani et al. [12]	PLA	Loctite EA 9466	Surfaces washed and dried	SLJ, Tensile test
Yap et al. [13]	ASA and NCF	Loctite E-20HP, ZAP Slo-Zap	Isopropyl alcohol	SLJ
Frascio et al. [14]	ABS	Loctite EA 9466, Teroson 9225, Teroson MS 9399	Loctite SF7063, 320 SiC emery paper, APP, LPP	SLJ
Frascio et al. [16]	PLA, A conductive PLA	Teroson PU9225	Isopropyl alcohol	SLJ
Leicht et al. [17]	Polyamide 12 (PA 12)	Delo Duopox SJ8665, SikaPower 1277	As received, Atmospheric plasma	The Adhesion Analyzer LUMiFrac (mechanical test method)
Kanani et al. [18]	PPA (Polyphthalamide)	Loctite EA 9497, Araldite 2015	60-grit sandpaper	SLJ
Kariz et al. [19]	ABS	Mitopur E45, Dorus KS 217, Bison Power	Acetone vapors to smoothen the surface	SLJ, Tensile test
Bürenhaus et al. [9]	Ultem 9085 (copolymer of PEI), PC	Weldyx Polyplast Delo-Pur 9694, Jowat 690.00, Loctite M-21 HP, UHU Endfest 300, Delo Duopax AD848	Mechanical roughening, Plasma activation	SLJ
Spaggiari and Favali [20]	ABS	Henkel Loctite 401	Loctite 7030	SLJ, Joggle lap joint, Double lap joint

Adhesive joints can be subjected to four types of stress. These are tensile, shear, cleavage and peel stresses [21, 22]. Loads acting on the connections (mechanical loading or thermal stresses) cause stress on the joints. It is important to know these loads in order to decide on the right design for adhesive joints. There are some joint types that are most commonly used: single-lap joints (SLJ), double-lap joints (DLJ), stepped-lap joints, and scarf joints [23–26]. Due to the ease of production and practicality, single lap joints are mostly used in adhesive joint works. Knowing the stresses on a joint and its value is helpful in determining if the joint is secure [27].

Adhesive bonding failures can cause undesirable results, causing the entire system to fail. Problems in bonding joints may be due to material (inappropriate adhesive-joint material pair selection, insufficient surface treatments, material manufacturing defects, adhesive thickness) or due to non-specialized personnel. For a successful application in adhesive joints, the materials and process must be defined.

7.2 MATERIAL AND METHODS

7.2.1 SAMPLE PROCESSING

Three different thermoplastics were used in the study, namely PLA (Polylactic Acid), PLA Wood (wood-reinforced PLA composite), and ABS+ (Acrylonitrile Butadiene Styrene). PLA is a thermoplastic derived from organic sources like corn starch and sugarcane. It is environmentally friendly, posing no harm to human health, and can naturally biodegrade. PLA Wood is a composite PLA with a 30% wood content. Due to the real wood reinforcement, it has higher heat resistance, less visible layer lines, and a matte wood appearance. ABS+ is a petroleum-based thermoplastic with high strength, hardness, and rigidity. The mechanical characteristics of the filament utilized in the study are presented in Table 7.2, while the printing parameters are given in Table 7.3.The Ultimaker and Esun brand filaments were printed using the Ultimaker 5S, and the Filameon brand filament was printed using the Creality 3S 3D printer.

7.2.2 ADHESIVE PROPERTIES

In the study, three different adhesives were used. The first adhesive was the epoxy polymer-based Loctite EA-9466 manufactured by Henkel Company. Loctite EA-9466

TABLE 7.2
Mechanical characteristics of Filament Materials [28–30].

Mechanical properties	PLA	ABS+	PLA Wood
Diameter size (mm)	2.85	2.85	1.75
Brand	Ultimaker	eSUN	Filameon
Color	Black	Green	Brown
Ultimate strength (MPa)	52.5	40	47
Percent elongation (%)	7.8	30	4
Density(g/cm³)	1.24	1.06	1.13

TABLE 7.3
Printing parameters of filaments.

Materials		PLA	ABS+	Wood PLA
Infill percentage (%)		100	100	100
LT (mm)		0.1-0.2-0.3	0.1-0.2-0.3	0.1-0.2-0.3
Temperature (°C)	Nozzle	210	250	220
	Bed	55	110	60
3D Print speed (mm/dk)		50	50	50

TABLE 7.4
Mechanical characteristics of Kwik Weld and Loctite EA 9466.

Kwik Weld [33]		Loctite EA 9466 [14, 32]	
Fixing time (minute)	6	Fixing time (minute)	180
Curing time (hours) 22 °C	4–6	Curing time (hours) 22 ° C	24
Ultimate strength (MPa)	14.51	E (MPa)	1718
Temperature resistance	150 °C	Tg (°C)	62
Mixing ratio	1:1	Tensile strength (MPa)	32
Density (g/cm³)	1.93	Density (g/cm³)	1.0
		Elongation %	3
		Shore hardness (Durometer D)	60

TABLE 7.5
Mechanical characteristics of Araldite 2015 [34–36].

E (MPa)	1850
Poisson ratio	0.33
Specific gravity (g/cm³)	1.43
Yield point strength (MPa)	12.63
Ultimate strength (MPa)	21.63
Shear modulus (MPa)	560
Shear yield strength (MPa)	14.6
Shear strength (MPa)	17.9

is a multi-purpose adhesive with a long working time and high bonding strength, used in metal, ceramic, and most plastic applications [31]. The second adhesive was the Kwik Weld adhesive produced by JB Weld, used for bonding metal, plastic, ceramic, wood, and glass [32]. The third adhesive was Araldite 2015 which is a two-component, multi-purpose resin and hardener adhesive used for bonding different materials with high corrosion resistance. Table 7.4 and Table 7.5 included the mechanical properties of the adhesives used.

7.2.3 The Design of Experiments (Taguchi Method)

The experimental design method was first proposed by Fisher [37]. Classic experimental design methods have disadvantages such as being complex and not user-friendly. Moreover, there is a necessary increase in the overall number of experiments, with an increase in the number of factors. Genichi Taguchi proposed the Taguchi approach which is a statistical method to solve this problem.

Taguchi included a three-stage in optimization of product or process. These stages are called system design, parameter design, and tolerance design [38]. The approach is based on employing a particular orthogonal array (OA) design, enabling the examination of the complete parameter space through a limited number of experiments. After completing the experimental phase, the collected results are converted into a S/N ratio. The S/N ratio is used as a measurement to evaluate the quality characteristics that deviate from the desired values. In the analysis of the S/N ratio, there are three types: "larger-is-better," "nominal-is-best," and "smaller-is-better." These classifications determine the desired direction of improvement for each specific quality characteristic. In addition, a statistical variance analysis (ANOVA) is conducted to identify the process parameters that exhibit significant statistical impact. The S/N and ANOVA analyses together aid in determining the optimal process parameters. Finally, a validation experiment is performed to verify the optimum process parameters [38–40]. Therefore, there are three objectives in the Taguchi Method (TM) for parameter design: 1) determining the optimum design parameters; (2) estimating the contribution of one by one design parameter to the quality; and (3) prediction the quality properties based on the optimum design parameters. In this study, the TM was selected for experimental design. Three factors were determined to examine the adhesion properties of the parts printed in 3D printers. These factors are filament material, layer thickness, and adhesive type. The factors which consist of three levels and their levels are given in Table 7.6. The shear strength values of the adhesive joints were utilized as the fundamental criteria for evaluating the factors.

Considering the factors in Table 7.6, the Taguchi L9 OA, which includes nine experimentals was chosen. Each experiment was repeated four times. Taguchi L9 experimental design determined which Minitab 15 is given in Table 7.7. In the study, S/N ratios were calculated based on maximum tensile stress using Equation 1.

TABLE 7.6
Factors and its levels used in the experimental.

Factor	Level I	Level II	Level III	Units
	I	**II**	**III**	
Materials	PLA	ABS +	PLA Wood	None
Adhesive	Kwik Weld	Loctite 9466	Araldite 2015	None
LT	0.1	0.2	0.3	mm

TABLE 7.7
Taguchi L9 OA and experimental results.

Experimental Number	Materials	Adhesive	Layer Thickness	Shear Strength (MPa)
1	PLA	Kwik Weld	0.1	3.24
2	PLA	Loctite 9466	0.2	4.56
3	PLA	Araldite 2015	0.3	3.57
4	ABS +	Kwik Weld	0.2	2.65
5	ABS +	Loctite 9466	0.3	2.87
6	ABS +	Araldite 2015	0.1	1.71
7	PLA Wood	Kwik Weld	0.3	1.70
8	PLA Wood	Loctite 9466	0.1	2.28
9	PLA Wood	Araldite 2015	0.2	3.28

Larger the better: It is used when the maximum results is desired [41].

$$\frac{S}{N} = -10\log\frac{1}{n}\left(\Sigma\frac{1}{yi^2}\right) \tag{1}$$

n: the number of tests
yi: the experimental result

In conclusion, ANOVA was conducted to determine the process parameters that had significant statistical impact. The optimal of process parameters could be determined using S/N and ANOVA analyses. Lastly, the optimum process parameters derived from the parameter design were validated through a separate experimental verification.

7.2.4 HARDNESS AND TENSILE EXPERIMENTS

To determine the mechanical characteristics of the filaments used in the study, dog-bone specimens (Figure 7.1a) were printed at 100% infill rates with different layer thicknesses according to ASTM D638-10 [42] standards. For the adhesive joints, 25x100x1.5 mm plates (as shown in Figure 7.1b) were manufactured. The tests were carried out at room temperature using a 5 KN capacity WDW-5 model universal tensile device, with a tensile speed of 1 mm/min.

The shore hardness measurement method was used for the hardness measurements. Shore hardness, which is the strength of a material to permanent deformation, is commonly used for quality control and quick and simple mechanical property evaluation in the elastomer industry [43]. To conduct the hardness tests, the ASTM D2240-15 standard was adhered to. The experiments utilized a LOYKA D-type shore hardness durometer. Measurements were repeated at five different points for each sample and the average of these measurements was reported. In addition, surface imaging was

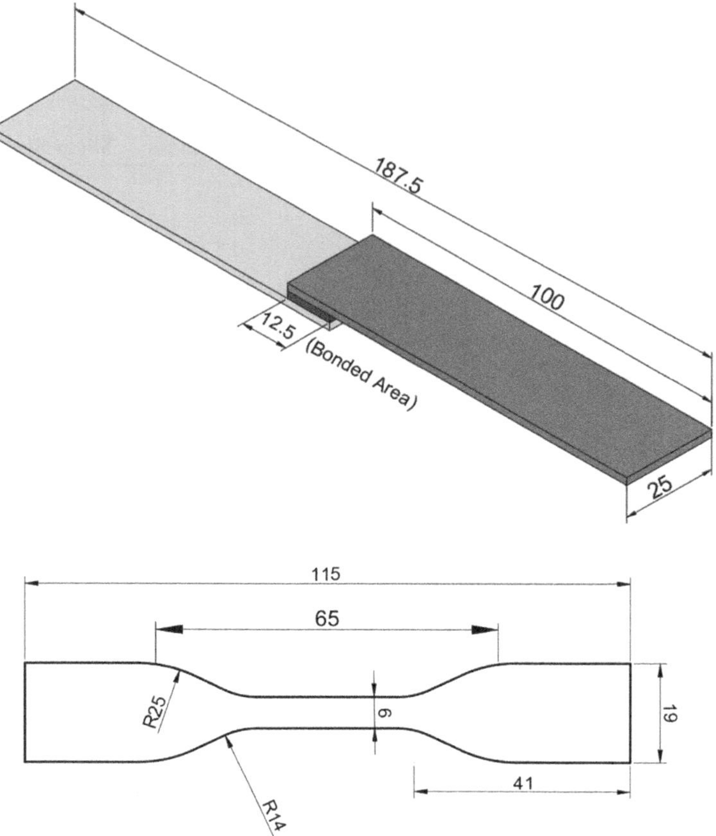

FIGURE 7.1 Dog-bone specimen and single-lap joint type dimensions (mm).

performed after 3D printing using a digital microscope. The damage types were discussed by taking images of the rupture surfaces of the adhesive joints.

7.3 RESULTS AND DISCUSSION

7.3.1 TENSILE TEST AND HARDNESS RESULTS

It is clear from the results that the tensile strength decreases with increasing layer thickness (Figure 7.2). This result is consistent with the literature [44–46]. The highest tensile strength values in all layer thicknesses were obtained in PLA, PLA Wood and ABS+, respectively. When the strength values for 0.2 mm and 0.3 mm layer thicknesses in PLA Wood and ABS+ are examined, it is seen that the strength values are close to each other, while the strength values of PLA are close to LT (0.1 and 0.2 mm). The lower strength observed in PLA Wood can be attributed to factors such as high porosity, inadequate compaction, and adhesion defects between the layers of the filament [47].

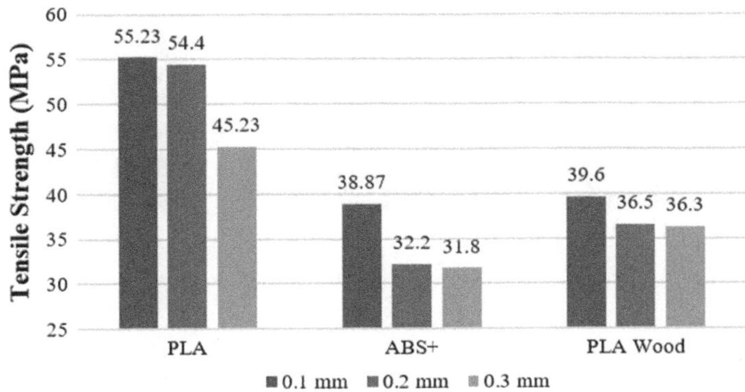

FIGURE 7.2   Tensile test results.

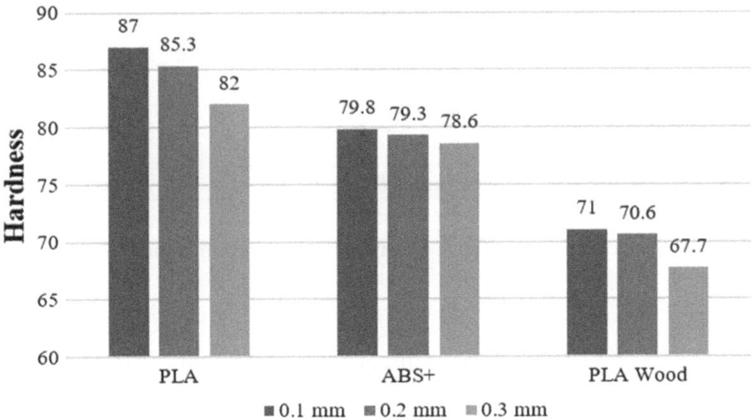

FIGURE 7.3   Hardness results.

When the hardness results are analyzed, it is seen that the hardness values change very little with the increase of layer thickness (Figure 7.3). As with the tensile strength, the highest hardness values were measured in PLA in all layer thicknesses. The reason for the low hardness of PLA Wood compared to PLA is due to the high porosity and adhesion defects between layers caused by the wood dusts contained in it. On the other hand, it was seen that the increase in layer thickness had little effect on the hardness values in ABS+. It was determined that this was due to the high printing temperature of ABS+, which causes the layers to interlock and fuse better.

7.3.2   Microstructure and Fracture Surface Images

Figure 7.4 shows surface images of PLA, ABS+, and PLA Wood in different layer thicknesses taken by a digital microscope after printing. The results obtained from tensile test samples of ABS+ with a layer thickness of 0.1 mm were found to be

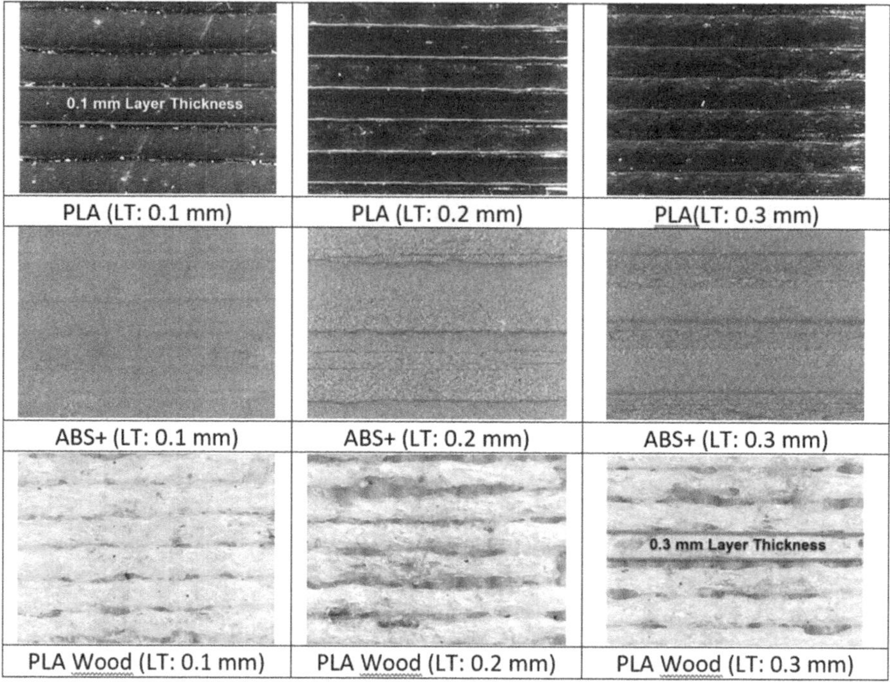

FIGURE 7.4 Surface images after 3D printing (Digital microscope/ 100x, LT: Layer thickness).

higher than those of ABS+ with layer thicknesses of 0.2 mm and 0.3 mm. The lowest tensile strength was obtained at 0.3 mm layer thickness. It was determined that the fusion between the layers was better in ABS+ parts with a layer thickness of 0.1 mm (Figure 7.4). Layer lines were blurred at 0.1 mm layer thickness. It was observed that the melting was relatively good at a layer thickness of 0.2 mm. At 0.3 mm thickness layer, the layers started to become more visible. In PLA, the layer thickness was visible and uniform, while in PLA Wood, the layers were not uniform. It is seen that wood dust, which is additive in PLA Wood, adversely affected the structure and caused anomalies during melting and solidification after melting. It was observed that the gaps between the layers increased with the increase in PLA Wood layer thicknesses. In addition, it was determined that the layers interpenetrated and fused better as the layer thickness decreased [48]. As a result, PLA has better printing formability than ABS and wood-based PLA [47]. After the tensile test, the fracture surfaces of the samples were analyzed in detail. The image of the fracture surfaces is given in Figure 7.5 and the evaluation of the fracture surface defects is given in Table 7.8.

The experiments were conducted at least three times under the same conditions. The adhesion ability obtained by using Loctite 9466 and Araldite 2015 in PLA and PLA Wood materials was higher than the part strength. For PLA and PLA based PLA Wood material, Kwik Weld adhesive showed adhesive failure. Adhesive failure was

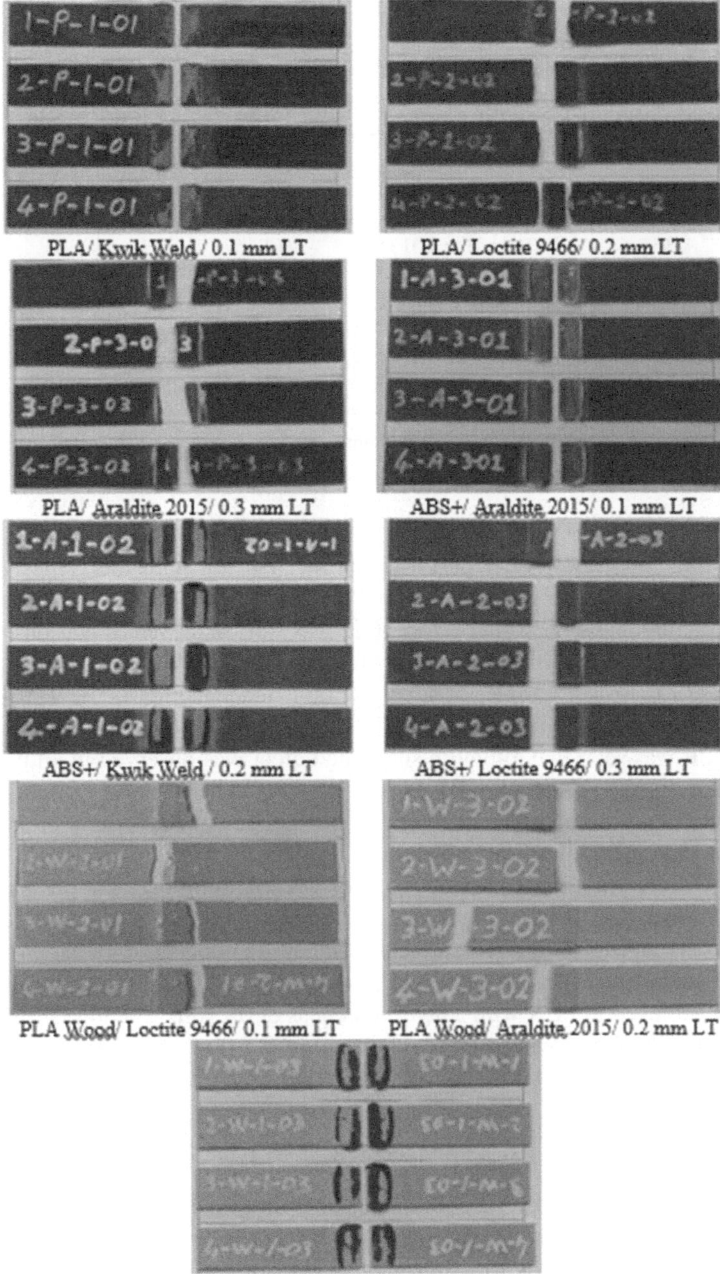

FIGURE 7.5 Fracture surface view of adhesive bond for all samples. (LT: Layer Thickness).

TABLE 7.8
Types of damage to adhesive joints.

Material	Adhesive	LT (mm)	Failure mode
PLA	Kwik Weld	0.1	AF
PLA	Loctite 9466	0.2	SF
PLA	Araldite 2015	0.3	SF
ABS+	Araldite 2015	0.1	AF
ABS+	Kwik Weld	0.2	AF
ABS+	Loctite 9466	0.3	SF
PLA Wood	Loctite 9466	0.1	SF
PLA Wood	Araldite 2015	0.2	SF
PLA Wood	Kwik Weld	0.3	AF

AF: Adhesive failure, SF: Substrate failure.

observed in all samples using Kwik Weld adhesive (Table 7.8). The surfaces of the materials did not show proper performance in adhesion with this adhesive. In the adhesive joints made using ABS+ material and Loctite 9466 adhesive, first the main material fractured, and the joint performance was fully achieved. It was observed that there was an adhesive failure in the joints made using Araldite 2015.

In Figure 7.6, three examples of adhesive joints made using Loctite 9466, which is one of the test samples with a substrate defect, are given. When the damage analysis was made, it was seen that although substrate defect was observed in all these parts, the way of rupture in the rupture areas was different. The joints in PLA broke in the lap area of the adhesive, broke at the end of the lap area in ABS+, and broke near the lap area in PLA Wood but from adherend.

To know how the joint is broken is useful in explaining the relationship between the adhesive and the bond strength value. It has been understood that the closer the rupture takes place to the lap joint area, the higher the bond strength. Upon analyzing the shear strength values of these joints, it was observed that the highest value (4.56 MPa) was achieved in PLA with a layer thickness of 0.2 mm. The lowest values were obtained in PLA Wood with a layer thickness of 0.1 mm (2.28 MPa) and in ABS+ with a layer thickness of 0.3 mm (2.87 MPa).

7.3.3 ANALYSIS OF THE FACTORS

The TM was used to analyze the effect of each control factor (material, adhesive and layer thickness) on the shear stress of the adhesive joints. First, the S/N response table was created. The S/N ratio for each level of control factors and the change from level 1 to level 3 depending on the levels of control factors are presented in Table 7.7.

As the S/N or Means difference between the levels of the factors increases, the effect ratio of the level increases. A higher difference means that the control factor has more influence on product control. The values of the factor effects were formulated

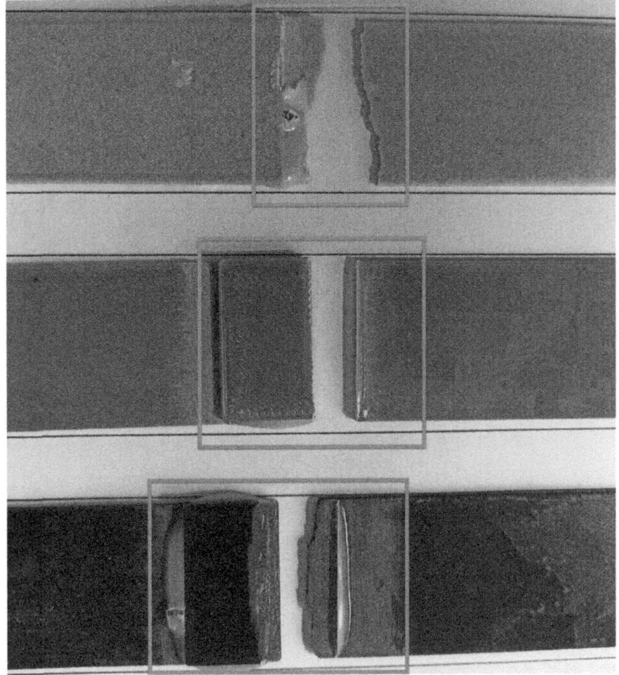

FIGURE 7.6   Some samples with substrate failure.

and calculated for each parameter and level using the TM. The S/N and Means graphs in Table 7.7 and Figure 7.7 were used to find the best combination and S/N result.

The most important criterion in the evaluation of experimental data in the TM is the S/N ratio [41]. In the study, according to the TM, the S/N ratio should have a maximum value in order to obtain the optimum shear stress. When the S/N ratios in Figure 11.6 are examined, it is seen that the optimum parameters are materials (PLA), adhesive (Loctite 9466) and layer thickness (0.2mm) (1 2 2 OA). When the S/N ratios for optimum shear strength were examined, the materials (PLA), adhesive (Loctite 9466) and layer thickness (0.2mm) (1 2 2 OA) were determined from Figure 11.7. In determining the adhesion conditions to be made under the same conditions, an evaluation can be made according to the level values of the material, adhesive and layer thickness factors. The mean S/N ratio for each level is calculated according to the tensile test results after bonding (Table 7.7).

Materials is the factor with the highest S/N ratio. The difference between high (11,488) and low (7.376) values was calculated as 4.112 (Table 7.9). The difference between the highest and lowest levels for adhesive was determined as 2.076 and as 3.288 for layer thickness. According to Taguchi's prediction, the larger the difference between the S/N ratio values, the more significant effect it has on shear stress. Accordingly, it is concluded that the change of filament material will significantly affect the shear stress. The reason for this is that there is a rupture from the material before the adhesive reaches its maximum value. This means that the mechanical

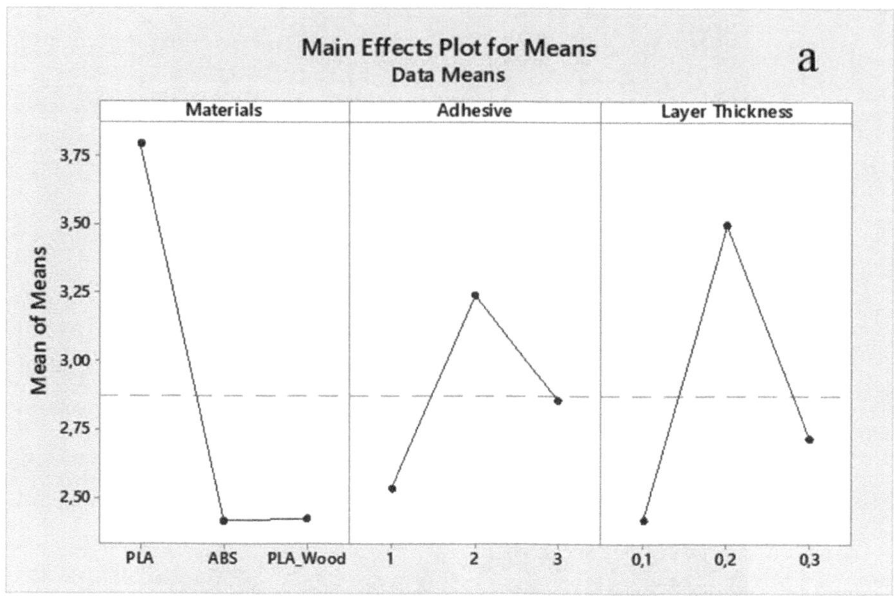

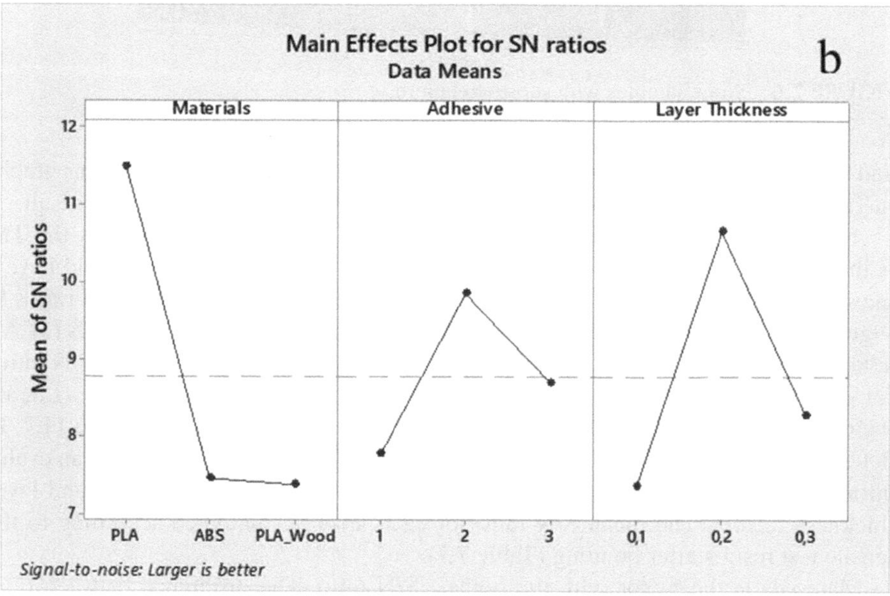

FIGURE 7.7   Means (a) and S/N (b) for shear stress.

properties of the material are decisive in the adhesive joint. The second effective factor is the layer thickness. When the tensile test results are examined, it is seen that the material strength changes depending on the layer thickness. This situation directly affects the adhesion strength. Adhesive type has the least effect on bond strength in adhesive joints.

TABLE 7.9
Response for S/N ratios and means.

Level	S/N Ratios			Means		
	Materials	Adhesive	LT (mm)	Materials	Adhesive	LT (mm)
1	11.488	7.772	7.370	3.793	2.533	2.416
2	7.452	9.848	10.658	2.416	3.242	3.499
3	7.376	8.696	8.287	2.423	2.858	2.717
Delta	4.112	2.077	3.288	1.377	0.709	1.082
Rank	1	3	2	1	3	2

TABLE 7.10
Results of ANOVA.

Source	DF	Seq SS	Contribution	Adj SS	Adj MS	F-Value	P-Value
Materials	2	3.7720	55.96%	3.7720	1.8860	11.08	0.083
Adhesive	2	0.7562	11.22%	0.7562	0.3781	2.22	0.310
Layer Thickness	2	1.8720	27.77%	1.8720	0.9360	5.50	0.154
Error	2	0,3404	5,05%	0,3404	0,1702		
Total	8	6,7406	100,00%				

Data analysis results obtained from S/N ratio and ANOVA are shown in Table 7.10 (where df is degrees of freedom, F is variance ratio and P is significant factor). Confidence level is 94.5%. P value is less than 0.05 for materials. This shows that there is a significant relationship between the shear stress in the adhesive joints of the materials. Table 7.8 shows the interaction between shear stress and determined factors in the adhesive joint. When evaluated according to the impact rates, the material has the highest impact with 55.96% impact rate, followed by layer thickness with 27.77% and the adhesive with 11.22% effect rate.

7.3.4 CONFIRMATION TESTS

After determining the optimum level of shear strength in adhesive joints, a verification test is required. If the optimum conditions suggested by Taguchi are in the L9 OA, direct calculation can be made. However, if the optimum conditions are not in the OA, a final verification test is required. The initial parameters were determined as $A_1B_1C_1$ according to the L9 OA, and the material, adhesive and layer thickness were determined as PLA, Kwik Weld and 0.1 mm, respectively. The results of the confirmation test according to the experimental and prediction data are shown Table 7.11. The suggested optimum parameters are PLA, Loctite 9466 and 0.2 mm layer thickness ($A_1B_2C_2$). According to the initial parameters, an improvement of 140.7% was achieved in the shear strength value. When the experimental results obtained with

TABLE 7.11
Results of the confirmation test.

	Initial adhesive parameter	Optimum adhesive parameters	
		Prediction	Experimental
Factor levels	$A_1B_1C_1$	$A_1B_2C_2$	$A_1B_2C_2$
Shear strength (MPa)	3.24278	4.77850	4.56298
S/N ratio (dB)	10.2184	14.4508	13.1850

the optimum value and the prediction results were compared, it was calculated that there was an acceptable difference of 4.5%.

7.4 CONCLUSIONS

The increase in the variety of materials used in accordance with human needs has led to the emergence of different techniques for joining or repairing materials. Although there are joining various techniques such as welding, mechanical fastening, soldering, etc. in the present day, adhesive process is often preferred by consumers due to its practicality and economy.

Adhesives have a wide variety of materials (natural adhesives, synthetic adhesives) that can appeal to almost every sector. The success of the adhesion process varies depending on the operation parameters and the environmental conditions in which the adhesion is conducted. Therefore, having knowledge of parameters such as the parts to be bonded, the adhesive to be used, and the applied load, etc. is extremely important. The rapid development of technology has affected both user requests and production techniques. Nowadays, 3D printers are very popular. In this technology, the number of PLA and ABS thermoplastic materials used as raw materials and their derivatives has increased considerably. While there are many resources available on manufacturing parts from 3D printers, data on joining these materials is relatively limited.

In the study, information is presented about the adhesive joints and the factors affecting the bond strength (surface preparation, material type, adhesive, etc.). In addition, studies on the methods used in the bonding of plastics are briefly summarized. Then, the plates produced from PLA, ABS+ and PLA Wood in three different layer thicknesses were bonded using three different adhesives. Taguchi experimental design was used to figure out the response of material, adhesive and layer thickness on shear strength in adhesive joints. The results are summarized below.

PLA was found as the material with the highest tensile strength and hardness value for all layer thicknesses. Tensile strength and hardness values decreased with increasing layer thickness for all materials.

When the surfaces of the plates after printing were examined, it was observed that the layers were well fused (joined) especially at low layer thicknesses due to the high nozzle and table temperature during ABS+ printing. In PLA Wood, it was determined that high porosity and interlayer adhesion defects increased with increasing layer thickness. In PLA, on the other hand, the layers were visible, uniform and there were no spaces between the layers.

When all the samples in the experiment were examined, it was observed that Kwik Weld adhesive gave adhesive failure and Loctite 9466 adhesive gave substrate failure. It is thought that the bonding performance of JB Kwik Weld adhesive can be improved when used with different surface preparations. Araldite 2015 adhesive provided the desired bond strength for PLA and PLA Wood, but it was understood that it was not appropriate for use with ABS+ material.

In this study, L9 OA was used to analyze the shear stress in the Taguchi experimental design. It was determined by statistical results (99.5% confidence level) that the shear strength effect on adhesive bond strength was obtained as 5.96%, 11.22% and 27.77% for materials (A), adhesive (B) and layer thickness (C), respectively. The optimum levels of the factors determined for shear strength in adhesive joints were determined as PLA, Loctite 6694, and 0.2 mm layer thickness ($A_1B_2C_2$). Accordingly, it was found that there was a 4.5% difference between the shear strength obtained from the experimental results of the validation experiments and the estimated shear strength.

REFERENCES

[1] Fay, P.A., A history of adhesive bonding, in: Robert D. Adams (ed), *Adhesive Bonding*, 2nd edn. 2021: Elsevier, United Kingdom. p. 3–40.

[2] Brockmann, W., P.L. Geiß, J. Klingen, and K.B. Schröder *Adhesive bonding: Materials, applications and technology*. 2008: John Wiley & Sons.

[3] Steuben, J., D.L., Van Bossuyt, and C. Turner. Design for fused filament fabrication additive manufacturing, Paper presented at *International design engineering technical conferences, and computers and information in engineering conferences*. 2-5 August 2015: Boston, Massachusetts, USA. p. 1-14.

[4] All3DP. *Best 3D Printer Filament: The Main Types in 2023*. 2023 Dec 26, 2022; Available from: https://all3dp.com/1/3d-printer-filament-types-3d-printing-3d-filament/.

[5] Harper, C.A., *Modern plastics handbook*. 2000: McGraw-Hill Education.

[6] Wahab, M.A., *Joining composites with adhesives: Theory and applications*. 2015: DEStech Publications, Inc.

[7] Ashcroft, W.R., *Industrial polymer applications: Essential chemistry and technology*. 2019: Royal Society of Chemistry.

[8] Pocius, A.V., *Adhesion and adhesives technology: An introduction*. 2021: Carl Hanser Verlag GmbH Co KG.

[9] Bürenhaus, F., E. Moritzer, and A. Hirsch, Adhesive bonding of FDM-manufactured parts made of ULTEM 9085 considering surface treatment, surface structure, and joint design. *Welding in the World*, 2019. 63: p. 1819–1832.

[10] Özenç, M., T. Tezel, and V. Kovan, Investigation into impact properties of adhesively bonded 3D printed polymers. *International Journal of Adhesion and Adhesives*, 2022. 118: p. 103222.

[11] Khosravani, M.R., P. Soltani, and T. Reinicke, Fracture and structural performance of adhesively bonded 3D-printed PETG single lap joints under different printing parameters. *Theoretical and Applied Fracture Mechanics*, 2021. 116: p. 103087.

[12] Khosravani, M.R., P. Soltani, K. Weinberg and T. Reinicke, Structural integrity of adhesively bonded 3D-printed joints. *Polymer Testing*, 2021. 100: p. 107262.

[13] Yap, Y.L., W. Toh, R. Konera, R. Lin, K.I. Chan, H. Guang, W.Y.B. Chan, S.S. Teong, G. Zheng, and T.Y. Ng, Evaluation of structural epoxy and cyanoacrylate adhesives on jointed 3D printed polymeric materials. *International Journal of Adhesion and Adhesives*, 2020. 100: p. 102602.

[14] Frascio, M., C. Mandolfino, F. Moroni, M. Jilich, A. Lagazzo, M. Pizzorni, L. Bergonzi, C. Morano, M. Alfano, and M. Avalle, Appraisal of surface preparation in adhesive bonding of additive manufactured substrates. *International Journal of Adhesion and Adhesives*, 2021. 106: p. 102802.

[15] Gilliam, M., Polymer surface treatment and coating technologies, in: Nee, A. (ed), *Handbook of manufacturing engineering and technology*. 2015: Springer. p. 99–124.

[16] Frascio, M., F. Moroni, E. Marques, R. Carbas, M. Reis, M. Monti, M. Avalle and L.F.M. Silva, Feasibility study on hybrid weld-bonded joints using additive manufacturing and conductive thermoplastic filament. *Journal of Advanced Joining Processes*, 2021. 3: p. 100046.

[17] Leicht, H., L. Orf, J. Hesselbach, H. Vudugula and E. Kraus, Adhesive bonding of 3D-printed plastic components. Journal of Adhesion, 2020. 96(1–4): p. 48–63.

[18] Kanani, A.Y., X-E. Wang, X. Hou, E.W. Reenie and J. Ye, Analysis of failure mechanisms of adhesive joints modified by a novel additive manufacturing-assisted method. *Engineering Structures*, 2023. 277: p. 115428.

[19] Kariz, M., M.K. Kuzman, and M. Sernek, Adhesive bonding of 3D-printed ABS parts and wood. *Journal of adhesion science and Technology*, 2017. 31(15): p. 1683–1690.

[20] Spaggiari, A. and F. Favali, Evaluation of polymeric 3D printed adhesively bonded joints: Effect of joint morphology and mechanical interlocking. *Rapid Prototyping Journal*, 2022. 28(8): p. 1437–1451.

[21] Cardoso, J.M.V., *Design and Analysis of the Mechanical Behaviour of Adhesively-Bonded CFRP T-Joints*. Dissertations. University of Beira Interior (Portugal) ProQuest. 2018.

[22] Adams, R.D., *Adhesive bonding: science, technology and applications*. 2021: Woodhead Publishing.

[23] Barbosa, N., R.D.S.G. Campilho, F.J.G. Silva and R.D.F. Moreira, Comparison of different adhesively-bonded joint types for mechanical structures. *Applied Adhesion Science*, 2018. 6(1): p. 1–19.

[24] Gürsel, A., Fundamentals in adhesive bonding design for complex structures and conditions: an overview. *Journal of Advanced Technology Sciences*, 2019. 8(1): p. 1–10.

[25] Banea, M.D. and L.F. da Silva, Adhesively bonded joints in composite materials: An overview. *Proceedings of the Institution of Mechanical Engineers, Part L: Journal of Materials: Design and Applications*, 2009. 223(1): p. 1–18.

[26] Comyn, J., *Adhesion science*. 2021: Royal Society of Chemistry.

[27] Silva, T. and L. Nunes, A new experimental approach for the estimation of bending moments in adhesively bonded single lap joints. *International Journal of Adhesion and Adhesives*, 2014. 54: p. 13–20.

[28] Ultimaker, *Ultimaker PLA Proterties*. 2023 4 Feb 2023; Available from: https://makerbot.my.salesforce.com/sfc/p/#j0000000HOnW/a/5b000004UiRV/ lt4XCkl0KOSLfPMcyG06mKKbES33WnYiFrMsG8bFGhw.

[29] Esun, *ABS+ Properties*. 4 Feb 2023; Available from: www.esun3d.com/abs-pro-product/.

[30] Filameon, *PLA Wood Natural*. Cited 1 Apr 2022; Available from: www.filameon. com/urun/filameon-pla-wood-filament/.

[31] Rahmani, A. and N. Choupani, Experimental and numerical analysis of fracture parameters of adhesively bonded joints at low temperatures. *Engineering Fracture Mechanics*, 2019. 207: p. 222–236.

[32] Weld, J.-B., *Kwikweld™ Twin Tube – 2 OZ.* 2022 10.09.2022; Available from: www. jbweld.com/product/kwikweld-twin-tube.

[33] Şentürk, B., K. Çetin, S.N. Ürküt, N. Anaç and O. Koçar, Jig design and manufacturing for adhesive thickness control in adhesive joints. *Journal of Materials and Manufacturing*, 2022. 1(2): p. 17–23.

[34] Ri, J.-H., M.-H. Kim, and H.-S. Hong, A mixed mode elasto-plastic damage model for prediction of failure in single lap joint. *International Journal of Adhesion and Adhesives*, 2022. 116: p. 103134.

[35] Pesquet, G., L.F. da Silva, and C. Sato, The use of thermally expandable microcapsules for increasing the toughness and heal structural adhesives. *Frattura ed Integrità Strutturale*, 2011. 5(16): p. 18–27.

[36] Hu, P., X. Han, W.D. Li, L. Li, and Q. Shao, Research on the static strength performance of adhesive single lap joints subjected to extreme temperature environment for automotive industry. *International Journal of Adhesion and Adhesives*, 2013. 41: p. 119–126.

[37] Fisher, R.A., *Statistical methods for research workers*. 1992: Springer.

[38] Yang, W.p. and Y. Tarng, Design optimization of cutting parameters for turning operations based on the Taguchi method. *Journal of Materials Processing Technology*, 1998. 84(1–3): p. 122–129.

[39] Ghani, J.A., I. Choudhury, and H. Hassan, Application of Taguchi method in the optimization of end milling parameters. *Journal of Materials Processing Technology*, 2004. 145(1): p. 84–92.

[40] Karna, S.K. and R. Sahai, An overview on Taguchi method. *International Journal of Engineering and Mathematical Sciences*, 2012. 1(1): p. 1–7.

[41] Abdulkadir, L.N., K. Abou-Al-Hossein, A.M. Abioye, M.M. Liman, Y-C, Cheng, A.A.S. Abbas, Process parameter selection for optical silicon considering both experimental and AE results using Taguchi L9 orthogonal design. *International Journal of Advanced Manufacturing Technology*, 2019. 103: p. 4355–4367.

[42] Polat, N., A. Nergizhan, and M. Faruk, Investigation of the effect of bonding parameters on the mechanical strength of the adhesive bonding PLA parts produced by additive manufacturing. *Politeknik Dergisi*, 2021. 26(3): p. 1145–1154.

[43] Qi, H., K. Joyce, and M. Boyce, Durometer hardness and the stress-strain behavior of elastomeric materials. *Rubber Chemistry and Technology*, 2003. 76(2): p. 419–435.

[44] Ayrilmis, N., M. Kariz, J.H. Kwon, M.K. Kuzman, Effect of printing layer thickness on water absorption and mechanical properties of 3D-printed wood/PLA composite materials. *International Journal of Advanced Manufacturing Technology*, 2019. 102: p. 2195–2200.

[45] Kovan, V., G. Altan, and E.S. Topal, Effect of layer thickness and print orientation on strength of 3D printed and adhesively bonded single lap joints. *Journal of Mechanical Science and Technology*, 2017. 31: p. 2197–2201.

[46] Shubham, P., A. Sikidar, and T. Chand, The influence of layer thickness on mechanical properties of the 3D printed ABS polymer by fused deposition modeling, *Key Engineering Materials*. 2016: Trans Tech. 706: p. 63–67.

[47] Liu, Z., Q. Lei, and S. Xing, Mechanical characteristics of wood, ceramic, metal and carbon fiber-based PLA composites fabricated by FDM. *Journal of Materials Research and Technology*, 2019. 8(5): p. 3741–3751.

[48] Narlıoğlu, N., Printing of furniture fasteners from wood-PLA composite filament using a 3D printer and investigating the effect of layer thicknesses on mechanical properties. *Furniture and Wooden Material Research Journal*, 2021. 4(2): p. 183–192.

8 Friction Stir Welding, Friction Stir Spot Welding and Spin Friction Welding of 3D Printed Components

Vivek Kumar Tiwary, Arunkumar Padmakumar and Vinayak R. Malik

8.1 INTRODUCTION

Digital Manufacturing (DM) is the set of technologies that can reduce the product lead time and price, also capable of addressing the need for customization, increased product quality, and reacting faster to the market [1]. A manufacturing technology can be called Digital manufacturing if it is centered around a computer system. Additive Manufacturing (AM) technologies perfectly fits into this criterion.

"Additive manufacturing" or "3D Printing (3DP)" refers to the layer-by-layer fabrication of three-dimensional physical models directly from computer-aided design (CAD). It is an automated method of producing objects (parts, prototypes, tools, and even assemblies) straight from CAD models without the use of cutters, tools, or fixtures customized to the product geometry [2],[3],[4]. In comparison to conventional methods, this technology shortens the design-to-manufacturing cycle time, lowers production costs by eliminating costly fixtures and tools, and satisfies the volatile market's rising demand [5],[6].

The main advantages of this technique over traditional production are its low buy-to-fly ratio, short time to market, design freedom, ability to manufacture customised products, low material waste, and comparatively low energy requirements [7],[8]. According to Fortune Business Insights, the AM printing industry is anticipated to grow from USD 18.33 to USD 83.9 billion, with a Compound Annual Growth Rate (CAGR) of 24.3% from 2022 to 2029 [9]. The details are depicted in the Figure 8.1 below.

Amongst the 32 OEM's offering 60 diverse AM systems globally, Fused Deposition Modeling (FDM) is one of the well-known techniques, capturing a majority market share due to its ease of operation, low-cost machinery, simple fabrication process and wide material compatibility [10].

DOI: 10.1201/9781032665351-10

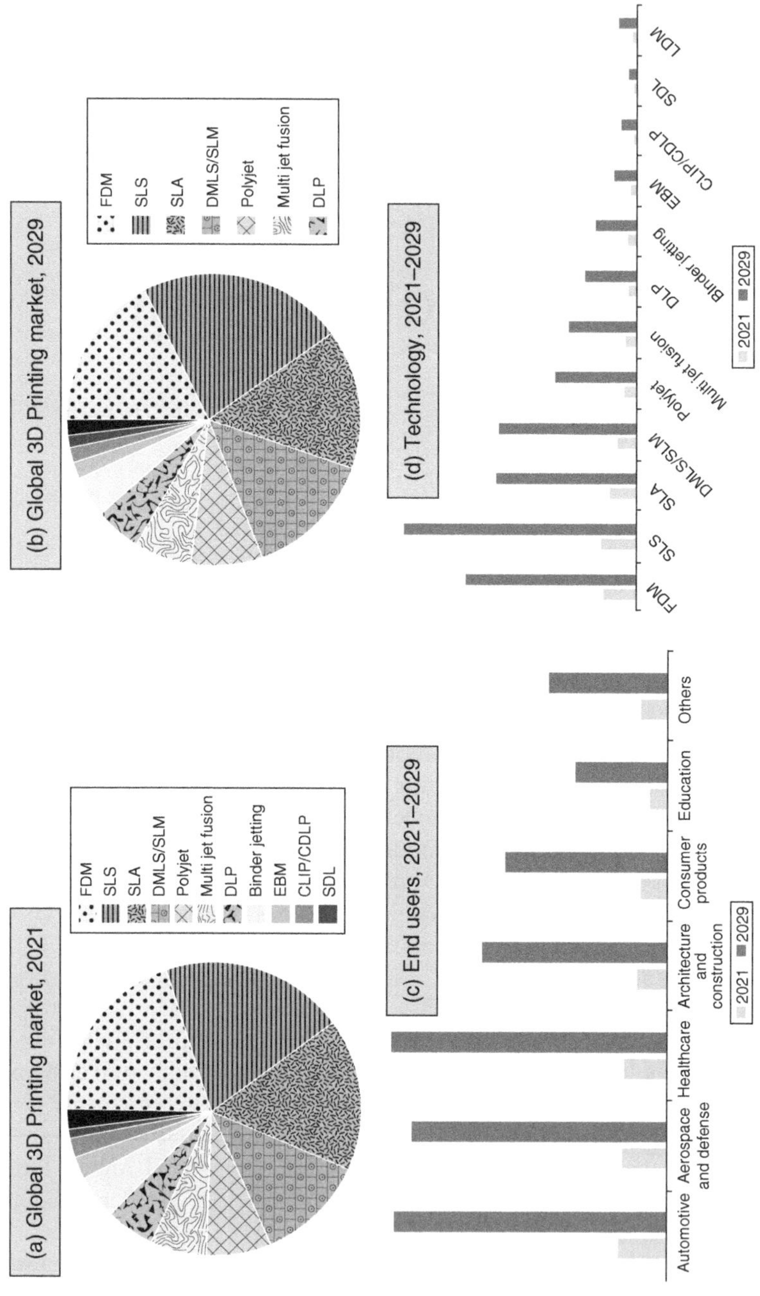

FIGURE 8.1 (a), (b), (c) and (d) Comparison of global 3D printing market share by technologies between 2021 & 2029 based on technologies and end users (Business Fortunes)

Prior to FDM-3D printing, an STL file is created by CAD software and divided into layers with the thickness of each layer determined according to the specifications. A heated nozzle and pressure from a driving gear and grooved bearing are used to expel a thermoplastic polymer filament onto a print bed during the 3D printing process. Based on the cross section of the finished component to be produced, software regulates the nozzle movement. After printing a layer, the printer either moves the nozzle or the bed by a unit layer height to make room for the next layer. Every layer is layered on top of the one before it. This cycle continues until the 3D solid component is finished [11]. The working of an FDM-3D printer is shown in the Figure 8.2 below.

Although the advantages of FDM-3DP are well known, there are a number of issues the technology confronts that prevent its full potential from being realized. This is depicted in Figure 8.3 below. FDM-3D printed parts suffer from inferior surface finish due to the staircase effect, inaccuracy in dimensions due to un proper cooling, low mechanical strengths as it is made up of basically thermoplastics, anisotropy due to its layer by layer fabrication method and low speed of fabrication [12],[13],[14],[15].

Among the several limitations mentioned above, the machine's bed size, which means it can only make objects up to the size of its bed, has been a key constraint of 3D printing that has gone largely unaddressed for a long time [16],[17]. Due to their low weight, 3D printed parts are increasingly widely employed in the aerospace, automotive, architectural, and military industries. However, this small bed size is a limitation that requires attention.

It is observed that, as the size of 3D printed component increases, it leads to instability, distortions, and warpages, further, also leading to an exponential increment in the cost [18]. A straightforward solution to evade the above issue would be to procure a bigger bed FDM-3D printer. However, this above consideration is discouraged when a cost and energy consumption comparison is made between a bigger and a commercially available FDM-3D printer.

Another logical approach to get around the problem is to divide the model into smaller pieces and then subsequently combine or weld the 3D printed components to create a lightweight structure with the appropriate geometry and the defined

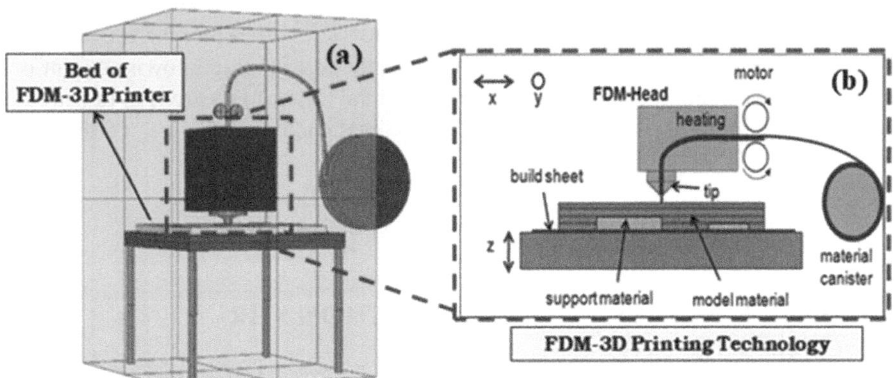

FIGURE 8.2   (a) FDM 3D printers showing the bed (b) working of FDM machines

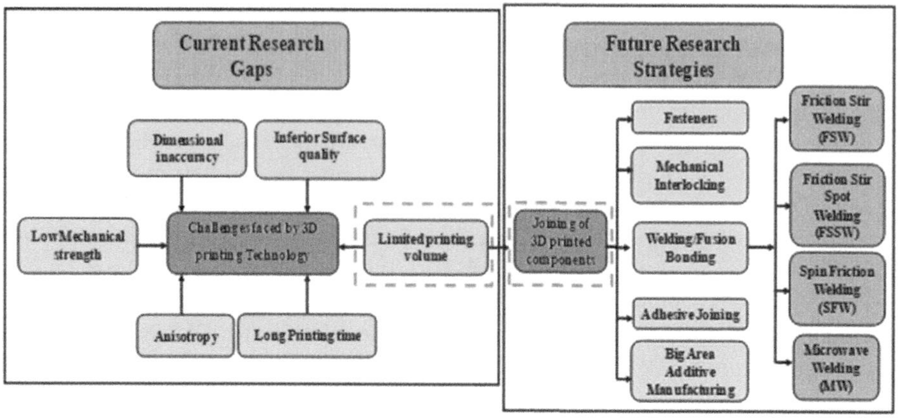

FIGURE 8.3 Challenges in FDM-3D Printing technology

dimensions [19]. Reviewing research articles related to welding of thermoplastics revealed that not much investigation has been carried out on joining or welding of 3D printed parts. Hence, the current chapter discusses the methods for welding 3D printed thermoplastics using friction stir, friction stir spot, and spin friction welding techniques, taking into account that this is a critical gap for 3D printing. Additionally, crucial process parameters in each of the joining processes are described for the benefit of practitioners. At last, case studies related to the welding techniques are also demonstrated.

8.2 EFFECTIVENESS OF THE WELD

The effectiveness of any weld refers to its ability to meet the required standards and perform its intended function in a specific application. The quality of a weld is crucial, as it directly affects the structural soundness and safety of the welded component. To determine the effectiveness of a weld, several tests including visual inspection, non-destructive testing and destructive testing can be employed. One of the regular tests being performed is the tensile test wherein the weld's strength is measured by applying a force to the welded joint until it breaks. From the test a value known as joint efficiency is determined. "Joint Efficiency" refers to the ratio of the strength of a welded joint to the strength of the base material being joined and ranges between 0 to 1.

$$\eta = \frac{\text{Peak strength acheived}}{\text{Base material strength}}$$

8.3 PROCESS DESCRIPTION FOR FRICTION STIR WELDING (FSW)

Friction Stir Welding (FSW) is a solid-state joining process developed by The Welding Institute (TWI), used to join materials, primarily metals, by creating a strong

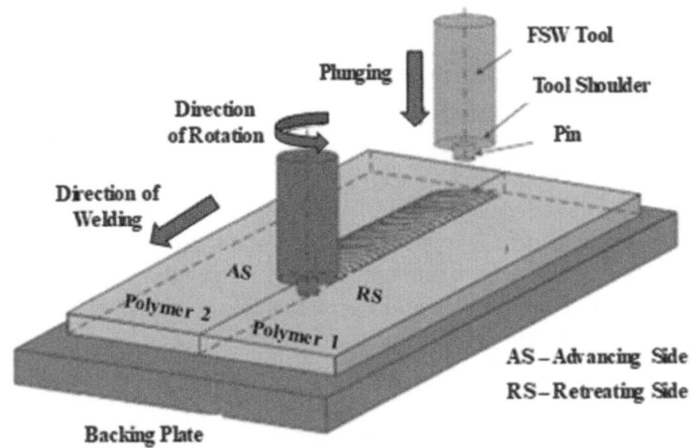

FIGURE 8.4 The Friction Stir Welding (FSW) Process

and defect-free bond between the two pieces. However, FSW can also be applied to thermoplastics, offering several advantages over traditional welding methods for these materials. This process was developed as a variant of conventional friction welding, which was originally used for metals.

Key features and advantages of Friction Stir Welding of thermoplastics include lower energy input, strong weld integrity, reduced part distortion and environmentally friendliness. Applications of Friction Stir Welding in thermoplastics include industries such as automotive, aerospace, marine, and construction, where high-strength and lightweight structures are required [20].

The process of FSW employs a tool having a shoulder and a profiled pin. The polymers to be joined are fixed on the platform of the Vertical Milling Machine (VMC) using a fixture, held by a backing plate. The tool is rotated at a pre-determined speed and steadily plunged into the joint line of the polymers. Next, the tool is moved ahead, ensuring a rigid contact between the tool and the polymers by an applied constant load. This mechanism creates a sufficiently high frictional heat resulting in stirring, softening of polymers, and finally a solid-state weld is created between the two polymers [21], [22], [23]. Friction Stir Welding being carried out on 3D printed thermoplastics is depicted in Figure 8.4 below.

8.4 PROCESS DESCRIPTION FOR SPIN FRICTION WELDING (SFW)

Spin friction welding is a solid-state welding process employed to join two cylindrical parts made of similar or dissimilar polymers or metals. The technique involves rotating one of the parts while applying pressure against the other, causing frictional heat generation at the interface. The heat softens the material, and when the rotation is stopped, the parts are pressed together, allowing them to forge together to form a strong bond. Below is a step-by-step explanation of how spin friction welding is done:

Step 1: Preparation: Before the actual welding process, the two parts to be joined are prepared. They are typically 3D printed to precise dimensions, ensuring a tight fit between the parts. The surfaces that will be in contact during welding are usually cleaned to remove any contaminants or oxides that might hinder the welding process.

Step 2: Clamping: The two parts are firmly clamped in a specialized machine designed for spin friction welding. While one of the components is kept fixed, the other is attached to a spindle that can rotate quickly. The clamping mechanism ensures proper alignment and stability during the welding process.

Step 3: Spin-Up: The spindle holding one of the parts is set in motion, and it starts to rotate at high speed. The rotational speed can vary depending on the materials being welded, but it is generally several thousand revolutions per minute (RPM).

Step 4: Applying Pressure: Once the required rotational speed is achieved, axial pressure is applied to the stationary part, pushing it towards the rotating part. The pressure ensures intimate contact between the two parts, maximizing frictional heat generation at their interface [24], [25].

Step 5: Frictional Heat Generation: As the rotating part rubs against the stationary part, frictional heat is generated at the contact surface. This heat softens the material at the interface, leading to localized plastic deformation.

Step 6: Spin-Down: After a specified amount of time or after reaching the desired temperature, the rotation is stopped. The frictional heat generated during the spinning phase allows the softened material to fuse together.

Step 7: Forge Welding: While the rotation is still stopped, the axial pressure is maintained, forcing the softened material to forge together and create a strong bond between the two parts. This process resembles a forging operation but occurs at a smaller scale.

Step 8: Cooling and Finishing: Once the welding process is complete, the welded assembly is allowed to cool down. Depending on the material and application, additional finishing processes may be performed to ensure the desired surface characteristics and dimensions of the welded joint [26].

Spin friction welding offers several advantages, such as the ability to join dissimilar polymers, high structural integrity, and minimal material loss. It is commonly used in various industries, including aerospace, automotive, and oil and gas, where high-quality and durable welded joints are required. Spin Friction Welding being carried out on 3D printed thermoplastics is depicted in Figure 8.5 below.

8.5 PROCESS DESCRIPTION FOR FRICTION STIR SPOT WELDING (FSSW)

Friction Stir Spot Welding is a solid-state welding process used to join two pieces of metal without melting them. Developed in the 1990s, FSSW has gained popularity in various industries due to its ability to create high-strength, high-quality welds with low thermal distortion. The FSSW process involves a rotating tool that is plunged into the workpieces to be joined. The tool is usually made of a hard, wear-resistant

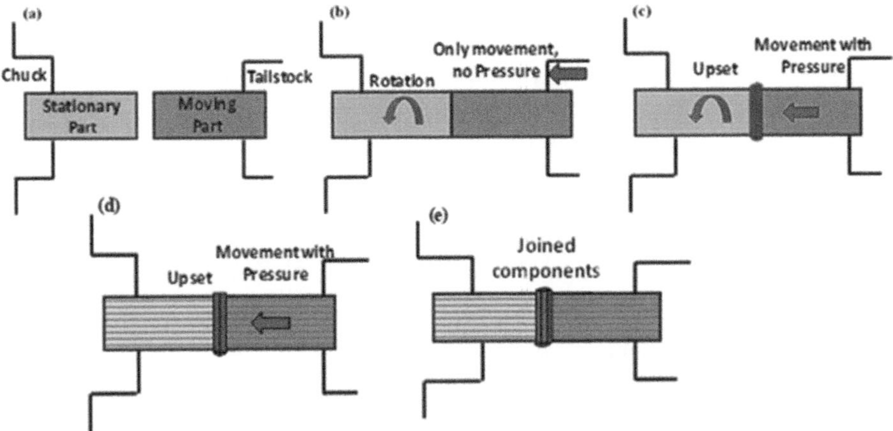

FIGURE 8.5 The Spin Friction Welding (SFW) Process (a) Clamping (b) Spin-up (c) Applying pressure (d) Forge Welding (e) Cooling down

material such as (WC) tungsten carbide or polycrystalline diamond. Frictional heat is produced at the tool-to-workpiece interface as a result of the rotating action of the tool. A step-by-step explanation of the FSSW process is as below:

1. Preparation: The two 3D printed thermoplastics to be joined are placed on a fixture or clamped together, with the surfaces to be welded in contact. The fixture helps to hold the workpieces securely during the welding process.
2. Tool placement: Above the joint and between the workpieces is where the revolving FSSW tool is situated.
3. Plunge phase: The rotating tool is plunged into the joint between the workpieces. The high rotational speed of the tool generates frictional heat at the interface between the tool and the workpieces. This heat causes the tool's surrounding area to become plasticized and soften them [27].
4. Stirring phase: The tool is moved along the joint, stirring the softened metal from both sides. This action creates a solid-state bond between the workpieces by mechanically intermixing and forging the metal. The process takes place at a temperature lower than the melting point of the polymer, preventing the formation of a liquid phase [28].
5. Weld formation: As the tool slides along the joint, it leaves behind a welded zone known as the "Exit Spot." The stirring action creates a strong and defect-free weld, ensuring a polymeric bond between the workpieces [19].
6. Cooling and solidification: After completing the stirring phase, the tool is withdrawn from the joint. The plasticized metal cools rapidly, solidifying the bond between the workpieces [29].

FSSW has several advantages over traditional welding methods, including reduced Heat-Affected Zone (HAZ), high joint strength, joining dissimilar thermoplastics and environmental friendliness.

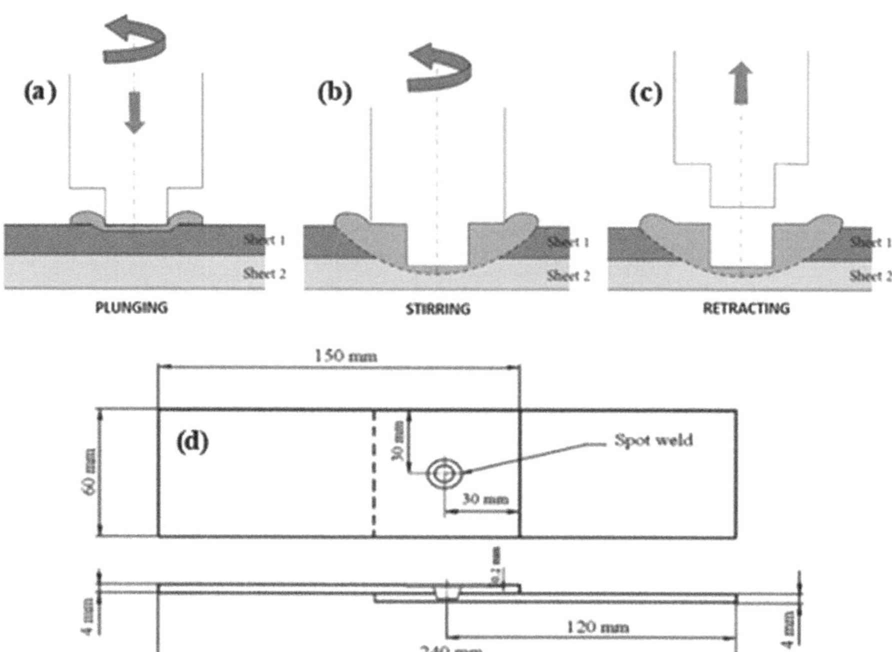

FIGURE 8.6   The Friction Stir Spot Welding (FSSW) Process (a) Plunging (b) Stirring (c) Retraction (d) The top view of the FSSW welded workpieces

8.6   IMPORTANT WELDING PARAMETERS FOR FRICTION STIR WELDING (FSW)

FSW is a practice that utilizes frictional energy generated to create a weld between two butt end configured work-pieces. Depending on the materials being welded and the desired result, several FSW process parameters may be used. However, here are some common process parameters that may be considered:

1. Tool rotational speed – The rotating speed of the tool is crucial during the FSW of thermoplastics. An ideal tool rotation would provide enough of a flow of material, proper stirring action, an ample amount of heat, and eventually a satisfactory weld. Numerous experts have acknowledged that this parameter has a significant impact on the weldability and strength of the weld [30],[31]. Usually the rotation speed can vary in between 800 to 1600 RPM.

2. Tool traverse speed – The speed at which the friction stir welding tool moves along the weld line is referred to as tool traverse speed. Numerous studies have documented how crucial tool traversal speed is to the FSW process [32]. The tool-workpiece contact time and heat input are both influenced by the ideal traverse speed, which also affects the crystallinity and subsequent characteristic features of the weld. Additionally, it has been found that this parameter

has an impact on the microstructure and surface quality of the welded pieces [33]. The usual range for the traverse speed is 30 to 50 mm/sec.

3. Tool geometry and its tilt angle – Material flow during welding and subsequent weld features are both impacted by the tool shape and tilt angle. Three main factors, including the tool geometry and tilt angle, influence the joint's formation in the following three ways: (1) heating the weld zone by friction and plastic deformation; (2) material extrusion from the front to the back of the pin; and (3) forging the plasticized materials by its shoulder [30], [34].

4. Number of passes – Researchers have reported that the number of passes also affects the output of the FSW process. It is usually seen that as the number of passes increases, mechanical and thermal loading increases, porosity level decreases, hardness as well as the strength level increases [35]. Further, in the case of Friction Stir Processing (FSP), the number of passes affects the dispersion of the nanoparticles at the joint. The usual range of passes are from 1 to 3 [36].

5. Taper pin tool angle – Tool pin taper angle, one of the possible tool geometries (tool pin length, pin diameter, pin profile, pin taper angle, shoulder concavity angle, and shoulder diameter), significantly influences the weld strength [37], [38]. A bigger pin taper angle causes a greater fracture force, which raises the stir zone thicknesses and expands the weld area. The usual range of the tool pin taper angle is $10°$ to $16°$.

8.7 IMPORTANT WELDING PARAMETERS FOR FRICTION STIR SPOT WELDING (FSSW)

In Friction Stir Spot Welding (FSSW), several process parameters display a crucial part in achieving successful welds and ensuring the desired weld quality. These parameters influence the heat generation, material flow, and joint formation during the process. Here are some of the important process parameters in FSSW:

1. Rotational Speed: The rotational speed of the FSSW tool directly affects the amount of frictional heat generated at the interface between the tool and the workpieces. Higher rotational speeds result in more significant heat generation, leading to better material softening and easier metal flow. However, excessively high rotational speeds may cause excessive wear on the tool or lead to flaws in the weld. The usual range is around 800 to 1500 RPM.

2. Plunge Depth: The depth of the plunge is the distance the FSSW tool pierces into the workpieces during the plunge phase. It determines the depth of the weld and affects the strength of the joint. A proper plunge depth is required to ensure complete penetration without causing excessive plastic deformation or over-penetration. The usual range of plunge depth is around 7 to 10 mm based on the thickness of the two polymers being welded [22].

3. Tool Geometry: The design of the FSSW tool, including its shoulder diameter and pin shape, plays a central role in determining the weld quality. The tool geometry affects the amount of frictional heat generation, the size of the stirred zone, and the mechanical intermixing of the materials [39].

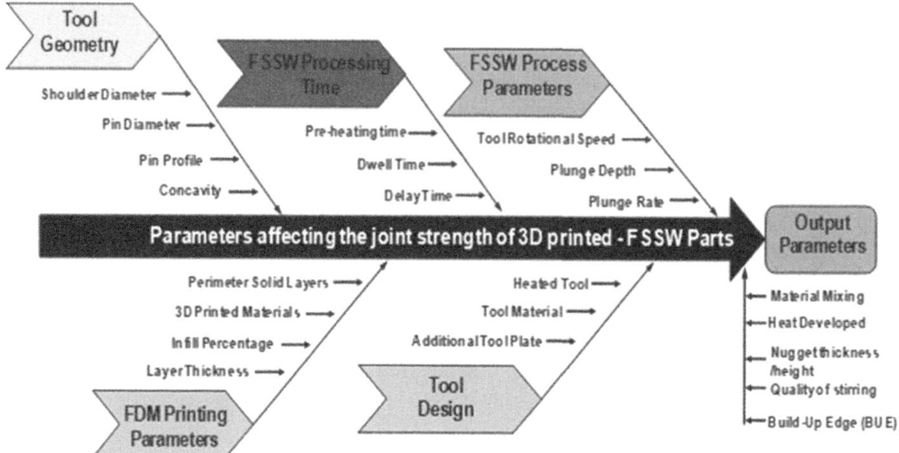

FIGURE 8.7 Important parameters affecting FSSW of 3D printed parts

4. Material Combination – The material type on which the welding process is done impacts the mechanical characteristics of the joint. It has been found that similar polymers join considerably more easily than the dissimilar ones. The reason for this is that they have different molecular weights, viscosities, melting points, MFIs, and glass transition temperatures (Tg).

5. Dwell Time – This is a crucial process parameter that refers to the duration during which the tool remains stationary at the bottom of the plunge depth before commencing the retraction movement. During this dwell period, no axial movement of the tool occurs [40]. It was observed that a high dwell time leads to more amount of material mixing, frictional heat and higher joint strength. The usual range is around 10 to 50 sec.

Optimizing these process parameters is essential to achieving high-quality FSSW joints with the desired mechanical properties and structural integrity. Adjusting these parameters based on the specific materials and application requirements is a common practice to ensure successful FSSW welds. Figure 8.7 below shows the important process parameters affecting the output of FSSWed joints.

8.8 IMPORTANT WELDING PARAMETERS FOR SPIN FRICTION WELDING (SFW)

In SFW, several process parameters play a crucial role in ensuring a successful and high-quality weld. These parameters need to be carefully controlled to achieve the desired weld characteristics and to prevent defects. Some of the important process parameters in SFW include:

1. Rotational speed: The rotating part's speed is a critical parameter as it directly affects the quantity of frictional heat generated at the interface. Higher

rotational speeds generally result in a significant heat generation, which softens the material for effective bonding. However, excessively high speeds might cause material degradation or create undesirable vibrations [41]. The usual range of rotational speed can be between 700 to 2000 RPM.

2. Axial Pressure: The axial pressure applied to the stationary part determines the contact force between the two parts during welding. Sufficient pressure is necessary to ensure intimate contact and promote effective heat generation at the interface. Too little pressure may result in incomplete bonding, while excessive pressure may lead to material flow issues and deformation.

3. Penetration depth: As reported by various researchers, the SFW's joint strength is affected by the temperature distribution in the welding region which in turn majorly depends upon two factors, the amount of heat developed as well as the amount of pressure applied [42]. The Specimens rotation speed, type of material, its infill would determine the amount of heat developed while the penetration depth and number of perimeter shells would determine the amount of pressure developed. Usually the penetration depth varies between 3 to 8 mm.

6. Material Combination: The mechanical characteristics of the joint are influenced by the material combination on which SFW is performed. It has been found that similar polymers join considerably more easily than the dissimilar ones. Again the reason for this is that they have different molecular weights, viscosities, melting points, MFIs, and glass transition temperatures (Tg) [43].

4. Perimeter Shells: In 3D printing, the perimeter layers refer to the outermost layers that form the shell or skin of a printed object. These layers define the outer boundary and surface of the printed model. It is observed that higher the number of perimeter shells, the more amount of material would be available during SFW and will result in better quality welds.

8.9 FRICTION STIR WELDING OF 3D PRINTED PARTS – A CASE STUDY

In the following section, a case study on welding two FDM-3D printed parts of an UAV wing having a wing span of 320 mm is presented. The 3D printer available in the institution had a platform dimension of maximum 170 mm, while the requirement of the customer was 320 mm length wing span. To address the issue, the CAD file was sectioned at mid-plane (160 mm) using NETFABB software and the sectioned CAD files were 3D printed as shown in Figure 8.8 (a). Later, the two 3D printed parts were FSWed on a Vertical Milling Machine (VMC) as shown in Figure 8.8 (b). The process parameters set were 1100 rev/min tool rotation speed, 30 mm/min traverse speed, and 14° tool taper pin angle. The 3D printing material chosen was PETG due to high tensile strength value. The printing settings were maintained at 0.2 mm layer thicknesses, 100% infill, 220°C extruder temperature. Figure 8.8 (c) depicts the FSWed components which showed good strength and integrity. The FSWed components were later painted and provided to the customer which is as shown in Figure 8.8 (d).

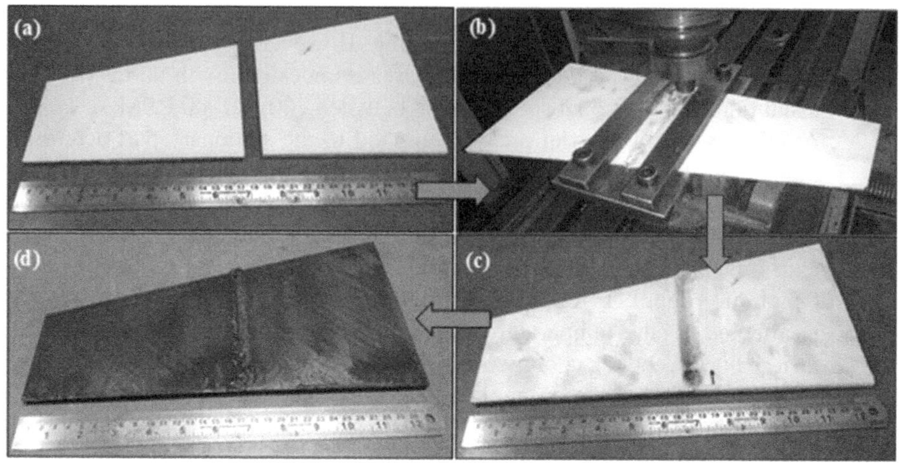

FIGURE 8.8 A case study on FSW on an UAV wing (a) 3D printed parts (b) FSW being carried out (c) After FSW (d) Final UAV wing

8.10 A CASE STUDY ON SPIN FRICTION WELDING OF 3D PRINTED PARTS

As a case study, the proposed technique of welding 3D printed parts by SFW was applied to a saddle point of pipelines. Figure 8.9 (a) shows a conventional service saddle point which comes in only fixed dimensions. Figure 8.9 (b) shows two 3D printed pipes placed at right angles to each other and the SFW process being performed. The process parameters were set as follows: rotation speed at 1400 RPM, penetration depth at 7 mm. While the 3D printing was done at 50% infill, with 9 mm perimeter shells and the material taken was ABS. ABS was chosen owing to its wide acceptability in the 3D printing domain. Figure 8.9 (c) shows the final saddle service point which can be used in place of the conventional pipeline. The process's benefit was the short lead time—just 120–180 seconds, or around 3 minutes—needed for a good weld, indicating the proposed method's practical importance.

8.11 COMPARISON OF DIFFERENT JOINING TECHNIQUES – A CASE STUDY

A comparison was attempted at our lab to study and compare the strength obtained with ABS/PLA 3D printed thermoplastics with different joining/welding techniques [44], [45], [43], [46]. The techniques under consideration were adhesive bonding, friction stir welding, friction stir spot welding, spin friction welding, and microwave welding. Figure 8.10 summarizes the maximum lap shear strength obtained for the ABS/PLA 3D printed thermoplastics. The strength obtained was in the order: Spin Friction Welding (60 MPa) > Friction Stir Welding (18 Mpa) > Microwave Welding (12 MPa) > Friction Stir Spot Welding (10 MPa) > Adhesive Bonding (4 MPa).

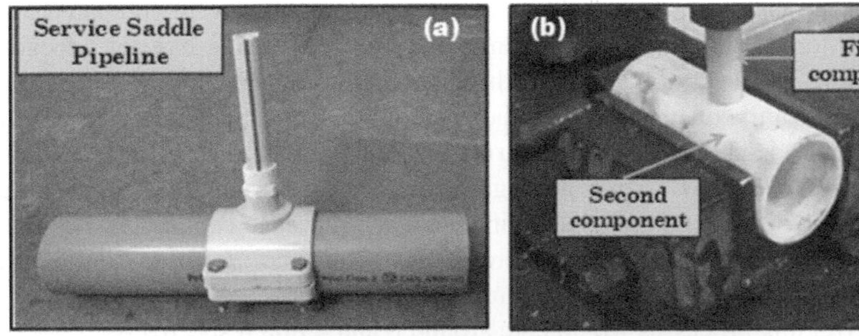

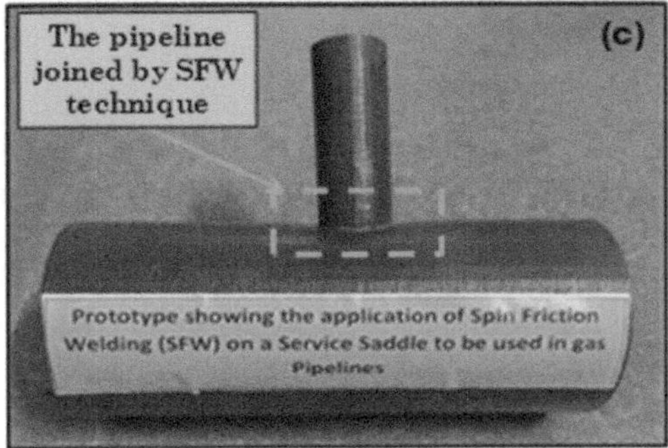

FIGURE 8.9 A case study on SFW on a service saddle point (a) Service saddle pipeline (b) 3D printed parts being SFWed (c) The final SFWed joints

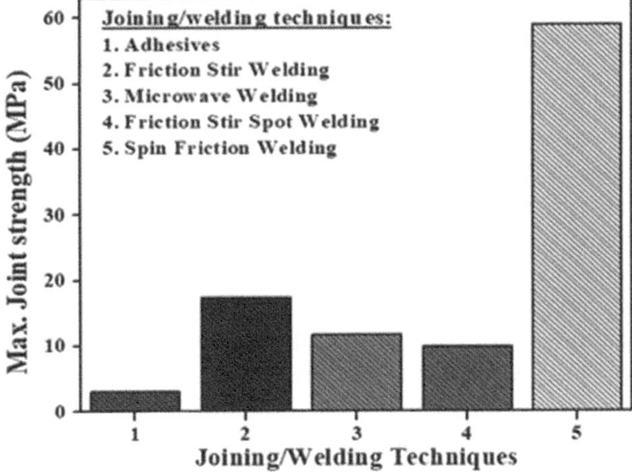

FIGURE 8.10 Comparison of the different joining techniques based on the max. strength achieved

In the present analysis, wherein different joining techniques were compared and evaluated, the longest duration of lead time was observed in the case of adhesive bonding (around 24 to 30 hrs), wherein the time consumed was during the surface preparation, application of adhesives and then loading the joint specimens with a load. The second highest lead time consumed during the joining process observed was in the case of SFW which involved fixing and aligning the work-pieces in the lathe machine, conclusion of the various stages and allowing the work-pieces to cool down. The total process would go around 15 min. The third highest lead time taken was in the case of MW which would go around up to 10 min. While Friction Stir Spot Welding and Friction Stir Welding processes would take around 8 and 4 min respectively.

8.12 CONCLUSIONS

3D printing empowers the imagination to materialize in intricate layers, transforming digital dreams into tangible reality. Despite receiving widespread attention, the limiting bed size of commercially available printers makes it difficult to use the technology to create massive objects. This might be solved simply but effectively by printing the large part in parts, and then welding them together using friction welding techniques like Stir Welding, Spin Welding and Stir spot welding famously known as FSW, FSSW, and SFW.

This chapter gives a general review of the technique of using FSW, FSSW, and SFW to weld FDM-3D printed pieces in order to get around the aforementioned problem. The chapter firstly describes the process to execute FSW, FSSW as well as SFW techniques to join FDM-3D printed parts. Secondly, critical process parameters of these welding techniques, whose judicial selection and optimization can yield better end results are presented and deliberated. Lastly, the chapter ends with a case study wherein UAV wings as well as service saddle pipelines are welded using these techniques.

Although FSW, FSSW, and SFW are very flexible techniques, the major limitation of these techniques is its inability to weld dissimilar thermoplastics. Secondly, welding of thermoplastics in itself seems to a big challenge owing to the thermosphysical difference in the properties. While SFW can be applied only to rotary work-pieces, FSSW leaves an unpleasing exit hole after the process. Further, a common drawback is related to equipment and setup cost which is common in all the three welding techniques, i.e., FSW, FSSW, and SFW.

It is felt that not enough experience has been gained for application of these techniques on an industrial scale yet to making it as common as other conventional welding technologies. Academicians, researchers, system manufactures as well as interested users are called upon to take the technologies to the next level.

ACKNOWLEDGMENTS

The authors would like to recognize the Centre of Excellence on Industrial Microwave Heating Applications-Enerzi Microwave Systems Pvt. Ltd., Additive Manufacturing and Reverse Engineering Lab, for the research facilities and KLS Gogte Institute of Technology, Belagavi for the research encouragement.

REFERENCES

[1] G. Chryssolouris, D. Mavrikios, N. Papakostas, D. Mourtzis, G. Michalos, and K. Georgoulias, "Digital manufacturing: History, perspectives, and outlook," *Proc. Inst. Mech. Eng. Part B J. Eng. Manuf.*, vol. 223, no. 5, pp. 451–462, 2009, doi: 10.1243/09544054JEM1241.

[2] J. Torres, E. Abo, and A. J. Sugar, "Effects of annealing and acetone vapor smoothing on the tensile properties and surface roughness of FDM printed ABS components," *Rapid Prototyp. J.*, vol. 29, no. 5, pp. 921–934, 2023, doi: 10.1108/RPJ-03-2022-0088.

[3] J. R. C. Dizon, C. C. L. Gache, H. M. S. Cascolan, L. T. Cancino, and R. C. Advincula, "Post-processing of 3D-printed polymers," *Technologies*, vol. 9, no. 3, pp. 1–37, 2021, doi: 10.3390/technologies9030061.

[4] A. Mathew, S. Ram Kishore, A. T. Tomy, M. Sugavaneswaran, S. G. Scholz, A. Elkaseer, V. H. Wilson, and A. J. Rajan, "Vapour polishing of fused deposition modelling (FDM) parts: a critical review of different techniques, and subsequent surface finish and mechanical properties of the post-processed 3D-printed parts," *Prog. Addit. Manuf.*, vol. 8, pp. 1161–1178, 2023, doi: 10.1007/s40964-022-00391-7.

[5] F. Ning, W. Cong, J. Qiu, J. Wei, and S. Wang, "Additive manufacturing of carbon fiber reinforced thermoplastic composites using fused deposition modeling," *Compos. Part B Eng.*, vol. 80, no. July, pp. 369–378, 2015, doi: 10.1016/j.compositesb.2015.06.013.

[6] H. Amelia, "The importance of 3D printing in Industry 4.0," *3D Natives*, 2021. www.3dnatives.com/en/3d-printing-in-industry-4-0-150220215/ (accessed Jun. 11, 2021).

[7] A. Vranić, N. Bogojevic, S. Ciric-kostic, and D. Croccolo, and G. Olmi, "Advantages and drawbacks of additive manufacturing," no. January, 2020, http://doi.org/10.5937/imk1702057v.

[8] Y. L. Yap, W. Toh, R. Koneru, R. Lin, K. I. Chan, H. Guang, W. Y. B. Chan, S. S. Teong, G. Zheng, T. Y. Ng, "Evaluation of structural epoxy and cyanoacrylate adhesives on jointed 3D printed polymeric materials," *Int. J. Adhes. Adhes.*, vol. 100, no. January, p. 102602, 2020, doi: 10.1016/j.ijadhadh.2020.102602.

[9] "3D Printing Market Size, Share, Business Fotune Insights," 2021. www.fortunebusinessinsights.com/industry-reports/3d-printing-market-101902 (accessed Jun. 11, 2021).

[10] V. K. Tiwary, A. Padmakumar, and V. R. Malik, "An overview on joining/welding as post-processing technique to circumvent the build volume limitation of an FDM-3D printer," *Rapid Prototyp. J.*, vol. 27, no. 4, pp. 808–821, 2021, doi: 10.1108/RPJ-10-2020-0265.

[11] Y. Tao, H. Wang, Z. Li, P. Li, and S. Q. Shi, "Development and application ofwood flour-filled polylactic acid composite filament for 3d printing," *Mater. (Basel).*, vol. 10, no. 4, pp. 1–6, 2017, doi: 10.3390/ma10040339.

[12] Z. Weng, J. Wang, T. Senthil, and L. Wu, "Mechanical and thermal properties of ABS/montmorillonite nanocomposites for fused deposition modeling 3D printing," *Mater. Des.*, vol. 102, pp. 276–283, 2016, doi: 10.1016/j.matdes.2016.04.045.

[13] T. Vivek, P. Arunkumar, A. S. Deshpande, M. Vinayak, R. M. Kulkarni, and A. Asif, "Development of polymer nano composite patterns using fused deposition modeling for rapid investment casting process," *AIP Conf. Proc.*, vol. 1943, 2018, doi: 10.1063/1.5029686.

[14] K. Kitsakis, P. Alabey, J. Kechagias, and N. Vaxevanidis, "A Study of the dimensional accuracy obtained by low cost 3D printing for possible application in medicine," *IOP*

Conf. Ser. Mater. Sci. Eng., vol. 161, no. 1, 2016, doi: 10.1088/1757-899X/161/1/012025.

[15] A. R. Torrado Perez, D. A. Roberson, and R. B. Wicker, "Fracture surface analysis of 3D-printed tensile specimens of novel ABS-based materials," *J. Fail. Anal. Prev.*, vol. 14, no. 3, pp. 343–353, 2014, doi: 10.1007/s11668-014-9803-9.

[16] L. Luo, I. Baran, S. Rusinkiewicz, and W. Matusik, "Chopper: Partitioning models into 3D-printable parts," *ACM Trans. Graph.*, vol. 31, no. 6, 2012, doi: 10.1145/2366145.2366148.

[17] P. Song, Z. Fu, L. Liu, and C. W. Fu, "Printing 3D objects with interlocking parts," *Comput. Aided Geom. Des.*, vol. 35–36, pp. 137–148, 2015, doi: 10.1016/j.cagd.2015.03.020.

[18] Michael Brooks, "RepRage – How much power does a 3D printer use?," *Reprage. Com*, 2017. https://reprage.com/post/39698552378/how-much-power-does-a-3d-printer-use (accessed Jun. 11, 2021).

[19] U. M. Dilberoglu, B. Gharehpapagh, U. Yaman, M. Dolen, E. Çanti, M. Aydin, and F. Yildirim, "Manufacturing and automotive industries," *Mater. Des.*, vol. 35, no. 4, pp. 1–8, 2018, doi: 10.1016/j.matpr.2021.05.140.

[20] V. Malik, N. K. Sanjeev, H. S. Hebbar, and S. V. Kailas, "Investigations on the effect of various tool pin profiles in friction stir welding using finite element simulations," *Procedia Eng.*, vol. 97, pp. 1060–1068, 2014, doi: 10.1016/j.proeng.2014.12.384.

[21] M. G. Kazen and P. Asadi, *Advances in Friction Stir Welding and Processing Related Titles:* Woodhead, 1981.

[22] F. Lambiase, A. Paoletti, and A. Di Ilio, "Effect of tool geometry on mechanical behavior of friction stir spot welds of polycarbonate sheets," *Int. J. Adv. Manuf. Technol.*, vol. 88, no. 9–12, pp. 3005–3016, Feb. 2017, doi: 10.1007/s00170-016-9017-2.

[23] V. Malik and S. V. Kailas, "Plasticine modeling of material mixing in friction stir welding," *J. Mater. Process. Technol.*, vol. 258, no. July 2017, pp. 80–88, 2018, doi: 10.1016/j.jmatprotec.2018.03.008.

[24] C. B. Lin and L. C. Wu, "Friction welding of similar and dissimilar materials: PMMA and PVC," *Polym. Eng. Sci.*, vol. 40, no. 8, pp. 1931–1941, 2000, doi: 10.1002/pen.11325.

[25] V. Malik, N. . Sanjeev, H. S. Hebbar, and S. V. Kailas, "Time efficient simulations of plunge and dwell phase of FSW and its significance in FSSW," *Procedia Mater. Sci.*, vol. 5, pp. 630–639, 2014, doi: 10.1016/j.mspro.2014.07.309.

[26] M. S. A. Parast, A. Bagheri, A. Kami, M. Azadi, and V. Asghari, "Bending fatigue behavior of fused filament fabrication 3D-printed ABS and PLA joints with rotary friction welding," *Prog. Addit. Manuf.*, vol. 7, no. 6, pp. 1345–1361, 2022, doi: 10.1007/s40964-022-00307-5.

[27] J. Sprovieri, "Friction Stir Spot Welding," 2016. www.assemblymag.com/articles/93337-friction-stir-spot-welding

[28] V. Malik, N. K. Sanjeev, H. S. Hebbar, and S. V. Kailas, "Finite element simulation of exit hole filling for friction stir spot welding – A modified technique to apply practically," *Procedia Eng.*, vol. 97, pp. 1265–1273, 2014, doi: 10.1016/j.proeng.2014.12.405.

[29] X. W. Yang, T. Fu, and W. Y. Li, "Friction stir spot welding: A review on joint macro- and microstructure, property, and process modelling," *Advances in Materials Science and Engineering*, vol. 2014, pp. 11, 2014. doi: 10.1155/2014/697170.

[30] H. A. Derazkola and M. Elyasi, "The influence of process parameters in friction stir welding of Al-Mg alloy and polycarbonate," *J. Manuf. Process.*, vol. 35, no. May, pp. 88–98, 2018, doi: 10.1016/j.jmapro.2018.07.021.

[31] A. Doniavi, S. Babazadeh, T. Azdast, and R. Hasanzadeh, "An investigation on the mechanical properties of friction stir welded polycarbonate/aluminium oxide nanocomposite sheets," *J. Elastomers Plast.*, vol. 49, no. 6, pp. 498–512, 2017, doi: 10.1177/0095244316674352.

[32] B. Vijendra and A. Sharma, "Induction heated tool assisted friction-stir welding (i-FSW): A novel hybrid process for joining of thermoplastics," *J. Manuf. Process.*, vol. 20, pp. 234–244, 2015, doi: 10.1016/j.jmapro.2015.07.005.

[33] U. M. Dilberoglu, B. Gharehpapagh, U. Yaman, M. Dolen, E. Çanti, M. Aydin, and F. Yildirim, "Manufacturing and automotive industries," *ACM Trans. Graph.*, vol. 100, no. 4, pp. 1–8, Jan. 2020, doi: 10.1201/b16763-29.

[34] V. Malik and S. V. Kailas, "Understanding the effect of tool geometrical aspects on intensity of mixing and void formation in friction stir process," in *Proceedings of the Institution of Mechanical Engineers, Part C: Journal of Mechanical Engineering Science*, 2020, vol. 235, no. 4, pp. 744–757. doi: 10.1177/0954406220938410.

[35] R. B. Azhiri, R. Mehdizad Tekiyeh, E. Zeynali, M. Ahmadnia, and F. Javidpour, "Measurement and evaluation of joint properties in friction stir welding of ABS sheets reinforced by nanosilica addition," *Meas. J. Int. Meas. Confed.*, vol. 127, pp. 198–204, 2018, doi: 10.1016/j.measurement.2018.05.005.

[36] P. A. Bajakke, S. C. Jambagi, V. R. Malik, and A. S. Deshpande, Friction Stir Processing: An Emerging Surface Engineering Technique, in: Gupta, K. (ed), *Surface Engineering of Modern Materials. Engineering Materials*, Springer, Cham, 2020. doi: 10.1007/978-3-030-43232-4_1.

[37] M. K. Bilici and A. I. Yukler, "Effects of welding parameters on friction stir spot welding of high density polyethylene sheets," *Mater. Des.*, vol. 33, no. 1, pp. 545–550, 2012, doi: 10.1016/j.matdes.2011.04.062.

[38] V. K. Stokes, "The effect of fillers on the vibration welding of poly(butylene terephthalate)," *Polymer (Guildf).*, vol. 34, no. 21, pp. 4445–4454, 1993, doi: 10.1016/0032-3861(93)90151-Y.

[39] W. S. Junior, U. A. Handge, J. F. dos Santos, V. Abetz, and S. T. Amancio-Filho, "Feasibility study of friction spot welding of dissimilar single-lap joint between poly(methyl methacrylate) and poly(methyl methacrylate)-SiO2 nanocomposite," *Mater. Des.*, vol. 64, pp. 246–250, 2014, doi: 10.1016/j.matdes.2014.07.050.

[40] D. Benyerou, O. C. El Bahri, H. Khellafi, H. M. Meddah, A. Benhamena, K. Hachelaf, and A. Lounis, "Parametric study of friction stir spot welding (FSSW) for polymer materials case of high density polyethylene sheets: Experimental and numerical study," *Frat. ed Integrita Strutt.*, vol. 15, no. 55, pp. 145–158, 2021, doi: 10.3221/IGF-ESIS.55.11.

[41] K. Prashantha and F. Roger, "Pure and applied chemistry multifunctional properties of 3D printed poly (lactic acid)/ graphene nanocomposites by fused deposition modeling," *J. Macromol. Sci., Part A Pure Appl. Chem.*, vol. 1325, no. December 2016, pp. 24–29, 2017, doi: 10.1080/10601325.2017.1250311.

[42] R. J. Wise and I. D. Froment, "Microwave welding of thermoplastics," *J. Mater. Sci.*, vol. 36, no. 24, pp. 5935–5954, 2001, doi: 10.1023/A:1012993113748.

[43] V. K. Tiwary, N. J. Ravi, P. Arunkumar, S. Shivakumar, A. S. Deshpande, and V. R. Malik, "Investigations on friction stir joining of 3D printed parts to overcome bed size limitation and enhance joint quality for unmanned aircraft systems," *Proc. Inst. Mech. Eng. Part C J. Mech. Eng. Sci.*, vol. 234, no. 24, pp. 4857–4871, 2020, doi: 10.1177/0954406220930049.

[44] V. K. Tiwary, A. Padmakumar, and V. Malik, "Adhesive bonding of similar/dissimilar three-dimensional printed parts (ABS/PLA) considering joint design, surface

treatments, and adhesive types," *Proc. Inst. Mech. Eng. Part C J. Mech. Eng. Sci.*, vol. 236, no. 16, pp. 8991–9002, Aug. 2022, doi: 10.1177/09544062221089849.

[45] V. K. Tiwary, A. Padmakumar, and V. R. Malik, "Investigations on FSW of nylon micro-particle enhanced 3D printed parts applied to a Clark-Y UAV wing," *Weld. Int.*, vol. 36, no. 8, pp. 474–488, 2022, doi: 10.1080/09507116.2022.2104141.

[46] V. K. Tiwary, P. Arunkumar, and P. M. Kulkarni, "Micro-particle grafted eco-friendly polymer filaments for 3D printing technology," *Mater. Today Proc.*, vol. 28, pp. 1980–1984, 2020, doi: 10.1016/j.matpr.2020.05.573.

9 Microwave and Ultrasonic Welding of FDM-3D Printed Components

*Vivek Kumar Tiwary, Mehmet Şükrü Adin,
Arunkumar Padmakumar and Vinayak R. Malik*

9.1 INTRODUCTION

Electromagnetic radiation refers to the energy that is propagated through space in the form of electromagnetic waves. These waves consist of oscillating electric and magnetic fields that are perpendicular to each other and travel at the speed of light (2.998 × 108 m/s) [1]. Electromagnetic radiation encompasses a wide range of wavelengths and frequencies, collectively known as the electromagnetic spectrum. The spectrum includes various types of radiation, each with distinct characteristics and applications [2].

The major types of electromagnetic radiations, in the order of increasing frequency and decreasing wavelength are radio waves, microwaves, infrared radiations, visible light, ultraviolet, X-rays and gamma rays. Figure 9.1 (a) depicts the manner in which electromagnetic waves propagates in vacuum while Figure 9.1 (b) shows the spread of various waves over the electromagnetic spectrum.

The phrase microwave (MW) refers to electromagnetic radiation with frequency spectrum from 300 MHz to 300 GHz and wavelength ranging from 1 mm to 1 m. The technique was initially created for communication, but it has since been widely used for food preparation, drying, material processing, vulcanizing rubber, and of late even for plastic welding [3].

Ultrasonic waves do not fall within the electromagnetic spectrum. While electromagnetic waves propagate through the interaction of electric and magnetic fields, ultrasonic waves are mechanical waves that require a medium (such as air, water, or solids) to travel [4]. Ultrasonic waves are characterized by high-frequency sound waves, typically above the upper limit of human hearing, which is approximately 20,000 Hz. These waves are produced by mechanical vibrations or oscillations, and they propagate through a medium by causing particles in the medium to vibrate, transmitting energy from one location to another [5].

DOI: 10.1201/9781032665351-11

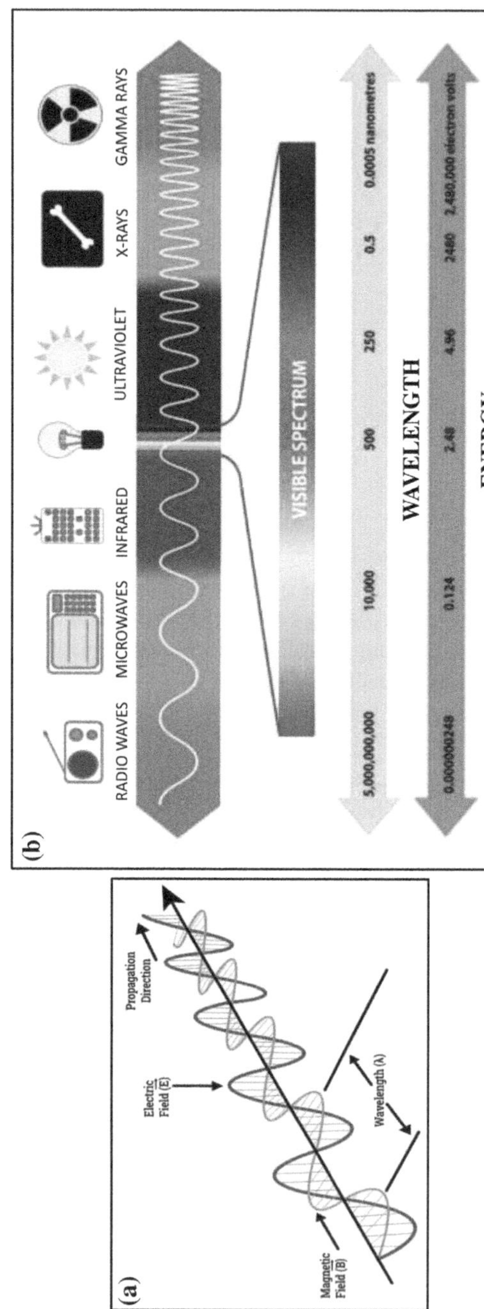

FIGURE 9.1 Figure depicting (a) propagation method of electromagnetic waves (b) spread of various waves over the electromagnetic spectrum

Ultrasonic waves find various applications in fields such as medical imaging (ultrasound), industrial testing and inspection, cleaning, distance measurement, and more recently even for polymer welding. However, unlike electromagnetic waves, which can travel through a vacuum, ultrasonic waves require a physical medium to propagate [6].

3D Printing/Additive Manufacturing (AM) is a revolutionary emerging technology that is expected to alter the century old approaches of design and manufacturing with intense geopolitical, economic, social, demographic, environmental, and security implications [7]. As a contrast to subtractive and formative manufacturing processes, the American Society for Testing of Materials (ASTM) describes 3D printing as a process of joining materials to build things from a 3D model data, generally "layer upon layer" [8]. The key benefits of the technology include design freedom, less setup cost, less time in getting the product, customization as well as almost zero wastages [9]. The applications of this technology can be observed in consumer products, medical, academic, aerospace, automotive, construction and almost all the sectors [10].

Like any other emerging technology that faces difficulties during its transforming stages, 3D printing also suffers from an inherent limitation of the maximum volume of the part it can fabricate [5]. This issue of the size limitation of 3D printers is represented in the Figure 9.2 below. On the other hand, if a bigger bed size 3D printer is procured, it consumes more energy as well as costs more compared to a smaller 3D printer. To overcome this problem, one of the possible solutions could be to split the model into smaller segments and further join/weld them by some technique. Ultrasonic Welding (UW) and Microwave Welding (MW) are such two methods which are still in their nascent stages. While UW is regarded as the greatest of its type

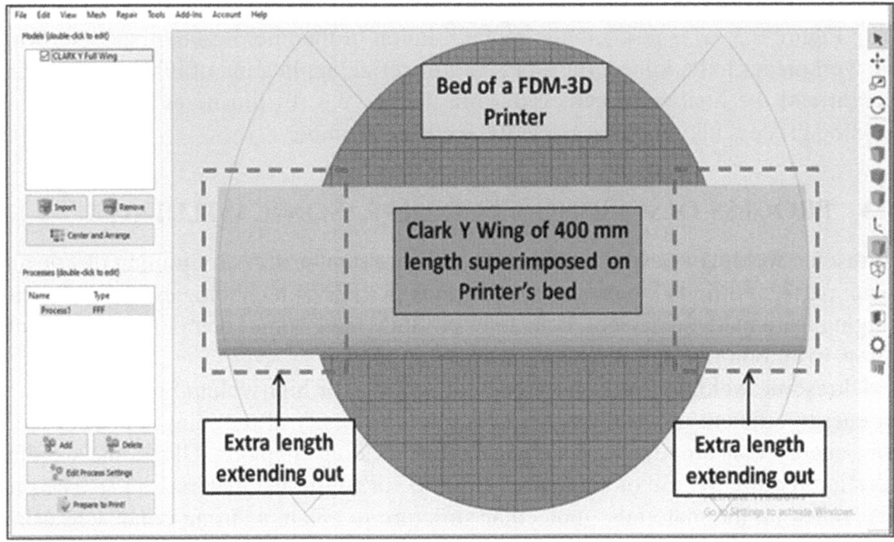

FIGURE 9.2   Bed size limitation in FDM-3D printers

and might be employed for the above purpose because of its "ease of automation" [4]. MW technology has enormous potential for development as a flexible assembly technique suited for use in large-scale manufacturing industries such as automobiles and aviation [11].

Reviewing research articles related to UW and MW of thermoplastics revealed that not much research has been done on joining of 3D printed parts by these techniques. Considering this as an important gap for 3D printing, the present chapter presents the techniques of UW and MW for joining 3D printed parts. Further, for the benefit of the practitioners, important parameters in each of the joining techniques are presented. At last, a case study of microwave being employed to attain a larger 3D printed component is also explained. In comparison to the valued literatures results, the current article is expected to provide a strong viewpoint, since the combined effect of 3D printing and microwave/ultrasonic welding technology is discussed at once.

9.2 PROCESS DESCRIPTION FOR MICROWAVE WELDING

MW is a relatively new and versatile welding process that has gained popularity due to its ability to provide fast and efficient welds for thermoplastics. It involves a volumetric based heating process employing a remote source and resulting in a quick and uniform temperature rise all the way through at the same time [12]. This technique seems to be the most flexible and cost effective method for mass production industries such as automobiles, able to weld intricate joint geometries with superior joint strength and quality [13], [14].

MW welding involves planting an absorbent material amid the thermoplastics to be joined which heats up under MW radiation and passes the heat developed on to the thermoplastics through thermal conduction. This makes the absorbent material being displaced into the weld bead creating a molten sub-layer. Finally, a light pressure is applied at the interface squeezing the molten layers together resulting in a new weld [15]. Figure 9.3 (a) is a schematic representation of the microwave process showing the workpieces to be joined, the absorbent material, application of microwaves, dead weight and the final weld achieved while Figure 9.3 (b) illustrates the microwave welding process taking place inside the welding chamber.

9.3 PROCESS DESCRIPTION FOR ULTRASONIC WELDING

Ultrasonic welding is a well-established and important process for joining plastic and metal parts, and finds increasing applications in various technologies and industries ranging from medical devices, consumer products, electronics components and packaging, to automotive and aerospace assembly [16].

Ultrasonic welding is a rapid process, allowing for high-volume production. It is an energy-efficient process since heat is generated only at the joint interface, reducing energy consumption compared to other welding methods. Ultrasonic welding does not require the use of adhesives, solvents, or additional materials [17]. The process relies on the materials' molecular structure to create a strong bond. Ultrasonic welding can be used to join a wide range of materials, including plastics, metals, and even dissimilar materials. The process does not produce smoke, fumes, or sparks. It

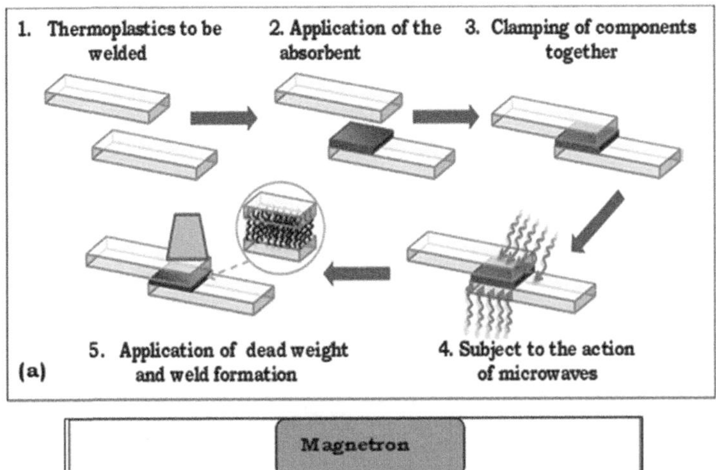

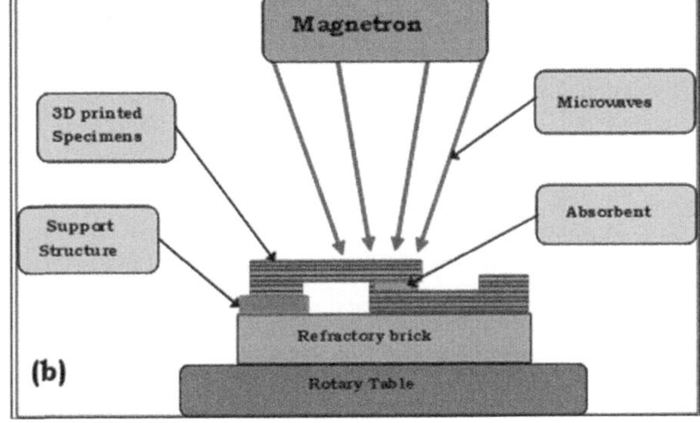

FIGURE 9.3 (a) Microwave welding process for 3D printed thermoplastics (b) Inside a microwave chamber

also allows for precise control over the welding parameters, resulting in consistent and high-quality welds [18]. Ultrasonic welding finds applications in various industries, such as automotive, electronics, medical devices, packaging, textiles, and more. It is commonly used for applications like sealing, staking, inserting, spot welding, and wire splicing [19].

The ultrasonic welding process involves preparing the surfaces to be welded thoroughly by cleaning, removing any contaminants or debris that could affect the quality of the weld. Next, the 3D printed thermoplastics to be welded are securely clamped together using a fixture or a horn (sonotrode) that applies pressure on the joint area. The horn or sonotrode generates high-frequency mechanical vibrations, typically in the range of 15 kHz to 300 kHz, but more typically 20, 30, or 40 kHz, transmitted through a booster, possibly amplifying the vibration, then to a horn (sonotrode) [20]. These vibrations are transferred to the joint interface, creating localized friction and heat. The high-frequency vibrations cause the material at the joint interface to soften and become more pliable. This softening occurs due to the mechanical energy converting into heat energy. As the material softens, the pressure applied by the sonotrode forces the materials together, allowing them to fuse and form a strong bond [21].

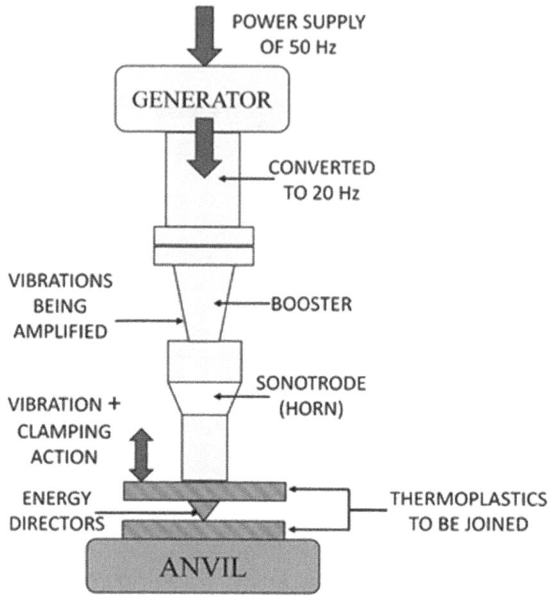

FIGURE 9.4   Schematic representation of ultrasonic welding of 3D printed parts

The process typically takes a few seconds. After the ultrasonic vibrations are stopped, the softened material solidifies, resulting in a solid and durable weld. A schematic representation of ultrasonic welding is depicted in Figure 9.4 below while Figure 9.5 shows the four stages involved in the process of ultrasonic welding.

9.4   IMPORTANT WELDING PARAMETERS FOR MICROWAVE WELDING

Microwave welding is a process that utilizes microwave energy to generate heat and create welds between two or more materials. The specific process parameters for microwave welding can vary depending on the materials being welded and the desired outcome. However, here are some common process parameters that may be considered:

1. Power level: The power level of the microwave energy is an important parameter. It determines the heating rate and temperature achieved during the welding process. The power level should be selected based on the material properties and thickness. Usually the power rating varies from 155 W to 6 kW in the case of industrial microwave setups [23].
2. Frequency: Microwave welding can be performed at different frequencies, such as 2.45 GHz or 915 MHz. The frequency choice depends on the material properties and the equipment available. The most preferred one is 2.45 GHz [24].

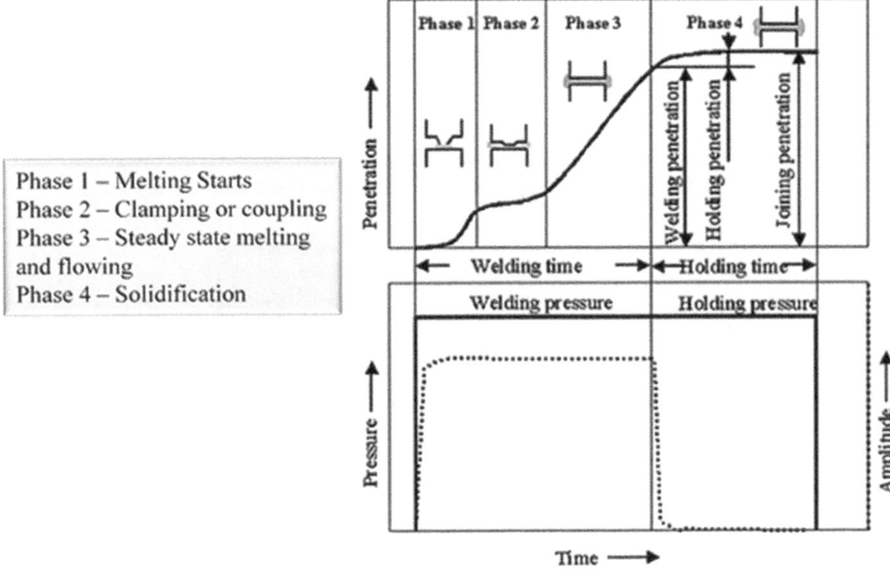

FIGURE 9.5 Different stages of Ultrasonic Welding (Lanxess [22])

3. Exposure Time: The welding time refers to the duration for which the microwave energy is applied to the materials. It is determined by factors such as the material type, thickness, and desired joint strength. Optimal time should be determined through experimentation and testing, that can vary between 1 to 15 min based on the material taken for welding [12], [25].

4. Welding Pressure: The application of pressure during microwave welding can help ensure good contact between the materials and promote intermolecular bonding. The pressure level depends on the material properties, joint configuration, and desired weld quality [26].

5. Preheating: In some cases, preheating the materials before applying microwave energy can improve the welding process. Preheating helps reduce thermal gradients, minimize stress, and enhance material flow during welding [27].

6. Implant material and its quantity: Microwave energy is absorbed by certain materials more readily than others. Adding or selecting an appropriate implant material can help concentrate the energy at the desired joint interface, promoting localized heating and better weld formation. Further, there is an optimum quantity of the implant material to be used that can be determined by a number of trial experiments. Example of the implant material include polypyrrol, polyaniline, ceramic materials (PbO_2, Cr_2O_3, MnO_2, Fe_2O_3), CNT, MWCNT, Graphene etc. [28], [29].

7. Polymers considered for welding: the thermoplastics considered during the welding process also determines the final strength achieved. It is observed that amorphous thermoplastics (ABS for example) responds better than crystalline

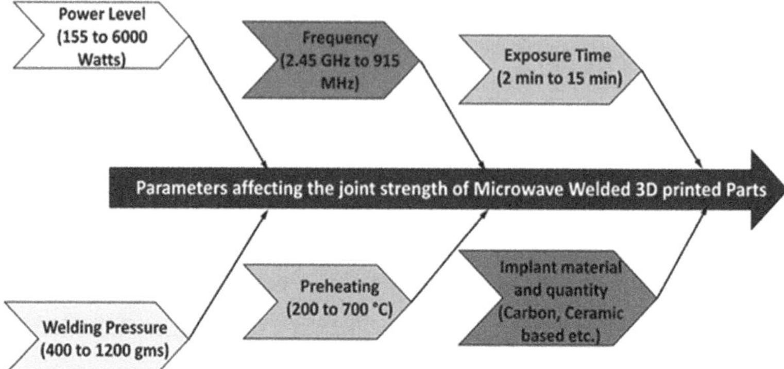

FIGURE 9.6 Factors affecting the strength of microwave welded 3D printed parts

ones (PLA). Further, joining dissimilar thermoplastics is a challenge which is still to be addressed [30].

It's important to note that the above parameters are general guidelines, and the optimal values may vary depending on the specific application and materials involved. It is recommended to consult the manufacturer's guidelines, perform trial welds, and conduct testing to determine the ideal process parameters for the specific welding needs. Figure 9.6.

9.5 MICROWAVE WELDING IMPLANTS

Implants play a very crucial role in the case of microwave welding. Here, in this section, some examples of implants that have been employed during microwave welding are presented:

1. Metal Mesh: A conductive metal mesh, such as stainless steel or copper, can be used as an implant between the surfaces to be welded. The mesh enhances heat distribution, improves bond strength, and provides structural reinforcement to the joint.
2. Carbon Fiber Mesh: Similar to metal mesh, carbon fiber mesh can be used as an implant to enhance the welding process. Carbon fiber offers high strength, lightweight properties, and good electrical conductivity, making it suitable for microwave welding applications.
3. Adhesive Films: Thermoplastic adhesive films, such as polyethylene or polypropylene films, can be used as implants to promote bonding between the surfaces being welded. These films soften and melt during the welding process, creating a strong bond between the parts.
4. Alignment Pins: Alignment pins or features can be incorporated into the design of the parts to aid in precise alignment during the welding process. These pins ensure accurate positioning of the components, facilitating proper alignment and improving joint quality [31].

5. Energy Directors: Geometric patterns or raised features, known as energy directors, can be designed into the parts as implants. These features concentrate the microwave energy, resulting in localized heating and improved welding efficiency.
6. Thermal Management Inserts: In cases where heat management is crucial, thermal management inserts can be employed. These inserts, typically made of thermally conductive materials like metals or ceramics, help dissipate heat and prevent thermal degradation or distortion of the parts during the welding process [32], [33].
7. Ceramic implants can be employed during microwave welding to enhance the welding process and achieve desired joint properties. Some of them include Alumina Ceramic Inserts, Zirconia Ceramic Pins, PbO_2, Cr_2O_3, MnO_2, Fe_2O_3 etc.

9.6 IMPORTANT WELDING PARAMETERS FOR ULTRASONIC WELDING

Ultrasonic welding is a solid-state welding process that uses high-frequency mechanical vibrations to create a weld between two materials. The process parameters for ultrasonic welding can vary depending on the materials being welded and the desired outcome. Here are some common parameters to consider:

1. Amplitude: Amplitude refers to the peak-to-peak displacement of the ultrasonic vibrations. It determines the energy applied to the materials and affects the weld strength. The amplitude is typically measured in micrometers (μm) or thousandths of an inch (mils) [21].
2. Frequency: Ultrasonic welding is typically performed at frequencies in the range of 15 kHz to 300 kHz, but more typically 20, 30, or 40 kHz. The frequency selection depends on the material properties, joint configuration, and equipment capabilities [20].
3. Pressure: Pressure is applied during ultrasonic welding to create sufficient contact and promote material bonding. The pressure level depends on the material properties, joint design, and desired weld quality. It is typically measured in pounds per square inch (psi) or newtons per square millimeter (N/mm²) [34].
4. Welding Time: The welding time refers to the duration for which the ultrasonic vibrations are applied to the materials. It is determined by factors such as material thickness, desired joint strength, and equipment capabilities. The welding time is typically measured in seconds [35].
5. Sonotrode Design: The sonotrode, also known as the horn or tooling, is a crucial component in ultrasonic welding. The design of the sonotrode, including its shape, material, and dimensions, can affect the welding process and joint quality. The specific sonotrode design should be chosen based on the materials being welded [36].
6. Clamping: Proper clamping of the materials is essential to ensure good contact and alignment during ultrasonic welding. The clamping force should

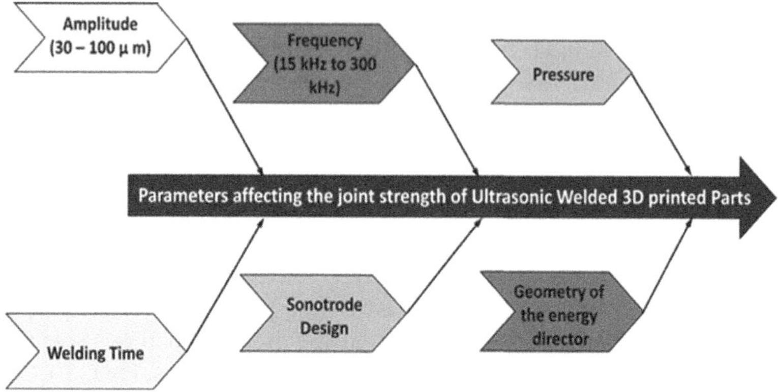

FIGURE 9.7 Important process parameters in ultrasonic welding of 3D printed parts

be sufficient to hold the materials together without distorting or damaging them [37].

7. Energy Control: Some ultrasonic welding machines provide the ability to control the energy delivered during the welding process. This control can be achieved through parameters such as energy limit, energy profile, or energy ramp-up [38].

8. Geometry of the energy director: An energy director is a raised triangular ridge of material molded on one of the joint surfaces. The apex of the energy director is under the greatest stress during welding and is forced into contact with the other part, generating friction, which causes it to melt. The molten energy director flows into the joint interface and forms a bond. Different designs like criss-cross energy director, cone energy director, interrupted energy directors can be employed [39].

It's important to note that the above parameters are general guidelines, and the optimal values may vary depending on the specific application and materials involved. It is recommended to consult the manufacturer's guidelines, perform trial welds, and conduct testing to determine the ideal process parameters for your specific ultrasonic welding needs. Figure 9.7 shows the important parameters involved in the process.

9.7 A CASE STUDY ON MICROWAVE WELDING OF 3D PRINTED PARTS

In the following section, a case study on joining two 3D printed rotor components by the microwave welding technique is demonstrated. Figure 9.8 a and b shows the top and the bottom view of the CAD model of the rotor component. Figure 9.8 c depicts the welding setup which was a household microwave oven, in this case. Figures 9.8 d and e show the final welded 3D printed rotor by the microwave technique.

Firstly, the two separate parts were 3D printed using ABS thermoplastic (Acrylonitrile Butadiene Styrene) polymer. ABS material was selected because of

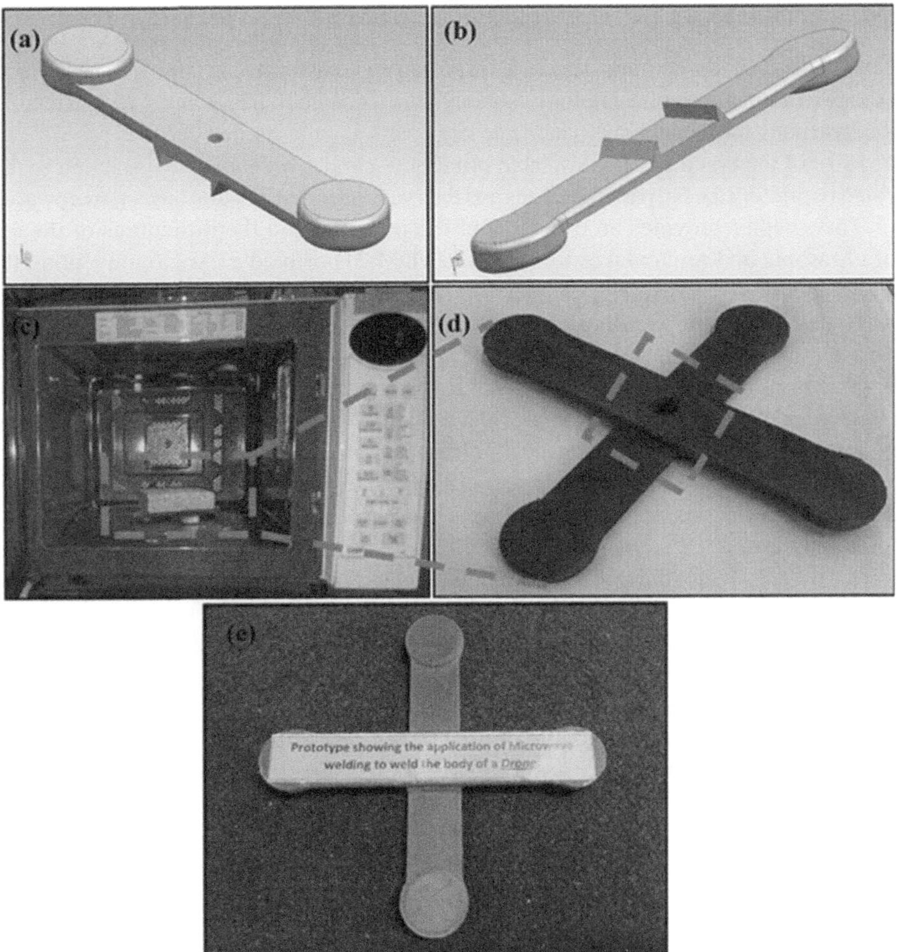

FIGURE 9.8 Part design CAD model of the drone rotor (a) Top View (b) Bottom View (c) Execution of the welding process (d) and (e) Final welded component

its universal acceptability in 3D printing technology. The printer settings while fabricating the parts were as follows: Layer thickness – 0.2 mm, Infill percentage – 35% with no support structure. The implant (susceptor, absorbent) material employed was 0.2 gms Fe_2O_3 (Ferric Oxide). Ferric Oxide was selected after a number of trial experiments using graphite and lead powders as susceptors.

The welding was done with a commercially available household microwave oven having a frequency of 2.45 GHz, while a 400 gms dead weight was used as a dead weight. The time of microwave exposure was around 90–120 seconds. The final microwave welded, as shown in the figure below, showed good strength and integrity. The present case study practically validated and confirmed that microwave welding technique can also be extended to servicing and repairing of broken 3D printed parts.

CONCLUSIONS

3D printing, in the coming decade is expected to transform itself from a niche technology to a mainstream technology. Although the technology has attained wide attention, implementing the technology to fabricate large parts remains limited due to the constricted bed size of the commercially available printers. A simple yet elaborative solution to this could be printing the big part in sections and then welding them by ultrasonic or microwaves.

This chapter provides an overview of the potential and the difficulties of the use of ultrasonic and microwaves to join/weld FDM-3D printed parts circumventing the above mentioned issue. The chapter firstly, describes the process to execute microwave and ultrasonic welding techniques to join FDM-3D printed parts. Secondly, critical process parameters of the welding techniques, whose judicial selection and optimization can yield better end results are presented and deliberated. Lastly, the chapter ends with a case study wherein a drone's rotor is being welded by microwave radiations using a commercially available household microwave oven.

Although microwave welding seems to be the most flexible and cost effective method for mass production industries such as automobiles, capable to weld intricate joint geometries with superior joint strength and quality, finding a suitable implant material to execute the process is a challenge. The second technique, ultrasonic welding, although regarded as the greatest of its type because of its "ease of automation", requirement of unique structural elements like raised ridges, tooling customization and possibility to weld only thin sections are its challenges.

As a concluding remark, both microwave and ultrasonic welding are beneficial as they are very quick and affordable option to weld 3D printed parts resulting in complex and high-strength components. However, it is also felt that not enough experience has been gained for application of these techniques on an industrial scale yet to making it as common as other conventional welding technologies. Academicians, researchers, system manufactures as well as interested users are called upon to take the technologies to the next level.

ACKNOWLEDGMENTS

The authors would like to acknowledge Centre of Excellence on Industrial Microwave Heating Applications-Enerzi Microwave Systems Pvt. Ltd., Additive Manufacturing and Reverse Engineering Lab, for the research facilities and KLS Gogte Institute of Technology, Belagavi for the research encouragement.

REFERENCES

[1] M. Q. Tran, "An introduction to classical electromagnetic radiation," *Nucl. Fusion*, vol. 38, no. 5, pp. 775–775, 1998, doi: 10.1088/0029-5515/38/5/701.

[2] R. Sinclair, "Addressing the telecommunications policy gaps," *Telecommun. J. Aust.*, vol. 61, no. 2, pp. 1307–1314, 2011, doi: 10.7790/tja.v61i2.213.

[3] S. Poyraz, L. Zhang, A. Schroder, and X. Zhang, "Ultrafast microwave welding/reinforcing approach at the interface of thermoplastic materials," *ACS Appl. Mater. Interfaces*, vol. 7, no. 40, pp. 22469–22477, 2015, doi: 10.1021/acsami.5b06484.

[4]   R. Kumar, R. Singh, and I. P. S. Ahuja, "Mechanical, thermal and micrographic investigations of friction stir welded: 3D printed melt flow compatible dissimilar thermoplastics," *J. Manuf. Process.*, vol. 38, no. November 2018, pp. 387–395, 2019, doi: 10.1016/j.jmapro.2019.01.043.

[5]   J. Favero, S. Belhabib, S. Guessasma, H. Nouri, and H. Nouri, "Optimal assembling of 3D printed features by modulation of interfacial behaviour," *Rapid Prototyp. J.*, vol. 25, no 1, pp. 13–21, 2018, doi: 10.1108/RPJ-11-2016-0178.

[6]   U. M. Dilberoglu, B. Gharehpapagh, U. Yaman, M. Dolen, E. Çanti, M. Aydin, F. Yildirim, "Manufacturing and automotive industries," *ACM Trans. Graph.*, vol. 100, no. 4, pp. 1–8, Jan. 2020, doi: 10.1201/b16763-29.

[7]   B. G. A. Thomas Campbell and C. Williams, "Strategic foresight report 2021," *Commun. from Comm.*, vol. COM/2021/7, no. Document 52021DC0750, p. 8, September, 2021.

[8]   "ASTM international and ISO unveil framework for global 3D printing standards," *ASTM Stand. News*, vol. 19, pp. 19–21, 2016, [Online]. Available: https://sn.astm.org/?q=outreach/astm-international-and-iso-unveil-framework-global-3d-printing-standards-nd16.html

[9]   V. K. Tiwary, A. Padmakumar, and V. R. Malik, "An overview on joining/welding as post-processing technique to circumvent the build volume limitation of an FDM-3D printer," *Rapid Prototyping Journal*, vol. 27, no. 4. Emerald Group Holdings Ltd., pp. 808–821, 2021. doi: 10.1108/RPJ-10-2020-0265.

[10] "3D printing market size, share, business fotune insights," 2021. www.fortunebusinessinsights.com/industry-reports/3d-printing-market-101902 (accessed Jun. 11, 2021).

[11] V. K. Tiwary, A. Padmakumar, and V. R. Malik, "Investigations on microwave-assisted welding of MEX additive manufactured parts to overcome the bed size limitation," *J. Adv. Join. Process.*, vol. 7, no. February, p. 100141, 2023, doi: 10.1016/j.jajp.2023.100141.

[12] P. Kathirgamanathan, "Polymer Commu Nications," vol. 34, no. 14, pp. 3105–3106, 1993.

[13] F. Castles, D. Isakov, A. Lui, Q. Lei, C. E. J. Dancer, Y. Wang, J. M. Janurudin, S. C. Speller, C. R. M. Grovenor, and P. S. Grant, "Microwave dielectric characterisation of 3D-printed BaTiO3/ABS polymer composites," *Sci. Rep.*, vol. 6, no. March, pp. 1–8, 2016, doi: 10.1038/srep22714.

[14] V. Malik and S. V. Kailas, "Plasticine modeling of material mixing in friction stir welding," *J. Mater. Process. Technol.*, vol. 258, no. March, pp. 80–88, 2018, doi: 10.1016/j.jmatprotec.2018.03.008.

[15] H. Potente, O. Karger, and G. Fiegler, "Laser and microwave welding – The applicability of new process principles," *Macromol. Mater. Eng.*, vol. 287, no. 11, pp. 734–744, 2002, https://doi.org/10.1002/mame.200290002.

[16] P. Yadav, M. G. Mauk, C. Ruiz, and R. Y. Chiou, "Manufacturing science laboratory: Robotic ultrasonic welding," *ASME Int. Mech. Eng. Congr. Expo. Proc.*, vol. 5, pp. 1–8, 2018, doi: 10.1115/IMECE2018-88437.

[17] J. Zhu, Y. Hu, Y. Tang, and B. Wang, "Effects of styrene–acrylonitrile contents on the properties of ABS/SAN blends for fused deposition modeling," *J. Appl. Polym. Sci.*, vol. 134, no. 7, pp. 1–5, 2017, doi: 10.1002/app.44477.

[18] F. Balle, G. Wagner, and D. Eifler, "Ultrasonic spot welding of aluminum sheet/carbon fiber reinforced polymer – Joints," *Materwiss. Werksttech.*, vol. 38, no. 11, pp. 934–938, 2007, doi: 10.1002/mawe.200700212.

[19] R. Y. Yeh and R. Q. Hsu, "Development of ultrasonic direct joining of thermoplastic to laser structured metal," *Int. J. Adhes. Adhes.*, vol. 65, pp. 28–32, 2016, doi: 10.1016/j.ijadhadh.2015.11.001.

[20] S. K. Bhudolia, G. Gohel, K. F. Leong, and A. Islam, "Advances in ultrasonic welding of thermoplastic composites: A review," *Mater. (Basel).*, 2020, [Online]. Available: www.mdpi.com/1996-1944/13/6/1284/htm

[21] I. Fernandez, D. Stavrov, and H. E. N. Bersee, "Ultrasonic welding of advanced thermoplastic composites: An investigation on energy directing surfaces," *ICCM Int. Conf. Compos. Mater.*, vol. 29, pp. 112–121, 2009.

[22] A. Blanc, "Component design," *Internal Components*, 2020. https://techcenter.lanxess.com/scp/americas/en/techServscp/77022/article.jsp?print=true&docId=86732

[23] P. Y. Foong, C. H. Voon, B. Y. Lim, *et al.*, "Feasibility study on microwave welding of thermoplastic using multiwalled carbon nanotubes as susceptor," *Nanomater. Nanotechnol.*, vol. 11, pp. 1–8, 2021, doi: 10.1177/18479804211002926.

[24] O. G. Kravchenko, V. S. Bonab, and I. Manas-zloczower, "Spray-Assisted Microwave Welding of Thermoplastics Using Carbon Nanostructures with Enabled Health Monitoring," pp. 21–24, 2019, doi: 10.1002/pen.25227.

[25] P. A. Bajakke, V. R. Malik, P. Mugali, and A. S. Deshpande, "Microwave Processing of Engineering Materials," pp. 31–55, 2021, doi: 10.1007/978-3-030-62163-6_2.

[26] T. Wu, Y. Pan, E. Liu, and L. Li, "Carbon Nanotube / Polypropylene Composite Particles for Microwave Welding," 2012, doi: 10.1002/app.

[27] F. Castles, D. Isakov, A. Lui, Q. Lei, C. E. J. Dancer, Y. Wang, J. M. Janurudin, S. C. Speller, C. R. M. Grovenor, and P. S. Grant, "Microwave dielectric characterisation of 3D-printed BaTiO3/ABS polymer composites," *Sci. Rep.*, vol. 6, no. March, pp. 1–8, 2016, doi: 10.1038/srep22714.

[28] R. J. Wise and I. D. Froment, "Microwave welding of thermoplastics," *J. Mater. Sci.*, vol. 36, no. 24, pp. 5935–5954, 2001, doi: 10.1023/A:1012993113748.

[29] V. Malik, N. K. Sanjeev, H. S. Hebbar, and S. V. Kailas, "Finite element simulation of exit hole filling for friction stir spot welding – A modified technique to apply practically," *Procedia Eng.*, vol. 97, pp. 1265–1273, 2014, doi: 10.1016/j.proeng.2014.12.405.

[30] V. K. Tiwary, A. Padmakumar, and V. Malik, "Adhesive bonding of similar / dissimilar three-dimensional printed parts (ABS / PLA) considering joint design, surface treatments, and adhesive types," vol. 236, no. 16, pp. 1–12, 2022, doi: 10.1177/09544062221089849.

[31] V. K. Tiwary, A. Padmakumar, and V. R. Malik, "Investigations on FSW of nylon micro-particle enhanced 3D printed parts applied to a Clark-Y UAV wing," *Weld. Int.*, vol. 36, no. 8, pp. 474–488, 2022, doi: 10.1080/09507116.2022.2104141.

[32] V. K. Tiwary, N. J. Ravi, P. Arunkumar, S. Shivakumar, A. S. Deshpande, and V. R. Malik, "Investigations on friction stir joining of 3D printed parts to overcome bed size limitation and enhance joint quality for unmanned aircraft systems," *Proc. Inst. Mech. Eng. Part C J. Mech. Eng. Sci.*, vol. 234, no. 24, pp. 1–15, 2020, doi: 10.1177/0954406220930049.

[33] V. Malik and S. V. Kailas, "Understanding the effect of tool geometrical aspects on intensity of mixing and void formation in friction stir process," in *Proc. Inst. Mech. Eng. Part C J. Mech. Eng. Sci.*, vol. 235, no. 4, pp. 1–14, 2020, doi: 10.1177/0954406220938410.

[34] A. Benatar, R. V. Eswaran, and S. K. Nayar, "Ultrasonic welding of thermoplastics in the near-field," *Polym. Eng. Sci.*, vol. 29, no. 23, pp. 1689–1698, 1989, doi: 10.1002/pen.760292311.

[35] Y. K. Chuah, L. H. Chien, B. C. Chang, and S. J. Liu, "Effects of the shape of the energy director on far-field ultrasonic welding of thermoplastics," *Polym. Eng. Sci.*, vol. 40, no. 1, pp. 157–167, 2000, doi: 10.1002/pen.11149.

[36] S. J. Liu, I. T. Chang, and S. W. Hung, "Factors affecting the joint strength of ultrasonically welded polypropylene composites," *Polym. Compos.*, vol. 22, no. 1, pp. 132–141, 2001, doi: 10.1002/pc.10525.

[37] P. J. Wolcott, A. Hehr, C. Pawlowski, and M. J. Dapino, "Process improvements and characterization of ultrasonic additive manufactured structures," *J. Mater. Process. Technol.*, vol. 233, pp. 44–52, 2016, doi: 10.1016/j.jmatprotec.2016.02.009.

[38] P. Rheology, "A study on the optimization of welding parameters for titanium drums using optimization algorithms," *J Weld Join.*, vol. 41, no. 03, pp. 185–194, 2023, Published online June 30, 2023, https://doi.org/10.5781/JWJ.2023.41.3.7.

[39] L. Langnau, "Ultrasonic welding 3D printed parts," *Make Parts Fast*, 2016. www.makepartsfast.com/how-to-ultrasonically-weld-3d-printed-parts/ (accessed Aug. 21, 2020).

Section III

Miscellaneous Topics

10 Big Area Additive Manufacturing (BAAM) of FDM Parts

Srikrishna B. Rao

10.1 INTRODUCTION

Additive Manufacturing (AM) is yet another method of making a product that will have a higher rate of mass production and materials of any shape and size can be obtained. It mainly refers to the use of 3D image which builds the component in layers by depositing one layer of material over the other. It eliminates many constraints imposed by a conventional manufacturing technique. It has increased applications in 3D objects in cross section, digital images in form of layers are sent and the recipient receives the layer images, that are further used by AM machines to fabricate the 3D object.

10.2 METHODS OF ADDITIVE MANUFACTURING

10.2.1 VAT Photo Polymerization

This method of manufacturing is also known as stereolithography. It makes use of a vat of liquid known as photopolymer resin. The laser beam is used to make a desired shape in the resin to create a layer, that provides a built-in platform for the resins to be lowered from top surface to the inner surfaces of the part produced. The average thickness of the layers formed is from 0.0025mm to 0.5mm. The controlled motor system is used to regulate the UV rays on the surface of the resin to harden the layer, during its formation. These steps are repeated to add the layers one after another [1]. This is shown in Figure 10.1.

10.2.2 Material Jetting

The deposition of material on the table is in the form of droplets and the print head is above the table surface during the process. Hundreds of micro-droplets are set on the table where the charged deflection plates are, and this helps to provide highly increased control to move the jet head and accuracy on the droplets deposited. These droplets get deposited and solidify, creating a constant layer on the table surface. The building up of layers one over the other is formed and the process is repeated

DOI: 10.1201/9781032665351-13

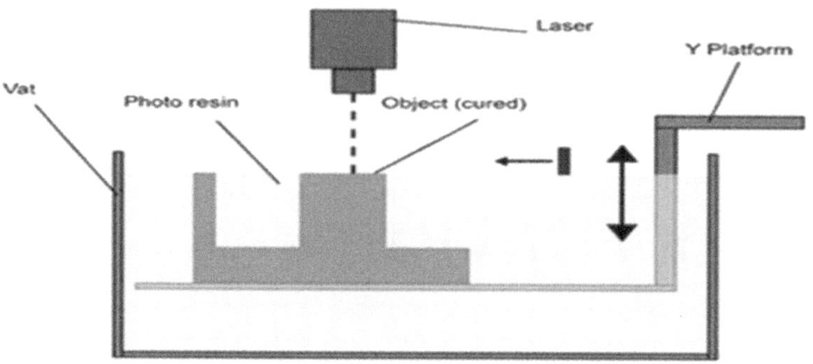

FIGURE 10.1   VAT Photo Polymerization process

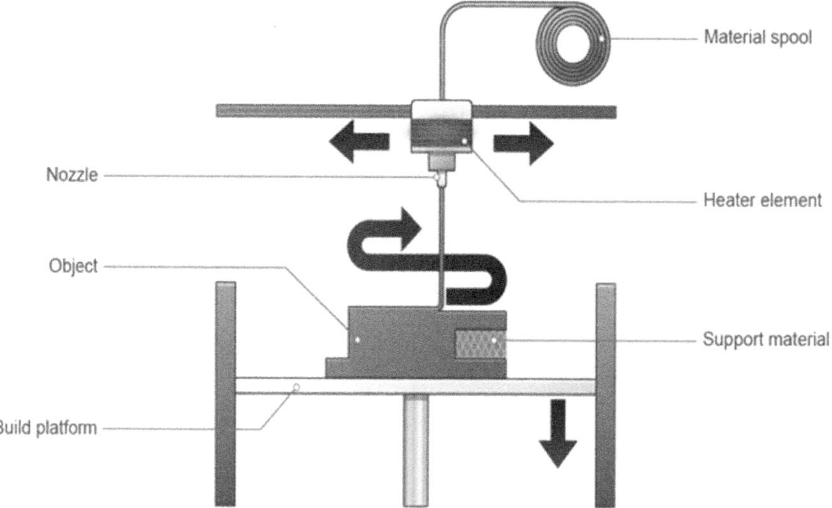

FIGURE 10.2   Material Jetting process

intermittently. With the drop on demand method, these droplets are distributed continuously or individually on the table surface [1]. This is shown in Figure 10.2.

10.2.3 BINDER JETTING

It uses a binder or a powder-based material for the process. This powder-based material is applied to build a platform that has a roller attached to it. The print head deposits the binder on top of the surface. The binder adheres the layers together and is usually in liquid form. Following a layer one after the other, the product is lowered onto the platform. This process is repeated to create layers one after the other until the product is finished. While using this process, different materials like polymers, ceramics, and metals can also be used. It is considered as one of the speediest and quickest of AM processes and allows for customization also. This is shown in Figure 10.3.

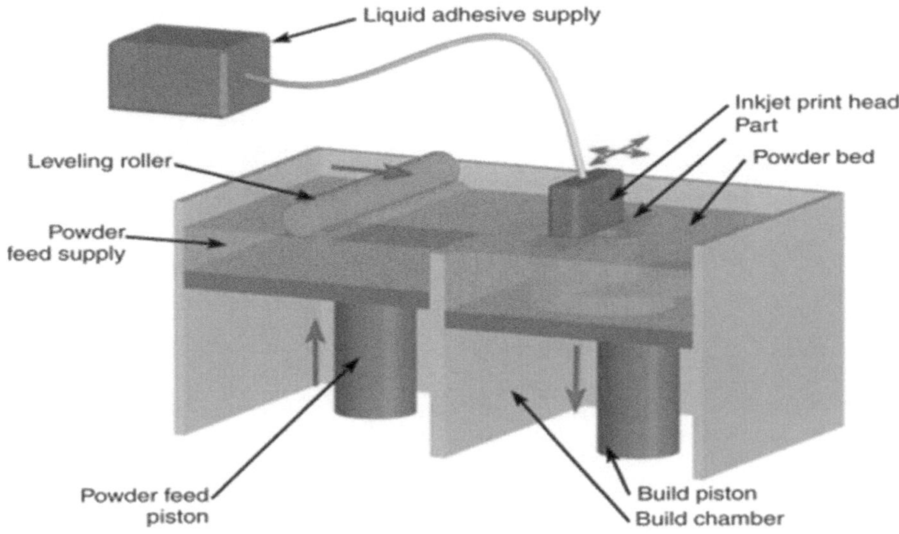

FIGURE 10.3	Binder Jetting process

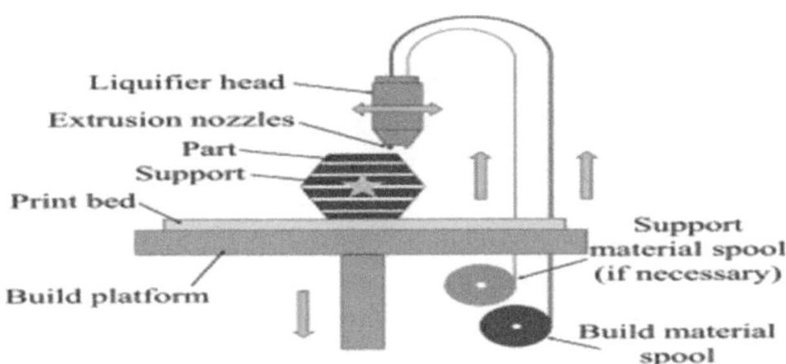

FIGURE 10.4	Material Extrusion process

10.2.4 MATERIAL EXTRUSION

It is an additive manufacturing process often used for home applications. The 3D printers are used where the material is drawn through a nozzle, heated, and then deposited in a continuous stream. This nozzle moves horizontally along the table surface, and the platform moves up, down and vertically to create layers along the surface. Since the material is heated before it is applied, it fuses to the previous layer quickly. Since the material extrusion is mostly seen in inexpensive models, it may have many capabilities due to which they provide strong structural support. But in some of the processes the accuracy is reduced due to the nozzle thickness. Many of the automotive companies making spare parts use the material jetting process to create various manufacturing devices used in assembly lines [2]. This is shown in Figure 10.4.

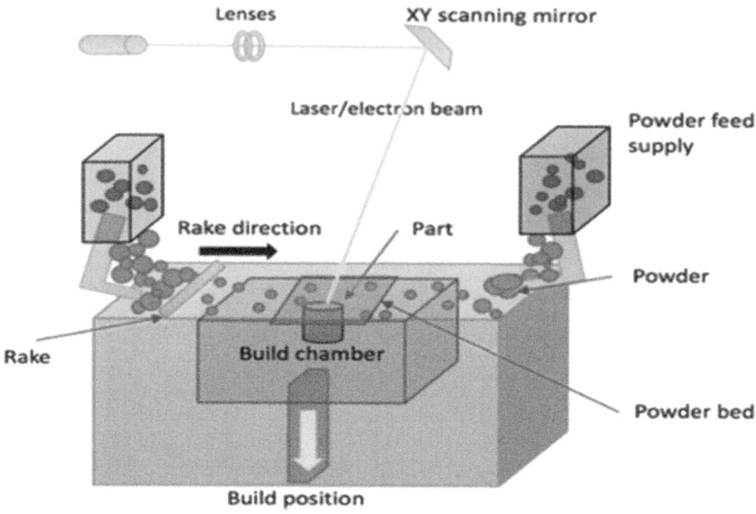

FIGURE 10.5   Powder Bed Fusion process

10.2.5   POWDER BED FUSION

A layer of powder is placed on the platform for the powder bed fusion additive manufacturing process. Before the second layer is applied to the platform with a roller or a blade, a thermal energy source, such as an electron or laser beam, fuses the powder perpendicular to the surface. And this layering process is repeated continuously. The process of the powder bed fusion process is carried out nearly in vaccum, with a pre-heated chamber with inert gas sealed inside. It is an appropriate process type for prototypes and visual models since it supports the structure with metals and polymer powder components [3]. This is shown in Figure 10.5.

10.2.6   SHEET LAMINATION

In this process, the layers are binded together using ultrasonic welding or the adhesive process. The process is carried out by placing the material on a cutting bed. A knife or a laser is used to cut off the appropriate shape once the layers have been placed and adhered to the substrate. This technique can also bind the different layers together, making it cost effective and a speedy process. When it comes to sheet lamination, the items could need post-processing as accuracy can occasionally be missing. It is often used to make prototypes [2]. This is shown in Figure 10.6.

10.2.7   DIRECTED ENERGY DEPOSITION PROCESS

It is the most complex types of AM processes. A four or five axis arm will move around, depositing the melted material around a fixed object. The material is melted by an electron beam or a laser which solidifies the deposited materials. Metal powders or wires are the most common materials used, but sometimes ceramics or polymers can also be used. Due to ability to repair the deposited layers a high degree of accuracy

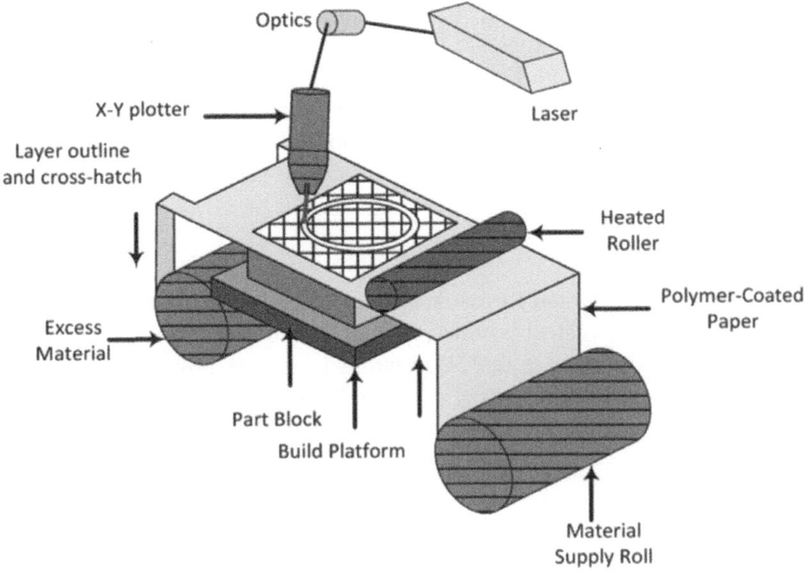

FIGURE 10.6   Sheet Lamination process

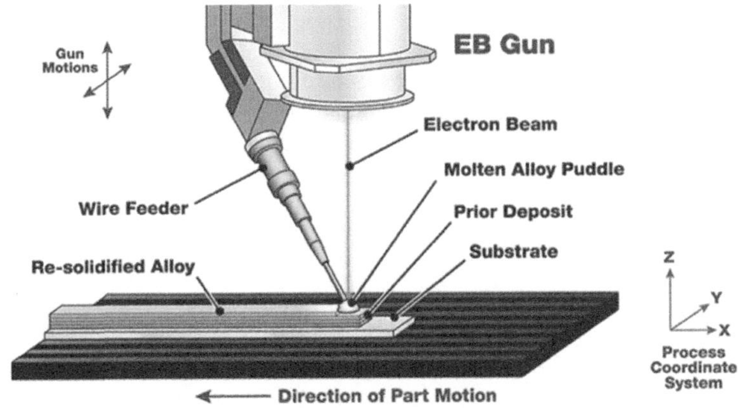

FIGURE 10.7   Directed energy deposition process

can be achieved, the grain size can also be effectively controlled. It is mainly used to repair fabricated parts [3]. This is shown in Figure 10.7.

10.3   MATERIALS USED IN ADDITIVE MANUFACTURING

10.3.1   Nylon

A synthetic thermoplastic material called nylon, also referred to as plastic, is occasionally used in place of silk. It is cheaper in rate and termed as one of the strongest thermoplastic materials available. To fit the newest technological advancements, complicated geometrical designs are produced using this method. Nylon is utilized extensively in a variety of additive manufacturing applications, including prototyping, modelling, and even in clothing and accessories, thanks to its great flexibility, durability, low friction, and corrosion-resistant properties. Nylon prints at a higher temperature than other materials, and it is more difficult to get it to adhere to the print bed.

10.3.2 STAINLESS STEEL

High strength and exceptional corrosion resistance are two characteristics of stainless steel. The use of stainless steel is widespread, from industrial to assistive technology. Fusion and laser sintering are the 3D printing processes utilized to create objects out of stainless steel. Stainless steel is printed using Direct Metal Laser Sintering (DMLS) and SML technologies, just like gold and silver.

10.3.3 TITANIUM

Titanium is mostly utilized for high-performance applications, such as spacecraft, aircraft, and the medical industry. It is a strong, lightweight, heat- and chemical-resistant material. It is quite difficult to machine with tools because of its tremendous strength. Because of this, it makes a fantastic material for additive manufacturing.

10.3.4 ALUMITE

The polyamide and aluminium powder mixture known as Alumite is used to make 3D printing material using the SLS technique. The metallic and porous surface of aluminumite has strong mechanical strength and can withstand temperatures of up to 340 °F. And because of this, the rapid prototyping and 3D printing industries employ it extensively. It is primarily utilized when producing small functional models that require great stiffness, complicated modelling, or designing.

10.3.5 HIGH IMPACT POLYSTYRENE (HIPS)

High impact polystyrene or high density polyethylene are examples of lightweight, flexible materials. It is often used in the fabrication of pipes, plastic bottles, and packaging materials because of its great molding ability. Since it can tolerate high temperatures and the majority of chemicals, with the exception of limonene, it is used as an alternative to ABS. To print, it needs a hot printing bed, and when the bed cools, it contracts and causes warpage.

10.4 CHALLENGES FACED IN ADDITIVE MANUFACTURING

10.4.1 VOID FORMATION

One of the major challenges for manufacturing in AM is void formation between the subsequent layers of AM [4]. It occurs mainly due to the reduced bonding strength in-between the layers of work pieces joined together. In some cases, void formation makes anisotropic mechanical behavior and delamination. In order to minimize the void formation effect, the nozzle geometry is changed in subsequent layers. The result of the study shows that the performance of rectangular nozzles is much better than cylindrical nozzles.

10.4.2 STAIR – STEPPING IN BAAM

The largest issue in AM is the appearance of the staircase effect or layering mistake on the surface of the manufacturing items. The external surfaces of the parts produced get substantially affected, as it doesn't matter if the internal fabricated surfaces get exposed to environment.

10.4.3 MECHANICAL AND MICROSTRUCTURE USED II ANISOTROPIC PROPERTIES

One more important challenge in AM is that when the layers are formed one above the other during their adhesion a small amount of disturbance is observed which may result in small fluctuations in the mechanical properties, which further changes the strength and hardness of the object. Due to these changes in the properties, it has in turn affected the microstructure of the surface. This results in development of anisotropic properties on the work piece surface.

10.4.4 SMALL BUILD VOLUME

Usually during the preparation of a prototype or an object the scale is reduced to a confined dimension, it adds time and effort on that particular part. Also it may not be feasible in all cases to scale down the part. When the parts are scaled down, the assembly may have less strength if adhesives are used, or it may become bulky if mechanical fasteners are utilized. For large-scale industrial applications, AM is not very effectively used, due to its complexity in use and manufacturing.

10.4.5 FABRICATION OF WEAPONS USED IN DEFENSE

Because AM can fabricate intricate structures, it can also be used to create defense-related weapons. But its application is limited to countries which have the scope for its future development.

10.4.6 FOOD AND DRUG ADMINISTRATION SAFETY STANDARD COMPLIANCE

In the field of medical applications many of the body part implants are made by AM. The food items are also prepared with AM, and are regulated and managed by the Food and Drug Administration (FDA). FDA is still working to a mandate to come out with the best ideas and process the activities so that the best result can be achieved.

10.5   BIG AREA CONCEPT IN ADDITIVE MANUFACTURING

As discussed earlier AM is the process of manufacturing a product layer by layer and is dependent upon the type of material or the additive used for making the parts in the whole assembly. In the concept of "Big Area", it the amount of space that will be taken into consideration for manufacturing the product. It is the area that will occupy the space for manufacturing the product, so that a larger dimension of work pieces can be made and if there is any complexity in the shape of the object that can be identified and made into a much simpler object. It primarily focuses on gaining a fundamental understanding of printability by carrying out a variety of polymer tests, in order to really discover the determinants for the number of components or, in the case of CNC machines, the number of axis of geometry needed for milling the work pieces.

10.6   RECENT TRENDS IN BIG AREA ADDITIVE MANUFACTURING

In the world of latest technologies, additive manufacturing has brought about a new revolution in the manufacturing industry, that fulfils the requirements of latest trends of light weight and lower cost products. It has become the new phase of the industrial development reaching out to the masses to a very large extent. In the case of Big Area Additive Manufacturing (BAAM), it has become prominent that a larger area of work space is to be covered, larger dimensions of work pieces are to be manufactured, greater complexity in the work pieces has to be handled, design based products with the influence of Artificial Intelligence (AI) and Machine Learning (ML) can be incorporated to match with the leading industrial developments and market trends.

The following are some of the latest and recent trends in the area of BAAM.

10.6.1   THE ADDITIVE SOFTWARE INNOVATION TECHNIQUE

The use of software in any manufacturing technique has been the backbone in the recent times. A very important phase in the AM workflow has been taken by the software, including the ordering stage, part design, production planning and monitoring, etc.

10.6.2   INCREASED FOCUS ON MACHINE CONNECTIVITY

AM's biggest trend is expected to be the integration of both the hardware and software, thus connecting it with the shop floor. Thus, the companies are rapping up with the AM operations, having the ability to seamlessly connect with the machines and software simultaneously. Thus many of the companies are rapidly shifting their focus to the systems which are open for integration with the factory shop floor.

10.6.3   CONVERGENCE OF AM AND AI

In today's competitive age artificial intelligence and machine learning have become the integral part of AM. The parts of AM has been a value chain process, that can benefit from AI, from material development, machine set up, part design sign and workflow automation.

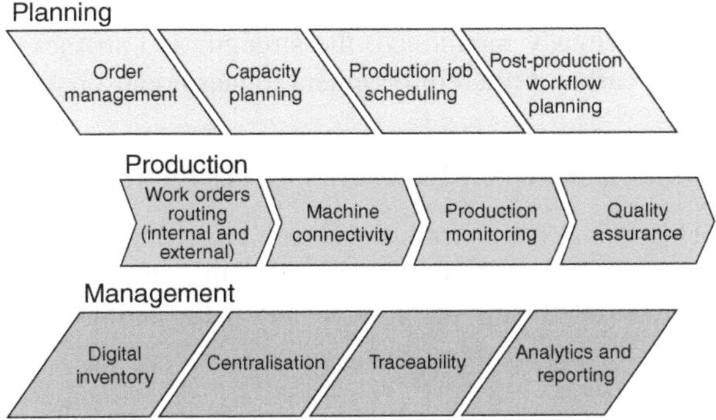

FIGURE 10.8 Chart showing BAAM process

1. AM will provide distributed manufacturing technique:
   The organizations are moving with the latest technology of supply chain, it allows the new technologies to be available at a lower cost or have more flexibility with the products manufactured. Thus, it needs greater flexibility that is influenced by the big area additive manufacturing, and agility that has given rise to a distributed, and localized manufacturing process for better applications and usage.
2. MES used in growth of additive manufacturing:
   The growth and scalability of AM across the organization has made the use of digital manufacturing technology the ultimate and only source in industries, one that requires the use of software as a foundation for various applications related to industrial development. Due to the use of required software for a particular application, there is an increased interest in Manufacturing Execution Systems (MES) for the additive manufacturing process. The process of order management, product scheduling and production planning and control, has made the use of MES software easier to make the workflow of AM more straightforward in order to automate the system [5–7]. This is shown in Figure 10.8.

10.7 AREAS OF APPLICATION OF BIG AREA ADDITIVE MANUFACTURING

10.7.1 EXTRUSION PROCESS IN BAAMS

Conventional fused deposition modelling is based on melting and extruding a filament of thermoplastic feedstock [8][9]. The flow rate of the filament is limited by the rate at which the filament can be melted. Big Area Additive Manufacturing is an extrusion process that uses injection molding material for the feedstock and a single screw extruder for melting and metering the flow rate. A gantry system, moves the extruder in x, y and z directions to build the part. The extruder is capable of delivering 100 kg/hour of thermoplastic materials from pellet feedstock. The gantry system can reach 200 /sec peak velocities with accelerations of 64.4 in/s2 and position precision of 0.002. As was previously mentioned, the strength and stiffness of the item are increased by adding carbon fibre reinforcement to thermoplastic resins.

10.7.2 Behavior of Material Properties While Printing

In the case of 3D printing, large sized objects are printed by taking into account the surface properties of the material and the texture of the surface that can be enhanced with the coating materials that gives good accuracy to the material and have better surface finish.

10.7.3 Material Surface Development with Advanced Composites

Extrusion deposition is quickly evolving as an energy-efficient method that drastically cuts the amount of time needed to manufacture new goods, such as vehicle constructions. There is now work being done to create reinforced thermoplastic materials that are specifically designed for the AM process and have the qualities required for engineering structures. Its primary goal is to establish a fundamental understanding of printability through a series of polymer characterization experiments, preliminary printability evaluations, and measurements of fundamental mechanical properties for various compounds.

10.7.4 Automotive Parts Made by BAAM

The goal of BAAM technology is to enable large-scale industrial applications for 3D printing. The machine chassis, motion system, and controls of the BAAM machines are all based on the laser platform from CI. However, an extruder and feeding mechanism have taken the role of the laser. Compared to current thermoplastic additive machines, BAAM machines operate at speeds 200–500 times quicker and on objects up to 10 times larger. Big Area Additive Manufacturing (BAAM) is one of the most significant 3D printing technologies now in use. BAAM made its debut at IMTS 2014 where it was utilized to 3D print an entire automobile. It was created by Oak Ridge National Laboratory (ORNL) in collaboration with Cincinnati Inc. This is shown in Figure 10.9.

10.7.5 BAAM Applied in Foundry Industry

By melting and pouring metal into cores and molds that have been specially designed, metal castings are produced. It by-passes the making of costly and time consuming patterns, cores and molds. The metal casting industry uses it as a tool to deal with the difficult shapes and intricate designs of the work components.

10.7.6 3D Printed Pre Cast Molds of Multi-Storied Buildings

In between 8 and 11 hours, the company was able to print each mold, and an additional 8 hours of machining was required to provide the desired surface polish. Carbon fiber-reinforced ABS, a popular thermoplastic combined with finely chopped

FIGURE 10.9 A 3D printed car fabricated by Local Motors by BAAM technology [7]

carbon fibers for added strength, was used to create the molds. In this instance, 3D printing was useful for making more sophisticated and long-lasting molds quickly. For the building sector, 3D printing huge parts has a number of advantages that give architects another practical precast mold manufacturing option.

10.7.7 INFRARED PREHEATING TO IMPROVE INTERLAYER STRENGTH OF COMPONENTS

It has the capacity to print structures that are several metres in length at high extrusion rates, potentially having a substantial impact on the many production industries. It has primarily attributed to poor bonding between the printed layers since the lower layers would cool down faster before the next layer is deposited. Thus, the potential of using infrared heating is considered for increasing the surface temperature of the printed layer just prior to the deposition of new layer to improve the interlayer strength of the components. It has made significant improvement in the bond strength for the deposition of acrylonitrile butadiene styrene (ABS) reinforced with 20% chopped carbon fibre when the surface temperature of the substrate material increased the surface temperature of the deposited layers using infrared heating. [7]

10.8 CASE STUDIES RELATED TO BIG AREA ADDITIVE MANUFACTURING (BAAM)

10.8.1 ADDITIVE MANUFACTURING OF DOWN HOLE ROTATION TOOL

The case study mainly focuses on the manufacturing of the rotor with a down hole rotation tool. The rotor of the tool was evaluated, designed, and fabricated using the AM technique. A standard master part was compared with an AM manufactured part with the same overall geometry of the component. Tests were planned to characterize the performance gains resulting from the AM process. The design targets were intended to match with the conventional parts. These targets included surface hardness, yield strength and dimensional tolerances and surface finish. A reduction in the rotational inertia was also achieved through topology optimization technique and power bed fusion fabrication process.

10.8.2 ADDITIVE MANUFACTURING IN THE SOUTH AFRICAN RAILWAY INDUSTRY

The case study deals with the making of the complex South African railway network which is utilized by freight and passenger locomotives. The Transnet Freight Rail (TFR) operates on the heavy haul railway routes utilizing the Big Area Additive manufacturing for expanding its organizational approach following the pillars of Industry 4.0 by taking advantage of the innovative global technologies to advance its measuring, manufacturing and maintenance operations. Its plan to switch from being an internal engineering service provider to becoming an original equipment manufacturer of wagons, passenger coaches, locomotives, and track equipment gives it a platform to research new potential manufacturing processes in order to apply the BAAM technique to the existing design and make it better.

10.8.3 BIG AREA ADDITIVE MANUFACTURING OF A SINGLE SEATER RACING CAR IN EUROPE

The main aim of the study is to design and build a single seater car which races on a course normally the reversed of Formula 1. It is a place where the innovation is very much at home, particularly when it comes to the new manufacturing and design engineering purposes. It mainly provides emphasis on the manufacturing of whole car through big area method, that can accompany each part of the car product, so that the required designing of the parts can be carried out and with respect to the manufacturing of the parts is made. Thus, the big area method is adopted for the entire car manufacturing process so that single seater racing car can be made.

10.8.4 USING THE BAAM TECHNIQUE REDUCES THE COUNT OF COMPONENTS

The study shows that the use of additive manufacturing technique has given another dimension to the production industry. The application of the big area method has made the design simpler and any complex shape of the product can be made without any objections and much hurdles in the manufacturing of component parts with respect to the given design can be made quickly. The design made with respect to BAAM is so complex that it overcomes the problems of assembly with dissimilar elements, and makes the most difficult assembly easier to produce, utilizing the maximum area of the work space available for assembly. [6]

10.8.5　Use of Selective Laser Sintering for BAAM

The core of the expanding trend in functional prototype and mass custom manufacturing is selective laser sintering technology. The manufacturing processes that are changing are those that use the proper additive technologies, materials, and finishes. The case study deals with the use of different types of lasers and varieties of laser shapes that provide the machining to the materials and mass production process in a very smooth manner so that it can be customized in a very smooth manner. The type of powder used also defines the type of additive used in manufacturing the product. The laser traces the pattern of each cross section layer by layer on the powder bed giving the clarity on the object to be prepared by AM. It also provides the guidelines on the different systems and materials available in the market, the workflow, its various applications over the traditional additive manufacturing process.

10.9　BIG AREA ADDITIVE MANUFACTURING USED IN THE FDM SYSTEM

Additive manufacturing utilizes a large amount of material for manufacturing a product. It mainly utilizes the use of complicated shapes to design the work pieces. There are software that make the 3D models as per the requirements of the customer, these model products are then transferred to the system with the use of a CAD process and then fed into the 3D machines. Then these machines will make the model with an additive manufacturing technique by using the fused deposition resins. The resins required for manufacturing are nylon, ABS, stainless steel, titanium, etc. [5], [10]

For a manufacturing process to qualify, it must involve the following three significant aspects:

10.9.1　Use of CAD for Creating 3D Models

The several CAD tools like CATIA, Solid works, AutoCAD, are used to develop the models and through computer technology, fed into it and develop complex models with the use of additive manufacturing technique. The amount of material to be extruded by the 3D printer and time taken to build the model is created by using G − code, which the printer can easily interpret.

10.9.2　Slicing and Generation of Tool Paths

Tool path is generated with the G−codes and fed into the CAD software to form layers, resin is fed through the nozzle making fusion deposition on the table surface.

10.9.3　Conversion of 3D Models into Real Time Product

The additive manufacturing machine such as 3D printer and laser convert the 3D model into actual product using engineering materials such as plastics, metal powders, composites, etc. The material is melted and allowed to flow according to the G-codes given, the model is developed by fusion deposition process.

10.10 CONCLUSION

This chapter shows the methods of operation of the additive manufacturing process, ways in which the 3D complex shaped objects are printed on the work table, the areas in which the scope of printing the products can be thought of. It shows the use of large areas to manufacture and develop new design technologies and process relations in which these products can be made. Further focus is given on the fusion deposition process that helps in making the best use of the materials/resins in developing the 3D products. Since fusion deposition is the fastest growing industry in this new revolution or generation, additive manufacturing process is getting greater scope in future generations to come.

REFERENCES

[1] Jichang Xie, Haifei Lu, Jinzhong Lu, Xinling Song, Shikai Wu, Jianbo Lei (2021). Additive manufacturing of tungsten using directed energy deposition for potential nuclear fusion application. *Journal of Surface and Coatings Technology*, Volume 409. https://doi.org/10.1016/j.surfcoat.2021.126884.

[2] Fu Hu, Jian Qin, Yixin Li, Ying Liu, Xianfang Sun (2021), Deep Fusion for Energy Consumption Prediction in Additive Manufacturing. *Journal of Procedia CIRP*, Volume 104, Pages 1878–1833. https://doi.org/10.1016/j.procir.2021.11.317.

[3] Panel Dea Jun Huang, Hua Li, (May 2021) ,A machine learning guided investigation of quality repeatability in metal laser powder bed fusion additive manufacturing. *Journal of Additive Manufacturing*, Volume 203, 109606, https://doi.org/10.1016/j.matdes.2021.109606.

[4] Systems Phillip Chesser, Brian Post, Alex Roschli, Charles Carnal, Randall Lind, Michael Borish, Lonnie Love, (August 2019). Extrusion Control for High Quality Printing on Big Area Additive Manufacturing (BAAM). *Journal of Additive Manufacturing*, Volume 28, Pages 445–455, https://doi.org/10.1016/j.addma.2019.05.020.

[5] Alex Roschli, Katherine T. Gaul, Alex M. Boulger, Brian K. Post, Phillip C. Chesser, Lonnie J. Love, and Fletcher Blue (2019). Designing for Big Area Additive Manufacturing. *Journal of Additive Manufacturing*, Volume 25, Pages 275–285. https://doi.org/10.1016/j.addma.2018.11.006.

[6] Matthew Korey, Mitchell L. Rencheck, Halil Tekinalp, Sanjita Wasti, Peter Wang. (May 2023) Recycling polymer composite granulate/regrind using big area additive manufacturing. *Journal of Material Composites Part B: Engineering*, Volume 256, 110652, https://doi.org/10.1016/j.compositesb.2023.110652.

[7] Vidya Kishorea, Christine Ajinjerua, Andrzej Nycz, Brian Post , John Lindahl, Vlastimil Kunc, Chad Duty. (March 2017) Infrared preheating to improve inter-layer strength of big area additive manufacturing (BAAM) components. *Journal of Additive Manufacturing*, Volume 14, Pages 7–12. https://doi.org/10.1016/j.addma.2016.11.008

[8] Joel C. Vasco, Additive manufacturing for the automotive industry. (May 2021) Handbook of Additive Manufacturing, Pages 505–530. https://doi.org/10.1016/B978-0-12-818411-0.00010-0.

[9] Wendy Triadji Nugroho, Yu Dong, Alokesh Pramanik. (June 2021). 3D printing composite materials: A comprehensive review. *Journal of Composite Materials. Manufacturing, Properties and Applications*, Pages 65–115. https://doi.org/10.1016/B978-0-12-820512-9.00013-7.

[10] Vidya Kishore, Ahmed Arabi Hassen. Polymer and composites additive manufacturing: Material extrusion processes. (May 2021). Journal of Additive Manufacturing. *Handbook in Advanced Manufacturing*, Pages 183–216. https://doi.org/10.1016/B978-0-12-818411-0.00021-5.

11 Machine Learning in Post-Processing of Fused Deposition Modelling Parts

Raman Kumar, Ranvijay Kumar and Jatinder Pal

11.1 INTRODUCTION

In the late 1980s, Stratasys invented the 3D printing method known as fused deposition modeling (FDM) (Chennakesava & Shivraj Narayan, 2014). It is a technique for turning computer designs into tangible items by piling materials on top of one another. FDM involves melting a thermoplastic filament and extruding it via a nozzle that travels along a predetermined route. The object is constructed layer by layer from the bottom by the nozzle as it deposits the melted filament onto a build platform (Banga et al., 2023). The material is immediately solidified upon deposition, enabling the subsequent layer to be placed on top. FDM is a frequently utilized and well-liked 3D printing method because of its adaptability, affordability, and simplicity (Zhang et al., 2020). It is used to make various items, including toys, models, and practical components for tools and machinery. The wide variety of sizes and materials that FDM printers can print with includes ABS, PLA, PETG, nylon, and TPU (HaghsefatLiu & Tingting, 2020). The capacity to print complicated geometries and unique patterns using FDM that could be challenging or impossible to produce using conventional manufacturing techniques is one of the technology's main benefits. By utilizing various filaments and altering the printing parameters, FDM may also produce pieces with different material characteristics, such as differing degrees of flexibility and strength. In the current decade, academicians and researchers have explored the FDM technique in the medical field applications for developing various parts using biocompatible materials (Singh et al., 2022; Singh et al., 2020; Banga et al., 2023). FDM is a versatile and widely-used 3D printing technology that has revolutionized how users design and manufacture objects in various fields.

11.1.1 OVERVIEW OF POST-PROCESSING TECHNIQUES FOR FDM PARTS

FDM 3D printing is a prevalent method for producing prototypes and valuable parts in the engineering and medical sectors {Formatting Citation}(Tilton et al., 2020). Further post-processing may occasionally be necessary to enhance the look, feel, or

DOI: 10.1201/9781032665351-14"""

mechanical qualities of FDM-produced items (Karakurt & Lin, 2020). The following are some typical post-processing methods for FDM parts.

11.1.1.1 Sanding and Polishing

FDM items frequently have rough surfaces and prominent layer lines. The surfaces can be smoothed out, and the part can look better visually with the help of sanding and polishing (Dizon et al., 2021). The sanding on the 3D printed parts offers the removal of the excessive edges. In recent years, researchers have been using many techniques successfully for the post-processing of FDM items, such as: HCM (Hot Cutter Machining), chemical post-processing treatment, Vibratory Bowl Abrasion Finishing, and Ultrasonic Abrasion Finishing (Kumbhar & Mulay, 2018; Galantucci et al., Percoco, 2009).

The sanding process is limited to only cleaning the edges of the external surfaces. The frictional forces are applied on the surface to remove the unwanted parts from the surface. In contrast, polishing results the cavities present as manufacturing defects. Figure 11.1 shows the concept of sanding.

11.1.1.2 Painting and Coating

An FDM part's surface can be painted or coated to enhance its look and defend against deterioration and moisture (Haidiezul et al., 2018). The coating and painting with the polymeric materials on the 3D printed surfaces ensure the improvement of dimensional acceptability as physical fitness (Žigon et al., 2020).

Figure 11.2 shows the coating of the unpolymerized resin on the 3D-printed parts (Kraemer Fernandez et al., 2020).

11.1.1.3 Vapor Smoothing

When a component is subjected to vapor smoothing, a solvent's vapor partly melts the surface, giving it a smoother finish. Figure 11.3 shows the pictorial view of the experimental setup used for the vapor smoothing process (Chohan et al., 2020). The work published by (Singh et al., 2017) has outlined that the exposure time of the

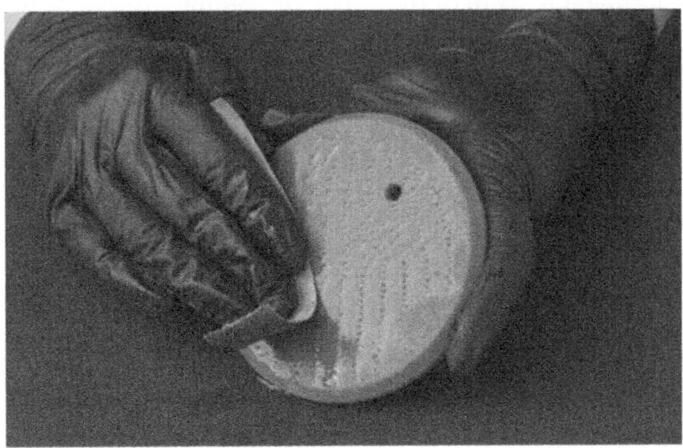

FIGURE 11.1 Post-processing of additively manufactured parts by sanding ("FormLabs Removing support marks," 2023)

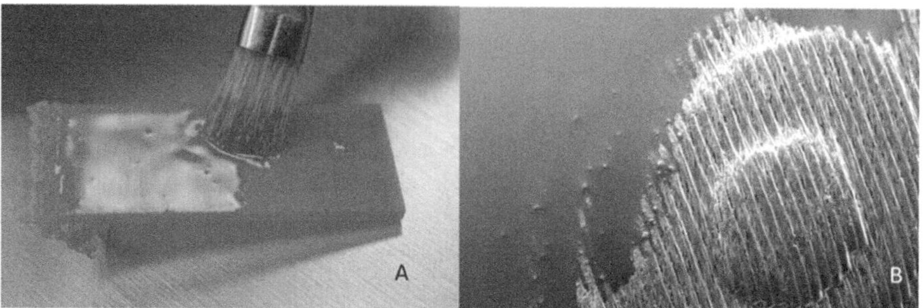

FIGURE 11.2 Coating of the unpolymerized resin on the 3D printed parts adapted from (Kraemer Fernandez et al., 2020) (CC BY 4.0)

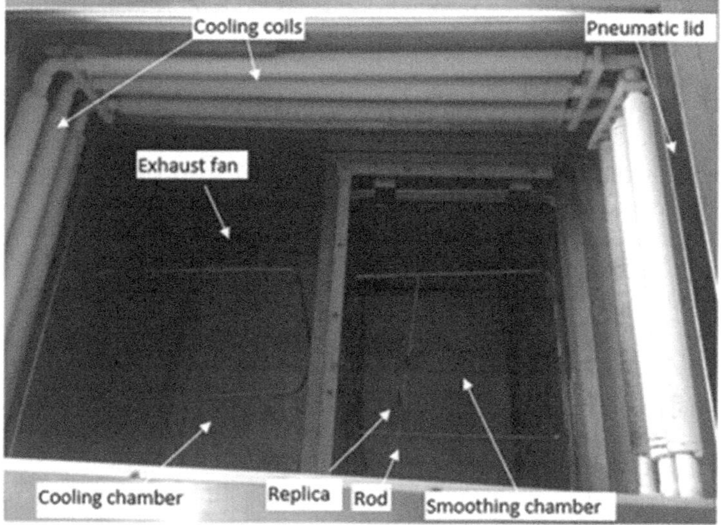

FIGURE 11.3 Experimental setup of the vapor smoothing process (Chohan et al., 2020)

acetone vapor has significantly affected the surface quality of the 3D printed parts and refer to Figure 11.4 (Singh et al., 2017).

11.1.1.4 Chemical Smoothing

FDM and chemical smoothing may be used in tandem to create inexpensive molds with excellent surface finishes (Ferretti et al., 2021). Chemical smoothing uses a solvent to melt and smooth the surface of the FDM part, just like vapor smoothing does. The previous study has reported satisfactory improvement of the surface clarity after exposure to the chemical on the 3D printed surface Figure 11.5 (Christian Leon-Cardenas et al., 2022).

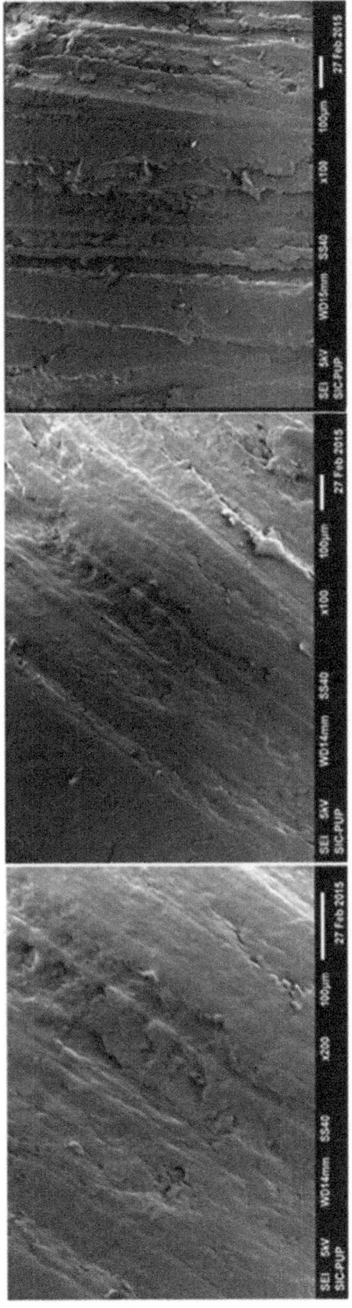

FIGURE 11.4 Micrographs of the 3D printed surface on acetone vapor exposure after (a) 25 sec,(b) 30 sec, and (c) 35 sec (R. Singh et al., 2017)

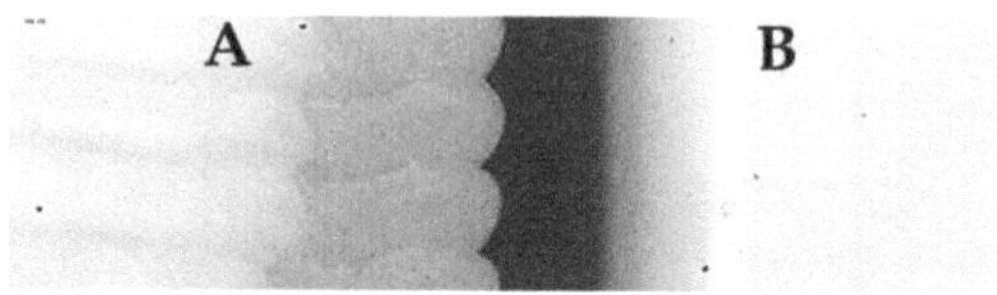

FIGURE 11.5 The 3D printed surface (a) before chemical smoothing and (b) after chemical smoothing (Christian Leon-Cardenas, 2022)

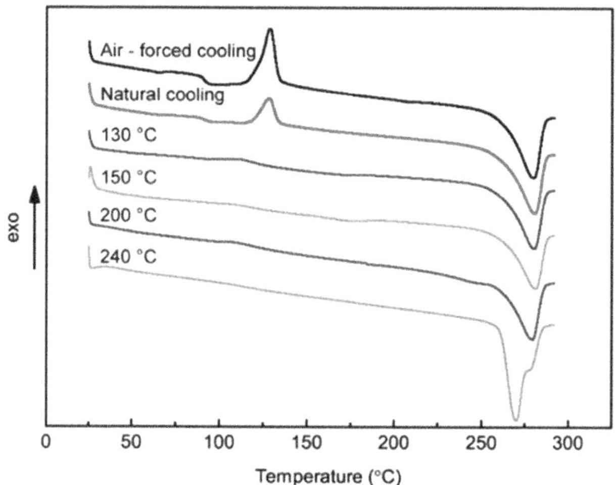

FIGURE 11.6 DSC plot of the PPS samples under different heating conditions (Geng et al., 2018)

11.1.1.5 Heat Treating

FDM parts can benefit from heat treatment to become stronger and more resilient. Enhancing the part's mechanical qualities entails heating it to a specified temperature for a predetermined time. The previous study reported that the crystallinity of the Polyphenylene sulfide (PPS) samples significantly improved if the heat treatment temperature increased (Geng et al., 2018) (Wach et al., 2018). Figure 11.6 and Table 11.1 show the PPS thermal properties under different heating conditions.

11.1.1.6 Support Removal

Support structures are frequently created with FDM components to assist them in keeping their form during printing. These support structures can be manually removed after printing or dissolved in water or a chemical bath. On the other hand, the support removal procedure might be challenging and time-consuming when there are no dissolving support materials accessible, as is the case with the outstanding durability thermoplastic Ultem TM 9085 (Chueca de Bruijn et al., 2020). The support removal in the 3D printing process is the first port process needed in

TABLE 11.1
Crystallinity of PPS samples under different heating conditions (Geng et al., 2018)

Sample	Crystallinity (%)
Air-forced cooling	19.13
Natural cooling	35.43
130°C	42.28
150°C	42.75
200°C	50.04
240°C	64.08

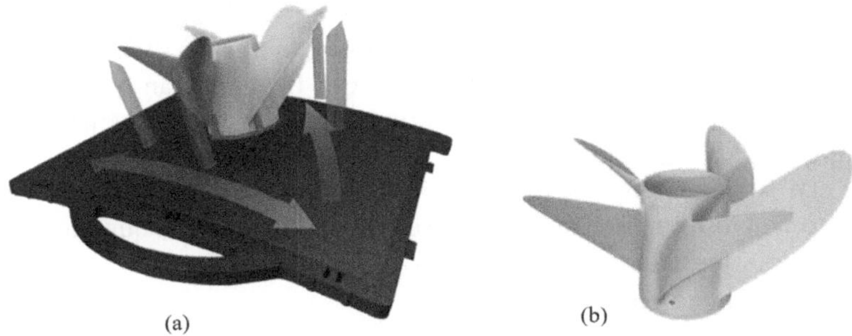

FIGURE 11.7 (a) Removal of the support from 3D printed part (b) 3D printed part after support removal (Hiemenz J., 2011)

almost every additive manufacturing process. Mechanical plucking is one of the most economical ways of support removal, but heating or cooling may also be applied if this does not work. If these processes cannot remove the part successfully, chemical treatment may be given to removing the support materials from the surfaces. Figure 11.7 shows the support removal process in the 3D printing process (Hiemenz J., 2011).

11.1.1.7 Insert Installation

Inserts like metal threads or pins may be included in FDM parts during design. After printing, these inserts can be added to the produced portion using specialized equipment. Figure 11.8 shows the 3D printed part after installation of the inserts as the post-processing shows the 3D printed part after installation of the inserts as the post-processing (Vasquez, 2023).

The insert installation may be executed when the mechanical/physical and surface properties are higher for 3D-printed parts. The item's intended usage and desired finish determine the post-processing methods for FDM parts. It could be necessary to use various ways to get the desired outcome.

FIGURE 11.8   3D printed parts after inset installation (Vasquez, 2023)

11.2 MOTIVATION FOR USING ML IN POST-PROCESSING

ML in post-processing for FDM parts is a growing well-liked field of study and development (Nasiri & Khosravani, 2021) (Ranjan et al., 2022). The component quality, dependability, and functionality can be significantly enhanced. ML algorithms can be trained to identify and fix typical flaws in FDM parts. Because each algorithm has distinct advantages for a given application, there is no one method that is suitable for all problems. For instance, regression, clustering, and classification algorithms are frequently used to predict material qualities (Nasiri & Khosravani, 2021). The primary reasons for employing ML in the post-processing of FDM parts are discussed briefly.

11.2.1 QUALITY CONTROL

The quality of FDM parts can be jeopardized by flaws like warping, cavities, and surface abnormalities, which can be detected by mechanical testing as well as using ML techniques (Kadam et al., 2021) (Baechle-Clayton et al., 2022). By doing so, producers may be able to increase the general quality of their items while requiring less manual examination.

11.2.2 DESIGN OPTIMIZATION

By determining the ideal orientation for printing and the most effective printing settings, including temperature and print speed, machine and deep learning can assist in improving the design of FDM parts (Sampedro et al., 2022) (Sharma et al., 2022). This can enhance the strength and mechanical qualities of the products while also cutting down on printing time and material waste (Zhang et al., 2019).

11.2.3 AUTOMATION

Many post-processing operations, like removing support structures, sanding, and polishing, can be automated using ML. This can decrease the need for manual labor and increase the manufacturing process efficiency (Kumar et al., 2023).

11.2.4 MATERIAL SELECTION

The choice of materials for FDM parts can be optimized using ML algorithms based on their mechanical characteristics, such as strength and flexibility (Cai et al., 2022). This can help to guarantee that components are made using the best materials for the purpose for which they are designed. The application of ML in post-processing for FDM components is a crucial area of research and development for the future of 3D printing, with the potential to significantly improve the process' quality, efficacy, and reliability (Nguyen & Lee, 2018).

11.3 ESSENTIAL STEPS OF ML

A type of artificial intelligence called ML employs statistical models and algorithms to help computers learn from data and improve at a given activity (Sarker, 2021). Collecting and preparing data, choosing and engineering features, choosing and training a model, assessing its performance, and deploying it into a production environment are the key phases of ML. Building precise and dependable machine-learning models to address real-world issues requires following these steps.

11.3.1 DATA COLLECTION AND PRE-PROCESSING

The data used in a study on ML for FDM post-processing will depend on the study's research question and methodology (Sharma et al., 2022). Some examples of the types of data that might be used:

- Quality control: Researchers may utilize a dataset of FDM components with known defects to train a ML system to detect problems in FDM parts (Goh et al., 2021). This dataset might have several flaws, including cavities, warping, and surface roughness. An optical or laser scanner might be used to scan the components and produce a 3D point cloud that could be used to train the algorithm.
- Design optimization: Researchers may utilize a dataset of parts with various orientations and support structures to optimize the orientation of FDM parts for printing. The dataset might include details about the pieces' mechanical characteristics and the printing settings used to make them. Based on the intended mechanical qualities of a new part, the data may train a ML system to forecast its best orientation and support structure.
- Automation: Researchers may use a dataset of FDM parts with known post-processing requirements, such as the position of support structures and regions that need sanding or polishing, to automate post-processing processes. Using this dataset, a ML system may be trained to spot these properties in fresh components and operate a robotic arm to perform the required post-processing operations (Ganguli et al., 2018).

In all cases, the data used to train the ML algorithm must represent the types of parts and defects the algorithm will detect or optimize (Sharma et al., 2022).

Additionally, the data should be collected consistently and reliably to ensure that the results are accurate and reproducible.

The initial phase gathers a sample dataset of FDM components and related information, such as scanning information, mechanical characteristics, and printing settings. To guarantee that the dataset is varied and contains a wide range of potential flaws and post-processing requirements, the parts should be printed in various settings and with various materials. To ensure the dataset is suitable for ML analysis once it has been gathered, it could be required to clean and pre-process the data. This may entail eliminating duplicates, reducing outliers, and normalizing or scaling the data to make the data consistent and comparable. It can be important to extract pertinent characteristics from the raw data before using it for ML analysis. This might entail extracting information like surface roughness, warping, or other faults using algorithms, and then choosing the most crucial features to include in the ML method.

The dataset may be divided into training and testing sets once the data has been pre-processed and the pertinent features have been retrieved and chosen. The testing set is used to gauge the ML algorithm's performance, while the training set is used to teach it. This may include utilizing methods like cross-validation to ensure the algorithm is reliable and adaptable to new data. The data collecting and preparation procedures may need to be iterated and improved to increase the accuracy and dependability of the algorithm, depending on the outcomes of the ML study. Additional data collection, feature extraction or selection procedure changes, or new ML algorithms or parameters may all be necessary to achieve this (Sandhu et al., 2020).

11.4 MACHINE LEARNING METHODS

11.4.1 SUPERVISED LEARNING

ML algorithm is trained on a labelled dataset, where the labels indicate the correct output for each input. In the context of FDM post-processing, supervised learning algorithms might detect defects in FDM parts or optimize the orientation and support structures for printing (refer to Figure 11.9a). Examples of supervised learning algorithms include decision trees, random forests, support vector machines, and neural networks (Barrios & Romero, 2019) (Farhan Khan et al., 2021) (Mahmood et al., 2020).

11.4.2 UNSUPERVISED LEARNING

ML algorithm is trained on an unlabeled dataset with no pre-defined labels for the inputs. In the context of FDM post-processing, unsupervised learning algorithms might cluster similar FDM parts or detect patterns in the data (refer to Figure 11.9b). Unsupervised learning algorithms include k-means clustering, principal component analysis (PCA), and autoencoders (Nam et al., 2020).

11.4.3 REINFORCEMENT LEARNING

The ML algorithm learns to make judgments based on input from its surroundings through reinforcement learning. In FDM post-processing, reinforcement learning

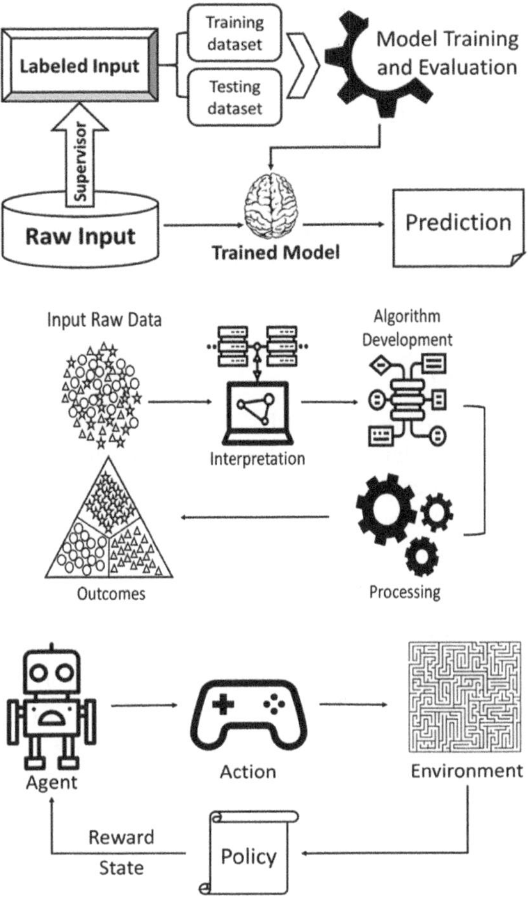

FIGURE 11.9 ML (a) Supervised learning, (b) Unsupervised learning, (c) Reinforcement learning adapted from Sagar et al. (2021) (CC BY 4.0)

algorithms might be used to control a robotic arm to perform post-processing tasks based on a camera's visual feedback (refer to Figure 11.9c). Reinforcement learning algorithms include Q-learning and policy gradient methods. In all cases, the choice of ML algorithm will depend on the specific research question and the dataset's characteristics, including the size, complexity, and structure. It is also essential to carefully evaluate and compare different ML algorithms to determine the most suitable task (Mishra & Jatti, 2023).

11.5 BRIEF EXPLANATION OF HOW EACH ALGORITHM WORKS AND WHY IT MAY BE CHOSEN

Without knowing the specific study conducted, it is not easy to provide a comprehensive list of the ML algorithms used, why they were chosen, and how they work. However, some common ML algorithms are used in the context of FDM post-processing.

11.5.1 Convolutional Neural Networks (CNNs)

CNN is a typical form of neural network for image identification and classification applications. Convolutional filters extract features from the input picture, passing through several layers to categorize the image. CNNs are frequently utilized in FDM post-processing studies to find flaws in printed items or improve the printing bed's structure and orientation (Farhan Khan et al., 2021).

11.5.2 Decision Trees

A method known as a decision tree divides the input space into regions and then fits a straightforward model (often a linear function) to each area. They are frequently utilized in FDM post-processing studies to make predictions about the qualities or behavior of printed items based on those parameters and other factors (Barrios & Romero, 2019; Sandhu et al., 2020).

11.5.3 Random Forests

The predictions of many decision trees are combined to form random forests, an ensemble of decision trees. They are frequently employed in FDM post-processing research to increase the reliability and accuracy of predictions or to determine the key characteristics for a particular assignment (Ranjan et al., 2022).

11.5.4 Support Vector Machines (SVMs)

SVMs are an algorithm that works by finding a hyperplane that separates the input space into two classes (Suniya & Gour, 2021). FDM post-processing studies often use them to classify printed parts based on their properties or behavior.

11.5.5 K-means Clustering

An unsupervised learning method called K-means organizes a dataset into a certain number of clusters based on how far apart the data points are from one another (Nasiri & Khosravani, 2021). FDM post-processing studies often use it to group similar printed parts based on their characteristics.

The particular research issue and the dataset's properties will determine the method used. If the research examines visual data, such as pictures of printed parts, CNNs could be selected. If the investigation includes predicting the qualities or behavior of printed components based on their printing settings and other factors, decision trees or random forests could be employed. The size and complexity of the dataset, the kind of issue, and the required performance qualities (such as accuracy, speed, and resilience), among others, will all influence the chosen method (Nasiri & Khosravani, 2021).

11.6 STRENGTHS AND LIMITATIONS OF ML ALGORITHM

The strengths and limitations of some commonly used ML algorithms regarding implementation in FDM post-processing are depicted in Table 11.2 (Goh et al., 2021; Nasiri & Khosravani, 2021). The CNNs can recognize and locate features within an image, learn sophisticated and hierarchical visual data representations, and are resilient to input variance. However, CNNs may be computationally costly, require a lot of labelled data to train properly, and learned representations can be tricky to read and comprehend. Decision trees manage categorical and continuous data, can handle missing data and outliers without pre-processing, and are simple to analyze and comprehend.

Decision trees are more prone to overfitting than random forests, which can handle high-dimensional data and feature interactions while being resilient to missing data and outliers (Barrios & Romero, 2019). However, Random Forests may not be as accurate as more complicated algorithms for particular jobs, as they are computationally costly and difficult to analyze and understand.

If the regularization parameter is calibrated correctly, SVMs can handle high-dimensional data and nonlinear decision boundaries, efficiently sort linearly separable classes, and resist overfitting. However, they can be computationally costly, are sensitive to the kernel function and its parameters, and might not be as effective as more complicated algorithms for a given task. K-means clustering is a straight-forward, quick, unsupervised learning approach that can handle huge datasets with numerous characteristics and is resilient to noisy and missing data. It is used to group data based on similarities. It depends on the algorithm's initialization, may converge to a poor solution, and is sensitive to the size and normalization of the features, all of which might impact the method's outcomes. When selecting the best machine-learning algorithm for a particular job, one should consider each method's advantages and disadvantages (Suniya & Gour, 2021).

11.7 ACCURACY AND USEFULNESS OF THE ML RESULTS

The statistical significance and quality of the data used, the suitability of the selected ML algorithms, and the degree to which the results can be generalized to other settings all affect how accurate and valuable the results of the ML analysis are. In the study on ML and FDM post-processing, the chosen algorithms can correctly forecast the quality of FDM components based on the post-processing characteristics that have been determined. By finding the most crucial variables for producing high-quality prints, ML may aid in optimizing FDM printing. Manufacturers and designers may decrease waste, increase productivity, and create more consistent and dependable components by using the detected factors to modify the printing process. It is crucial to remember that the representativeness and quality of the training data determine how accurate and generalizable the ML models are. A thorough comprehension of the underlying physical principles and mechanisms of the FDM printing process should complement the application of ML techniques. ML may be used to create predictions and find connections, but it cannot always explain the reasons behind observed patterns. It is critical to consider the study's limitations when interpreting

TABLE 11.2
Strengths and Limitations of some ML algorithms

Algorithm	Strengths	Limitations	Reference
Convolutional Neural Networks (CNNs)	• Can learn complex and hierarchical representations of visual data • Capable of detecting and localizing features within an image • Robust to variations in input, such as changes in lighting or orientation	• Require a large amount of labelled data to train effectively • It can be computationally expensive and require powerful hardware • Learned representations may be challenging to interpret and understand	(Suniya & Gour, 2021)
Decision Trees	• Easy to interpret and understand • Can handle both categorical and continuous data • Can handle missing data and outliers without requiring pre-processing	• Prone to overfitting if the tree is too deep or the data is noisy • Sensitive to small changes in data, which can lead to instability and overfitting • It may not be as accurate as more complex algorithms for certain tasks	(Barrios & Romero, 2019)
Random Forests	• They are less prone to overfitting than decision trees, as they aggregate the predictions of multiple trees • Can handle high-dimensional data and interactions between features • Robust to missing data and outliers	• Can be computationally expensive, especially for large datasets with many features – Predictions may be challenging to interpret and understand • It may not be as accurate as more complex algorithms for specific tasks	(Nasiri & Khosravani, 2021)
Support Vector Machines (SVMs)	• Effective for separating linearly separable classes • Can handle high-dimensional data and nonlinear decision boundaries through the use of kernel functions • Robust to overfitting if the regularization parameter is adequately tuned	• Sensitive to the choice of kernel function and its parameters, which can affect the performance of the algorithm • It can be computationally expensive, especially for large datasets with many features • It may not perform as well as more complex algorithms for specific tasks	(Goh et al., 2021)
K-means clustering	• A simple and fast unsupervised learning algorithm for clustering data based on similarities • Can handle large datasets with many features • Robust to noisy and missing data	• Requires the user to specify the number of clusters in advance, which may be challenging to determine for some datasets • May converge to a suboptimal solution, depending on the initialization of the algorithm • Sensitive to the scale and normalization of the features, which can affect the results of the algorithm	(Nasiri & Khosravani, 2021)

the results and to use caution when extrapolating the findings to other contexts (Cai et al., 2022).

11.7.1 COMPARISON OF THE ML OUTCOMES WITH TRADITIONAL POST-PROCESSING TECHNIQUES

For FDM components, ML has the potential to be more advantageous than conventional post-processing methods in several ways. Traditional post-processing methods frequently rely on labor-intensive, time-consuming manual or semi-automated methods that require high skill levels. Contrarily, ML algorithms can swiftly and reliably analyze enormous datasets, identifying correlations and predicting component quality based on various post-processing characteristics. Additionally, when the number of variables and factors involved in FDM printing rises, conventional post-processing methods might not always be efficient in optimizing the FDM printing process. ML algorithms may analyze numerous parameters at once to consider the intricate relationships between factors and provide a complete picture of the printing process.

Nevertheless, it is crucial to remember that ML algorithms are a complementary strategy rather than a replacement for conventional post-processing methods. However, ML cannot provide light on the fundamental physical mechanics of the FDM printing process. It can only find patterns and correlations in the data. It is essential to comprehend these mechanisms to create efficient post-processing methods and ensure that the outcomes of ML models are relevant and comprehensible. The complexity of the printing process and the specific application determine how ML and conventional post-processing methods compare. While traditional post-processing methods can offer a more thorough knowledge of the physical principles and mechanisms, ML may frequently help streamline the FDM printing process and enhance part quality. Combining these methods may make it feasible to create post-processing methods that are more efficient and produce FDM components that are of higher quality.

11.8 SUGGESTIONS FOR FUTURE RESEARCH

The chapter offers insightful knowledge on the subject of ML-based FDM post-processing. It is recommended that future studies confirm the efficacy of ML methods for FDM post-processing using more extensive and varied datasets. The dataset's generalizability might be increased by increasing the variety of FDM materials and printing settings. More complex algorithms can be created to increase the precision and effectiveness of FDM post-processing. Researchers might investigate more sophisticated algorithms like deep learning and reinforcement learning to discover more nuanced relationships between FDM printing settings and component quality. Even though ML can offer valuable insights into FDM post-processing, it cannot wholly replace physics-based modeling. ML and physics-based modeling may be used to determine the physical principles that underlie FDM printing and to develop even more precise and thorough post-processing methods. Assessing how well ML models perform in practical settings is crucial. The usefulness of ML models in many

FDM applications, such as aerospace, automotive, and medical devices, might be investigated in the future. Look at various 3D printing techniques' usage of ML.

11.9 POTENTIAL APPLICATIONS OF THE ML METHODS

There are several possible uses for the proposed ML techniques in FDM post-processing. The settings for FDM printing may be optimized using the generated ML models to raise the caliber of produced items. ML models can change the printing parameters in real-time to attain the target quality level by detecting the most crucial post-processing elements that affect component quality. Due to the relatively sluggish nature of FDM printing, mistakes or flaws can waste much material. Before a printed item is finalized, ML models can forecast its quality, enabling early identification of any problems that can result in material waste. Over time, this may save you much money. ML models can aid in boosting FDM printing efficiency by decreasing material waste and optimizing the printing settings. Increased manufacturing capacity and quicker turnaround times may result from this. The quality control capabilities of the ML models enable the speedy identification and remediation of any printing-related problems. This can lessen the need for expensive rework or scrap by ensuring that the parts satisfy the necessary quality requirements. More studies on FDM printing may be conducted using ML models to find novel relationships between post-processing variables and component quality. This may result in new post-processing methods that would enhance the effectiveness and caliber of FDM printing even more. ML techniques have the potential to considerably increase the effectiveness and quality of FDM printing and have a wide range of applications in a variety of FDM printing-using businesses.

11.10 CONCLUDING REMARKS

The chapter covered ML strategies for post-processing components from fused deposition modelling. The chapter briefly introduced FDM technology, outlining its salient features and significant difficulties. The constraints of conventional post-processing methods for FDM components, such as manual support structure removal and sanding, are explored. It is clear why ML should be used in FDM post-processing, including the requirement for increased precision, effectiveness, and automation. The pros and weaknesses of the ML algorithms, such as decision trees, random forests, and neural networks, are briefly discussed. The results demonstrate the potential for ML to enhance the accuracy and efficacy of FDM post-processing. Discussions are conducted regarding the effects of both conventional and ML post-processing methods.

11.10.1 FINAL THOUGHTS ON THE USE OF ML IN FDM POST-PROCESSING

Applying ML techniques might significantly enhance the quality and effectiveness of the FDM post-processing process. ML models can optimize FDM printing settings, decrease material waste, boost productivity, and improve quality control by identifying the most crucial post-processing factors that affect component quality. The models created may be used to optimize the post-processing settings and raise the quality of FDM parts. ML can forecast the quality of FDM components with a high degree of accuracy. ML can assist in automating the process and saving time and expenses by minimizing the reliance on conventional post-processing methods. ML models' performance depends on the training data's quality. To ensure the quality and generalizability of the models, rigorous selection and pre-processing of the data are essential.

11.10.2 RECOMMENDATIONS FOR PRACTITIONERS AND RESEARCHERS

High-quality data must be gathered and arranged by industry professionals. For ML models to be developed effectively, accurate and pertinent data is essential. As a result, it's crucial to gather and organize high-quality data while concentrating on the factors that affect part quality. It's vital to periodically assess the ML models' accuracy and efficiency and test and improve them using various datasets. When developing and applying ML models, it's critical to be mindful of their limits. For instance, ML models could not be successful at forecasting outcomes that fall beyond the area of the training data. For FDM post-processing, various ML algorithms may be utilized. Thus, it's critical to investigate new, more powerful or effective approaches. Conduct more tests and research to confirm the efficacy and applicability of ML models for FDM post-processing. Therefore, conducting more research and experiments that evaluate the models using various datasets and settings is crucial. Sharing information and findings can foster a more active and cooperative research community and hasten the advancement of the discipline. Therefore, it is vital to freely and cooperatively exchange data and findings.

REFERENCES

Baechle-Clayton, M., Loos, E., Taheri, M., & Taheri, H. (2022). Failures and flaws in Fused Deposition Modeling (FDM) additively manufactured polymers and composites. *Journal of Composites Science, 6*(7), 202. https://doi.org/10.3390/jcs6070202

Banga, H. K., Kumar, R., Channi, H. K., & Kaur, S. (2023). Parametric design and stress analysis of 3D printed prosthetic finger. In *Woodhead Publishing Reviews: Mechanical Engineering Series, Innovative Processes and Materials in Additive Manufacturing* (pp. 57–80). Woodhead Publishing. https://doi.org/10.1016/B978-0-323-86011-6.00014-3, ISBN 9780323860116

Barrios, J. M. & Romero, P. (2019). Decision tree methods for predicting surface roughness in fused deposition modeling parts. *Materials, 12*(16), 2574. https://doi.org/10.3390/ma1 2162574

Cai, R., Wang, K., Wen, W., Peng, Y., Baniassadi, M., & Ahzi, S. (2022). Application of machine learning methods on dynamic strength analysis for additive manufactured polypropylene-based composites. *Polymer Testing, 110*, 107580. https://doi.org/10.1016/j.polymertesting.2022.107580

Chennakesava, P., & Shivraj Narayan, Y. (2014). Fused Deposition Modeling – Insights. *International Conference on Advances in Design and Manufacturing (ICAD&M'14)* National Institute of Technology Tiruchirappalli, India, volume: III, 1345–1350.

Chohan, J. S., Singh, R., & Boparai, K. S. (2020). Vapor smoothing process for surface finishing of FDM replicas. *Materials Today: Proceedings*, *26*, 173–179. https://doi.org/10.1016/j.matpr.2019.09.013

Chueca de Bruijn, A., Gómez-Gras, G., & Pérez, M. A. (2020). Mechanical study on the impact of an effective solvent support-removal methodology for FDM Ultem 9085 parts. *Polymer Testing*, *85*, 106433. https://doi.org/10.1016/j.polymertesting.2020.106433

Dizon, J. R. C., Gache, C. C. L., Cascolan, H. M. S., Cancino, L. T., & Advincula, R. C. (2021). Post-processing of 3D-printed polymers. *Technologies*, *9*(3), 61. https://doi.org/10.3390/technologies9030061

Farhan Khan, M., Alam, A., Ateeb Siddiqui, M., Saad Alam, M., Rafat, Y., Salik, N., & Al-Saidan, I. (2021). Real-time defect detection in 3D printing using machine learning. *Materials Today: Proceedings*, *42*, 521–528. https://doi.org/10.1016/j.matpr.2020.10.482

Ferretti, P., Santi, G. M., Leon-Cardenas, C., Freddi, M., Donnici, G., Frizziero, L., & Liverani, A. (2021). Molds with advanced materials for carbon fiber manufacturing with 3D printing technology. *Polymers*, *13*(21), 3700. https://doi.org/10.3390/polym13213700

FormLabs Removing support marks. (2023). Retrieved from Form 3BL website: https://support.formlabs.com/s/article/Removing-Support-Marks?language=en_US

Galantucci, L. M., Lavecchia, F., & Percoco, G. (2009). Experimental study aiming to enhance the surface finish of fused deposition modeled parts. *CIRP Annals*, *58*(1), 189–192. https://doi.org/10.1016/j.cirp.2009.03.071

Ganguli, A., Pagan-Diaz, G. J., Grant, L., Cvetkovic, C., Bramlet, M., Vozenilek, J., Kesavadas, T., & Bashir, R. (2018). 3D printing for preoperative planning and surgical training: a review. *Biomedical Microdevices*, *20*(3), 1–24. https://doi.org/10.1007/s10544-018-0301-9

Geng, P., Zhao, J., Wu, W., Wang, Y., Wang, B., Wang, S., & Li, G. (2018). Effect of thermal processing and heat treatment condition on 3D printing PPS properties. *Polymers*, *10*(8), 875. https://doi.org/10.3390/polym10080875

Goh, G. D., Sing, S. L., & Yeong, W. Y. (2021). A review on machine learning in 3D printing: applications, potential, and challenges. *Artificial Intelligence Review*, *54*(1), 63–94. https://doi.org/10.1007/s10462-020-09876-9

HaghsefatLiu, K., & Tingting, L. (2020). Conference: ICIRES – International conference on innovation and research in engineering sciences at: Tbilisi – Georgia. *FDM 3D Printing Technology and Its Fundemental Properties*, 1–3. Tbilisi – Georgia.

Haidiezul, A., Aiman, A., & Bakar, B. (2018). Surface finish effects using coating method on 3D printing (FDM) parts. *IOP Conference Series: Materials Science and Engineering*, *318*, 012065. https://doi.org/10.1088/1757-899X/318/1/012065

Hiemenz, J. (2011). *3D printing with FDM: How it Works*. Stratasys Inc.

Kadam, V., Kumar, S., Bongale, A., Wazarkar, S., Kamat, P., & Patil, S. (2021). Enhancing surface fault detection using machine learning for 3D printed products. *Applied System Innovation*, *4*(2), 34. https://doi.org/10.3390/asi4020034

Karakurt, I., & Lin, L. (2020). 3D printing technologies: Techniques, materials, and post-processing. *Current Opinion in Chemical Engineering*, *28*, 134–143. https://doi.org/10.1016/j.coche.2020.04.001

Kraemer Fernandez, P., Unkovskiy, A., Benkendorff, V., Klink, A., & Spintzyk, S. (2020). Surface characteristics of milled and 3D printed denture base materials following polishing and coating: An in-vitro study. *Materials*, *13*(15), 3305. https://doi.org/10.3390/ma13153305

Kumar, S., Gopi, T., Harikeerthana, N., Gupta, M. K., Gaur, V., Krolczyk, G. M., & Wu, C. (2023). Machine learning techniques in additive manufacturing: a state of the art review on design, processes and production control. *Journal of Intelligent Manufacturing, 34*(1), 21–55. https://doi.org/10.1007/s10845-022-02029-5

Kumbhar, N. N., & Mulay, A. V. (2018). Post processing methods used to improve surface finish of products which are manufactured by additive manufacturing technologies: A review. *Journal of the Institution of Engineers (India): Series C, 99*(4), 481–487. https://doi.org/10.1007/s40032-016-0340-z

Leon-Cardenas, C., Patrich F., Gian M. S., Alessandri, G., & Mattia C. (2022). Replicability assessment of a 3D-printed mold with advanced polymers and chemical smoothing. *Proceedings of the International Conference on Industrial Engineering and Operations Management Istanbul, Turkey, March 7–10, 2022*, 1747–1755. Turkey.

Mahmood, M. A., Visan, A. I., Ristoscu, C., & Mihailescu, I. N. (2020). Artificial neural network algorithms for 3D printing. *Materials, 14*(1), 163. https://doi.org/10.3390/ma14010163

Mishra, A., & Jatti, V. S. (2023). Reinforcement learning based approach for the optimization of mechanical properties of additively manufactured specimens. *International Journal on Interactive Design and Manufacturing (IJIDeM), 17*, 2045–2053. https://doi.org/10.1007/s12008-023-01257-0

Nam, J., Jo, N., Kim, J. S., & Lee, S. W. (2020). Development of a health monitoring and diagnosis framework for fused deposition modeling process based on a machine learning algorithm. *Proceedings of the Institution of Mechanical Engineers, Part B: Journal of Engineering Manufacture, 234*(1–2), 324–332. https://doi.org/10.1177/0954405419855224

Nasiri, S., & Khosravani, M. R. (2021). Machine learning in predicting mechanical behavior of additively manufactured parts. *Journal of Materials Research and Technology, 14*, 1137–1153. https://doi.org/10.1016/j.jmrt.2021.07.004

Nguyen, T. K., & Lee, B.-K. (2018). Post-processing of FDM parts to improve surface and thermal properties. *Rapid Prototyping Journal, 24*(7), 1091–1100. https://doi.org/10.1108/RPJ-12-2016-0207

Ranjan, N., Kumar, R., Kumar, R., Kaur, R., & Singh, S. (2022). Investigation of fused filament fabrication-based manufacturing of ABS-Al composite structures: Prediction by machine learning and optimization. *Journal of Materials Engineering and Performance, 32*, 4555–4574. https://doi.org/10.1007/s11665-022-07431-x

Sagar, M. S. I., Ouassal, H., Omi, A. I., Wisniewska, A., Jalajamony, H. M., Fernandez, R. E., & Sekhar, P. K. (2021). Application of machine learning in electromagnetics: Mini-review. *Electronics, 10*(22), 2752. https://doi.org/10.3390/electronics10222752

Sampedro, G. A. R., Agron, D. J. S., Amaizu, G. C., Kim, D.-S., & Lee, J.-M. (2022). Design of an in-process quality monitoring strategy for FDM-Type 3D printer using deep learning. *Applied Sciences, 12*(17), 8753. https://doi.org/10.3390/app12178753

Sandhu, K., Singh, G., Singh, S., Kumar, R., Prakash, C., Ramakrishna, S., … Pruncu, C. I. (2020). Surface characteristics of machined polystyrene with 3D printed thermoplastic tool. *Materials, 13*(12), 2729. https://doi.org/10.3390/ma13122729

Sarker, I. H. (2021). Machine learning: Algorithms, real-world applications and research directions. *SN Computer Science, 2*(3), 160. https://doi.org/10.1007/s42979-021-00592-x

Sharma, P., Vaid, H., Vajpeyi, R., Shubham, P., Agarwal, K. M., & Bhatia, D. (2022). Predicting the dimensional variation of geometries produced through FDM 3D printing employing supervised machine learning. *Sensors International, 3*, 100194. https://doi.org/10.1016/j.sintl.2022.100194

Singh, G., Singh, S., Kumar, R., Parkash, C., Pruncu, C., & Ramakrishna, S. (2022). Tissues and organ printing: An evolution of technology and materials. *Proceedings of the Institution of Mechanical Engineers, Part H: Journal of Engineering in Medicine, 236*(12), 1695–1710. https://doi.org/10.1177/09544119221125084

Singh, G., Singh, S., Prakash, C., Kumar, R., Kumar, R., & Ramakrishna, S. (2020). Characterization of three-dimensional printed thermal-stimulus polylactic acid-hydroxyapatite-based shape memory scaffolds. *Polymer Composites, 41*(9), 3871–3891. https://doi.org/10.1002/pc.25683

Singh, R., Singh, S., Singh, I. P., Fabbrocino, F., & Fraternali, F. (2017). Investigation for surface finish improvement of FDM parts by vapor smoothing process. *Composites Part B: Engineering, 111*, 228–234. https://doi.org/10.1016/j.compositesb.2016.11.062

Suniya, N. K., & Gour, A. (2021). Applications of supervised machine learning in FDM manufacturing: A review. *Webology (ISSN: 1735-188X), Volume 18*(Number 6), 6199–6218.

Tilton, M., Lewis, G. S., Bok Wee, H., Armstrong, A., Hast, M. W., & Manogharan, G. (2020). Additive manufacturing of fracture fixation implants: Design, material characterization, biomechanical modeling and experimentation. *Additive Manufacturing, 33*(March 2020), 101137. https://doi.org/10.1016/j.addma.2020.101137

Vasquez, J. (2023). Threading 3d printed parts: How to use heat-set inserts. Retrieved from https://hackaday.com/2019/02/28/threading-3d-printed-parts-how-to-use-heat-set-inserts/

Wach, R. A., Wolszczak, P., & Adamus-Wlodarczyk, A. (2018). Enhancement of mechanical properties of FDM-PLA parts via thermal annealing. *Macromolecular Materials and Engineering, 303*(9), 1800169. https://doi.org/10.1002/mame.201800169

Zhang, J., Wang, P., & Gao, R. X. (2019). Deep learning-based tensile strength prediction in fused deposition modeling. *Computers in Industry, 107*, 11–21. https://doi.org/10.1016/j.compind.2019.01.011

Zhang, X., Fan, W., & Liu, T. (2020). Fused deposition modeling 3D printing of polyamide-based composites and its applications. *Composites Communications, 21*, 100413. https://doi.org/10.1016/j.coco.2020.100413

Žigon, J., Kariž, M., & Pavlič, M. (2020). Surface finishing of 3D-printed polymers with selected coatings. *Polymers, 12*(12), 2797. https://doi.org/10.3390/polym12122797

12 A Study on Additive Manufacturing Processes, Standards and Mechanical Properties

Mehmet Şükrü Adin

12.1 INTRODUCTION

Nowadays, fabrication methods recognized as three-dimensional (3D) printing or additive manufacturing, which are used to build 3D products layer upon layer, are very popular [1, 2]. Thanks to our digital industrial age, the potential of AM technologies appears to be revolutionary, as it removes the limits of manufacturing technologies [2, 3, 4, 5]. AM methods are a subject of intense research since they are a relatively new technology. Unlike traditional manufacturing methods, AM technologies use raw materials such as wires or powders to build 3D products. In this way, it enables much more complex and expensive products to be manufactured with cost-effective solutions (reduction in overall weight, increased functionality, complex geometries, and number of parts reduction). Additionally, it provides outstanding solutions to sectors (leading sectors) like aerospace and automotive, where low weight and high strength are crucial. [1, 3, 6, 7]. In AM printers, titanium, aluminium and steel alloys are used extensively as raw materials. There are many defects in the parts built with these raw materials. One of the basic reason for this is that the production processes of these production technologies have not yet been optimized at the desired level. Therefore, in recent years, when the researches in the literature are reviewed, it is felt that intensive research on this subject has started to be made. Furthermore, despite the importance of the mechanical properties of AM-produced products, little research has been done in this area [3, 8, 9, 10].

Therefore, in this chapter, important information is given about the usage areas of different standards. Further, consideration has been given to the hardness, tensile, compression, surface roughness, fracture toughness, and fatigue strength properties of AM materials. In this context, the results of different mechanical properties obtained from studies with materials such as Ti-6Al-4V, Al-8.5, SS316L, IN718, Al-12Si, and Fe-1.3V-1.7Si were compared. The most popular AM approaches and their distinctions are also highlighted within the context of the study.

DOI: 10.1201/9781032665351-15

12.2 ADDITIVE MANUFACTURING PROCESS (AM PROCESS)

Although the material of the manufactured parts, the technology used and the manufactured part's design are different, the AM process generally consists of eight separate production stages. These production stages are "CAD", "Converting to STL ", "File transfer to machine", "Machine setup", "Build", "Part removal and clean-up", "Post-processing" and "Application", respectively [3, 11, 12]. The production stages are depicted in Figure 12.1 [11].

12.2.1 COMPUTER AIDED MANUFACTURING (CAD)

At this stage, drawings of the parts whose conceptualization studies have been completed are made using 3-Dimensional (3D) CAD programs. 3D CAD programs (Figure 12.2 [13]) are the foundation of AM technology. Because these programs are the key to the transition from CAD to Computer Aided Manufacturing (CAM) [3, 11, 14].

12.2.2 CONVERTING TO STL (STEREOLITHOGRAPHY)

All AM technologies use the STL file format that describes a CAD model only in terms of geometry. Thanks to 3D CAD programs, this conversion (Figure 12.3) is usually done automatically. The term STL comes from Stereolitograhy (STL), an AM technology in the 1990s [3, 4, 14, 15].

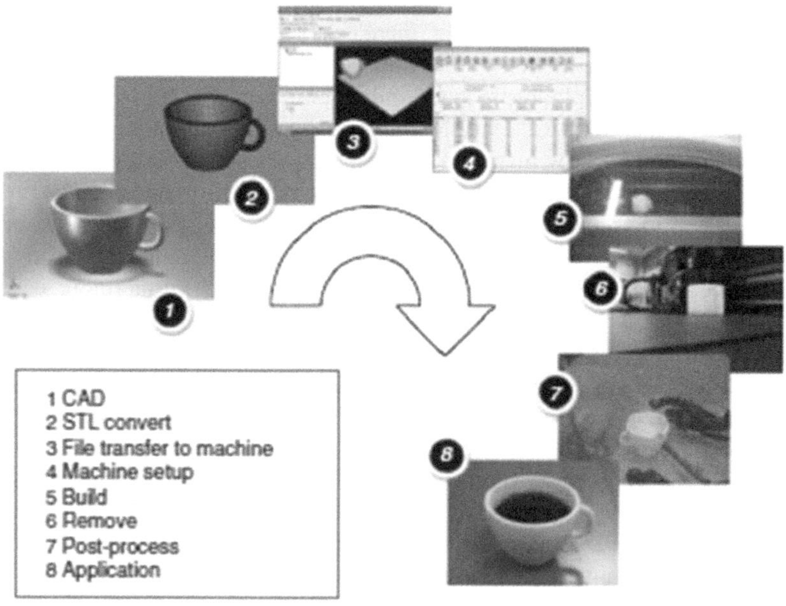

FIGURE 12.1   AM Process stages [11]

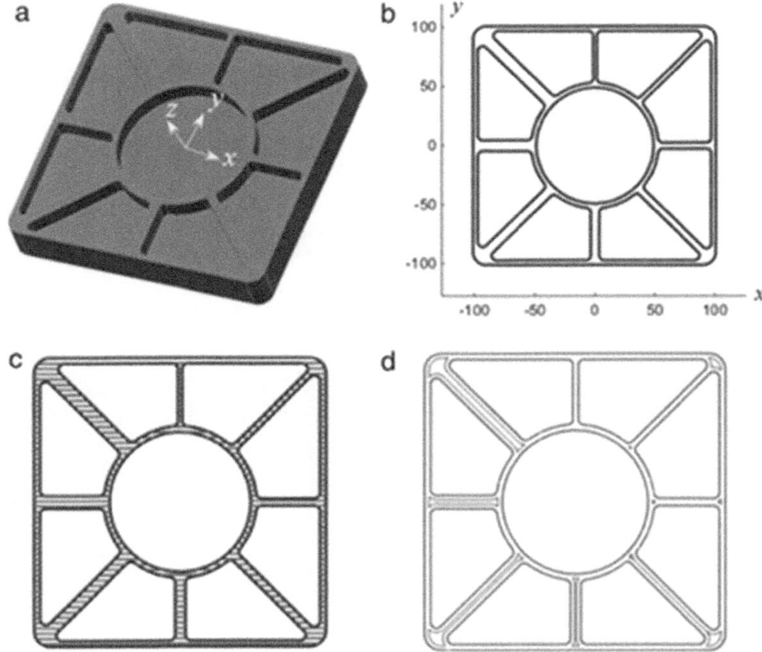

FIGURE 12.2 CAD program models [13]

FIGURE 12.3 CAD converting to STL [3]

12.2.3 File Transfer to Machine

The first step is to determine whether the part is in the correct dimensions. Since STL files are easily scalable, if it is found that there is an incorrect application, the dimensions in the file can be changed. In addition, AM machines allow the user to easily intervene in the resulting product. In this way, the direction and orientation of the product are changed by the user and brought to the appropriate format. What needs to be done will be to upload the STL files to the AM machine (Figure 12.4) and get the product as output from the machine [3, 11, 15, 16].

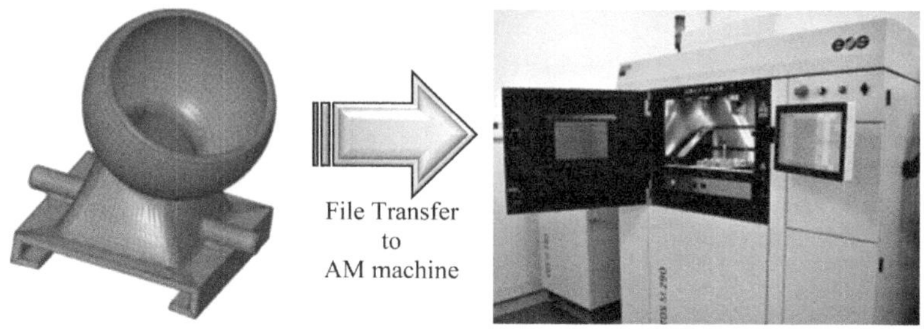

FIGURE 12.4　File Transfer to AM machine [3, 16]

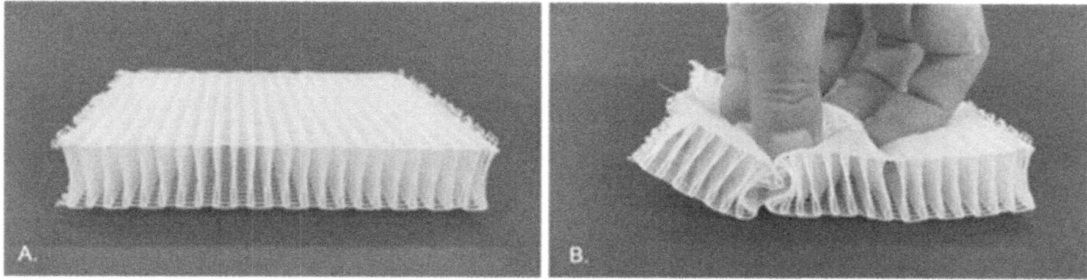

FIGURE 12.5　3D printed specimen [17]

12.2.4　MACHINE SETUP

The number of parameters may be different based on the type of AM machine utilized. Some AM machines can work with one or two materials. In this case, it is to be ensured that there will be not too many changes in the parameter settings. A complex product made from a number of different components would, however, require a number of parameter adjustments during manufacture. In this case, it is important to make the settings very convenient and carefully. Otherwise, unanticipated effects will inevitably occur [3, 9, 11, 15, 16].

12.2.5　BUILD

It is assumed that the entire AM process takes place automatically. However, partially automatic tasks that may need manual control, interaction, and decision-making must be completed before this procedure. Layer-based production begins with the computer-controlled construction phase of the processes. The processes are repeated until the product is completed or the raw material used is consumed. The device notifies the user when the product is finished or the material utilized is exhausted (Figure 12.5) [3, 9, 11, 16, 17, 18].

12.2.6　PART REMOVAL AND CLEAN-UP

It is anticipated that the finished product from the machine is removed from its location with ease. However, it is likely that material residues will form around some products. The extraction of these formed residues from the material requires skill and good manual dexterity. In this way, the product is brought to the structure it should be by using sensitive applications [3, 4, 11, 17, 18, 19, 20].

12.2.7 POST-PROCESSING

After the product clean-up process (Figure 12.6) has been performed, it is the stage where some operations are performed, which still need to be done mostly manually. Depending on the product's structure, some processes are completed with minimal changes, while others allow changes to be made that require attention due to the delicate and fragile structure of the final product [9, 11, 21, 22, 23].

12.2.8 APPLICATION

The parts whose production has been completed and the final operations have been completed have become available. Although the material used in the part is identical to the materials used in other manufacturing methods, it shouldn't be expected that the part obtained at the end of the AM process would have a standard property. The fabrication process of AM machines is being renewed day by day, and applications are increasing in order for the parts to have standard part characteristics [9, 11, 21, 22, 24, 25].

12.3 MOST USED AM METHODS

Because of the development of AM technology, complex structures can now be created with high precision. To achieve this, there are some development factors. The most important of these factors can be expressed as reducing defects (printing), being able to build large-sized parts, quickly (rapid) building prototypes, and improving the mechanical properties of the built parts. Polymer filaments are most typically employed in the popular 3D AM technique recognized as Fused Deposition Modelling (FDM). At the same time, AM techniques including stereolithography,

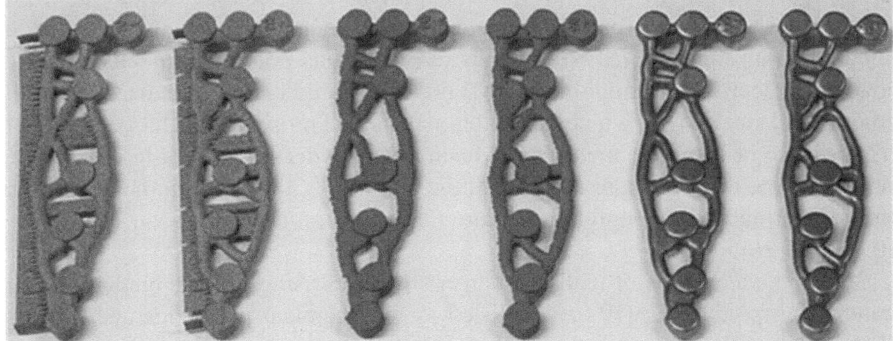

FIGURE 12.6 Different cleaning stages of post-processing [21]

direct energy deposition (DED), selective laser melting (SLM), and selective laser sintering (SLM) are also referred to as the key techniques. Two-photon polymerization (TPP), electrohydrodynamic printing (EHDP), projection micro stereolithography (P-SLA), and non-contact micro or nano printing techniques are some of the most recent AM technologies. Among all these, the most popular techniques for AM technologies include FDM, Powder bed fusion, SLA, Inkjet printing, DED and LOM [3, 9, 10, 11, 13, 14, 15, 16, 22].

12.3.1 FUSED DEPOSITION MODELLING (FDM)

The FDM technique uses thermoplastic polymer filament to print layers of materials in three directions. This technique ensures the filament's fluidity by heating it in the nozzle and turning it semi-liquid. On the printer bed or on pre-printed layers, liquid filament is extruded. The thermoplasticity of the plastic wire is the fundamental property that makes the filaments to join (fuse) together during printing in the extrusion process and to cool down at room temperature after the completion of printing. The critical factors that influences the mechanical properties of printed parts are the layer thicknesses, its width, direction of the filaments and the infill percentage. It had been found that the main cause of low strength of FDM 3D printed parts is the inter-layer distortion. Low cost, quick production, and simplicity of use are the benefits of FDM, whereas poor mechanical qualities, a staircase appearance, inferior surface finish, and a small selection of polymeric materials are its drawbacks. The foremost issues with composite parts are fibre alignment, bonding between matrix and fiber, and void formation, despite the fact that fibre reinforced composites have been designed to improve the mechanical properties of 3D printed components [3, 4, 10, 11, 26].

12.3.2 POWDER BED FUSION (PBF)

PBF techniques combine layer structures with a laser beam or a binder to create the final product. These layers are made up of tightly packed, finely dispersed particles on the bed of the printer. The preceding layers are rolled over by the powders to create the first 3D part, and then they are fused to create the final 3D part. Excess dust is removed using a vacuum. If needed, methods for coating, sintering, or percolation with advanced processing and detailed techniques are used. There are selective laser sintering (SLS) methods using various polymers, metals and alloy powders, selective laser melting (SLM) methods using certain metals such as aluminium and steel. SLM, which has better mechanical properties, achieves this property because of complete melting and fusing the powders used after laser scanning. While in SLS, the molecular merging of the granules is a result of both the powders' partial melting and the high temperature that exists on the grain's surface [3, 4, 9, 10, 13, 14, 26].

3DP is the technique of realizing bigger structures using a liquid binder. The outstanding properties of 3DP can be listed as the chemical properties and rheology of the binder, the size and shape of the dust particles, the precipitation rate, the interaction between the powder and decoupling, and post-processing techniques. The critical factors that influence the sintering process are the intensity of laser power as well as the speed of the scanning. Used in various manufacturing setups, for example in advanced applications such as tissue engineering, cages, scaffolds for aerospace and electronics, this method is used appropriately for printing intricate structures, due to the fact that it has a fine resolution and high print quality. The fundamental benefit of powder bed fusion is this. Another advantage is the use of the dust bed as a support allows overcoming the difficulty of removing the support material. The disadvantages are the slow progress of the process, the high costs when the powder is fused along with the binder, and the high porosity of the part [3, 10, 11, 22].

12.3.3 Contour Crafting with Inkjet Printer

The fabrication of intricate and sophisticated ceramic structures for uses like tissue engineering scaffolds uses inkjet printing, one of the primary techniques for ceramic additive manufacturing. Because of its quick and effective nature, the technology gives enough flexibility to the design and printing of complex structures. The distribution of ceramics' particle sizes, the viscosity and solid content of the ink, the extrusion speed, nozzle size, and printing speed are all variables that affect how effectively inkjet printing will turn out. The method's drawbacks, however, is the maintenance, rough decolonization, and lack of bonding among the layers [3, 11, 10, 26].

12.3.4 Stereolithography (SLA)

SLA is a method that uses UV light to ignite a series of reactions on an outer surface of resin or monomer solution. One of the early additive manufacturing methods, it can print high-quality items at a resolution less than 10 μm. Complex nanocomposites can be successfully manufactured via this method. As for the disadvantages, while it should be recorded that it is comparatively sluggish and costly compared to other methods, the materials used in printing are limited, the reaction kinetics and the hardening process are multifaceted [9, 10, 11, 22, 26].

12.3.5 Direct Energy Deposition (DED)

High-performance super alloys have been created using the DED process. A beam of electron (laser) is employed as the source of energy, while simultaneously melting the raw material (wire or powder). In addition, the fact that no powder bed is used in the method ensures that it has a property similar to FDM according to the properties separated from SLM and the raw materials used. The processing can be finished by combining the DED with conventional extraction techniques. This procedure is often used in aerospace applications using alloys like titanium, Inconel, stainless steel and aluminium. DED is categorized by exceptionally big working envelopes and a rapid working speed. As for disadvantages, this is because of the fact that it has a lower precision and inferior surface quality and can produce less intricate parts in comparison to SLM or SLS [4, 9, 10, 11, 22].

12.3.6 Laminated Object Manufacturing (LOM)

Laminated object manufacture (LOM), one of the oldest forms of AM, uses layer-by-layer removing and laminating of material sheets. This technique involves utilizing a mechanical cutter or laser to precisely cut successive layers. When cutting is complete, extra materials left behind is utilized as support before being removed and recycled. Electronics, foundries, paper manufacturing, smart constructions, and other industries all use LOM technology. The benefits of the technique include a reduction in tooling costs and manufacturing times. While it is considered to be one of the most preferred AM techniques; the disadvantages are that it can be because of the fact that it has an inferior surface quality without post-treatment and its dimensional accuracy is lower compared to powder bed methods. In addition, the removal of excess parts of the laminates after the fabrication of the component is not used for complex shapes because of the time-consumption compared to powder bed methods [9, 10, 11, 22].

12.4 MECHANICAL PROPERTIES OF AM MATERIALS

The primary purpose of using metal AM components in the AM practice is due to its high-density materials by examining the relationship between the structure and its mechanical properties (example: ductility, anisotropy and strength). Measurements of the performance of metal products produced by additive manufacturing (AM) methods under load are made according to internationally valid standards. In the test methods of the manufactured products, properties such as surface density, hardness, strength, compression strength, fatigue as well as degradation of printed metallic parts are examined and the congruence of the results with the international standard values are observed. As per the positive data obtained as a result of the observations, mass production is started using 3D fusion technology. When loading is performed on a product manufactured with 3D printing; a 3D printing test is performed to determine its resistance, quality and mechanical behavior against this load. Manufacturers determine whether products meet industry standards by applying ASTM-based tests. Mechanical tests are applied in research and development conducted to ensure the high quality of the production process or product [3, 10, 12, 14, 27, 28, 29, 30, 31]. In this context, we can list the standards and definitions used in mechanical testing of metal products as in **Table 12.1** [3, 4, 32].

12.4.1 Hardness Properties of AM Materials

Because the mechanical strength of the metal AM test components can be determined, hardness testing is a useful technique. In recent researches, it was revealed that the microstructural properties of AM titanium alloy are related to the microhardness of Vickers and its compatibility with the Hall-Petch relationship. In the research conducted on the influence of cross-sectional area on hardness values it has been noted that there is very no correlation between metal AM component size and microhardness, most likely as a result of inadequate thermal insulation. While it was reported in a recent study that the cross-sectional area depends on the hardness property, a decrease in micro-hardness due to microstructural coarsening has been

TABLE 12.1
Standards and definitions [4, 8, 32]

ISO/ASTM 52900:2015 establishes and defines terms used in additive manufacturing (AM) technology, which applies the additive shaping principle and thereby builds physical 3D geometries by successive addition of material.

ISO/ASTM 52921:2013 includes terms, definitions of terms, descriptions of terms, nomenclature, and acronyms associated with coordinate systems and testing methodologies for additive manufacturing (AM) technologies in an effort to standardize terminology used by AM users, producers, researchers, educators, press/media, and others, particularly when reporting results from testing of parts made on AM systems.

ISO 17296-2:2015 describes the process fundamentals of Additive Manufacturing (AM). ISO 17296-2:2015 explains how different process categories make use of different types of materials to shape a product's geometry. It also describes which type of material is used in different process categories.

VDI 3405 is aimed at users and producers of additive manufacturing processes, which i.e. are used to manufacture prototypes, tools and end products. VDI 3405 explains the principles of commercially available additive manufacturing processes and specifies their quality parameters. This facilitates a better assessment of different additive manufacturing processes depending on the application. The standard covers the principle considerations which apply to the design, fabrication and assessment of parts produced by additive manufacturing. It specifies terms and definitions and describes relevant quality parameters. It considers component testing and the draw-up of supply agreements. It also covers safety-related and environmental aspects.

ISO 17296-4:2014 covers the principal considerations which apply to data exchange for additive manufacturing. It specifies terms and definitions which enable information to be exchanged describing geometries or parts such that they can be additively manufactured. The data exchange method outlines file type, data enclosed formatting of such data and what this can be used for.

ISO/ASTM 52915:2016 provides the specification for the Additive Manufacturing File Format (AMF), and interchange format to address the current and future needs of additive manufacturing technology. The AMF may be prepared, displaced and transmitted provided the requirements of this specification are met.

ASTM F2924 covers additively manufactures titanium-6aluminium-4vanadium (Ti-6Al-4V) components using full-melt powder bed fusion such as electron beam melting and laser melting. It indicates the classifications of the components, the feedstock used to manufacture Class 1, 2, and 3 components, as well as the microstructure of the components. This specification also identifies the mechanical properties, chemical composition, and minimum tensile properties of the components.

ASTM F 3001-14 establishes the requirements for additively manufactured titanium-6aluminium-4vanadium with extra low interstitials (Ti-6Al-4V ELI) components using full-melt powder bed fusion such as electron beam melting and laser melting. The standard cover the classification of materials, ordering information, manufacturing plan, feedstock, process, chemical composition, microstructure, mechanical properties, thermal processing, hot isostatic pressing, dimensions and mass, permissible variations, retests, inspection, rejection, certification, product marking and packaging, and quality program requirements.

(continued)

TABLE 12.1 (Continued)

ASTM F3049–14 introduces the reader to techniques for metal powder characterization that may be useful for powder-based Additive Manufacturing processes including binder jetting, directed energy deposition and powder bed fusion. It refers the reader to other, existing that may be applicable for the characterisation of virgin and used metal powders processed in Additive Manufacturing systems.

The intention of this article is to provide purchasers, vendors, or producers of metal powder to be used in Additive Manufacturing processes with a reference for existing standards or variations of existing standards or variations of existing standards that may be used to characterize properties of metal powders used for Additive Manufacturing processes.

It will serve as a starting point for the future development of a suite of specific standard test methods that will address each individual property or property type that is important to the performance of metal-based Additive Manufacturing systems and the components produced by them. While the focus of this standard is on metal powder, some of the referenced methods may also be appropriate for non-metal powders.

ISO 17296-3:2014 covers the principal requirements applied to testing of parts manufactures by Additive Manufacturing processes. It specifies main quality characteristics of parts, specifies appropriate test procedures, and recommends the scope and content of test and supply agreements.

It is aimed at machine manufacturers, feedstock suppliers, machine users, part providers and customers to facilitate the communication on main quality characteristics. It applies wherever Additive Manufacturing processes are used.

ASTM F 3055–14a defines the requirements for additive manufacturing of nickel alloy (UNS N07718) using full-melt powder bed fusion such as electron beam melting and laser melting. The standard many be used by purchasers and producers of additively manufactures UNS N07718 components to specify the requirements and ensure component properties, and by users to obtain components that will satisfy the minimum acceptance requirements. The standard covers terminology and classification as well as the requirements with respect to ordering information, manufacturing plan, feedstock, thermal processing, chemical composition, microstructure, mechanical properties, hot isostatic pressing, dimensions and permissible variations, retests, inspection, rejection, certification, product marking and packaging, maintenance of a quality program, and the significance of numerical limits.

ASTM F3056–14e1 covers additively manufactures UNS N06625 components using full-melt powder bed fusion such as electron beam melting and laser melting. The components produced by these processes are used typically in applications that require mechanical properties similar to machined forgings and wrought products. Components manufactured to this specification are often, but not necessarily, post processed via machining, grinding, electrical discharge machining (EDM), polishing, and so forth to achieve desired surface finish and critical dimensions.

TABLE 12.1 (Continued)

VDI 3405 Part 2.1 contains material characteristic data for additively manufactured parts made from the aluminium alloy AlSi10Mg obtained in a round robin test. The test procedures and methods described in VDI 3405 Part 2 were used. Since all these procedures and methods correspond to recognised industry standards, it is possible to compare the characteristic values with those of conventional manufacturing process. Additive manufacturing processes, rapid manufacturing – Laser beam melting of metallic parts; Material data sheet aluminium alloy AlSi10Mg.

ASTM F2971–13 describes a standard procedure for reporting results by testing or evaluation of specimens produced by Additive Manufacturing. This practice provides a common format for presenting data for AM specimens, for two purposes; to establish further data reporting requirements and to provide information for the design of material property databases.

The standard was established because, due to variables unique to each AM process and piece of equipment, it is critical to standardize descriptions used to report the preparation, processing and post processing of specimens produced for tests or evaluation. The intent of this standard is to ensure the consistent documentation of the materials and processing history associated with each specimen undergoing test or evaluation. The level of detail for the documentation will match the application.

This practice establishes minimum data element requirements for reporting of material and process data for the purpose of:

- Standardising test specimen descriptions and test reports
- Assisting designers by standardising AM materials databases
- Aiding material traceability through testing and evaluation
- Capturing property-parameter-performance relationships of AM specimens to enable predictive modelling and other computational approaches.

shown with an increase in the cross-sectional area. Due to the thicker cross-sectional part, the larger temperature input to the part and the slower solidification rate cause microstructural coarsening. In this investigation, it was shown that another effect on microhardness is due to differences in 2D planar geometry due to variances in heat flow. Research conducted on the stiffness effect of building height have shown contradictory results. In the first of two different studies, the researchers showed that there were variations in the structure height and Vickers microhardness values; while the other researchers did not report any important differences in Vickers microhardness levels up to 25 mm from the base. Moreover, they showed that the decrease in the Vickers microhardness value was caused by an increase in the height of EBM fabricated Ti-6Al-4 V [14, 32, 33, 34].

12.4.2 TENSILE PROPERTIES OF AM MATERIALS

It is observed that the tensile strength of materials fabricated by the AM process often meets the special necessities of industries comparison to the tensile strength of existing steel grades. Grain thinning significantly improves yield and ultimate tensile strength. Elongation values according to the material used are in terms of ductility, lower porosity (0.1%) results in a ductile mode of fracture. Simultaneously,

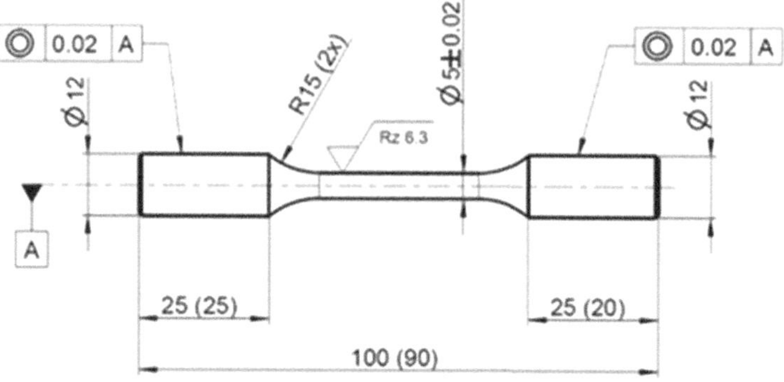

FIGURE 12.7   Tensile test sample [39]

flexible modes with significantly lower elasticity are caused by a porosity of more than 2.4%. Some specific ASTM procedures observe the expected material properties for 3D printed metal powder-based sintering applications. The mechanical properties obtained by dividing the elastic modulus stress by strain include the elongation at the moment of fracture, the final stress caused by tension, or the maximum stress (Figure 12.7). SLA materials are more brittle and tougher than comparable goods made with injection moulds. [3, 4, 14, 33, 34, 35, 36, 37, 38, 39].

12.4.3   AM Fabricated Parts – Compression Test

In research conducted the mechanical properties of metal AM components (tantalum amalgam combined with SLM method) were investigated using a pressure test (Figure 12.8). It has been found that the compressive yield strength is superior to the upward equal one due to the sliding of the crystallographic surfaces. Despite its anisotropic hardness, the mechanical properties of the tantalum fabricated by SLM method were discovered to be better than those of the product made using an electron pole heater or powder metallurgy. Porosity, which is considered an important defect in SLM materials, surprisingly folds, expands, mixes and disperses, although the idea of stacking under compression load closes the pores and is a defect, this is misleading. When the results are evaluated, it is reasoned that the effects of porosity on the stacking layer; the condition evaluated as a defect indicates that once the orientation of the layer resembles the stacking path and the direction of the layer is moved along these lines, it has a wider effect than if the orientation of the layer is reversed [3, 9, 14, 38].

12.4.4   AM Fabricated Parts – Surface Roughness

By affecting the surface properties of AM components, various inputs can also affect performance, leading to the growth of numerous noticeable and measurable output variables. These specifications encompass all aspects of production, including materials, component design, process sections, specification of the process, and post-processing. Improper melting from partly fused powder granules has been beneficial

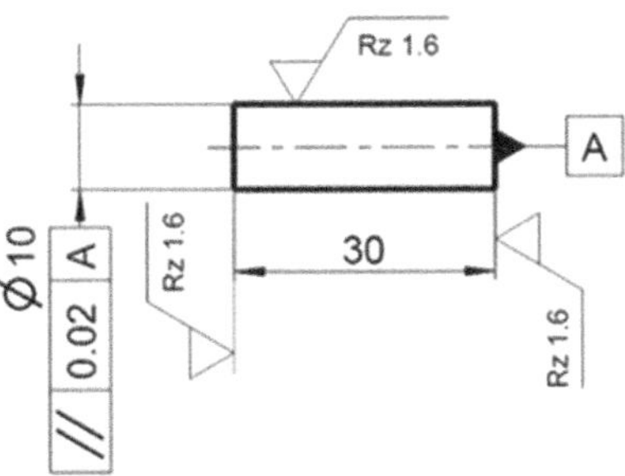

FIGURE 12.8   Compression test sample [39]

due to the absence of a fusion layer or detection pathways such as construction, fusion, or aggregation, such as striation. One of the most harmful influences on the performance and rotation fatigue of metal products is surface roughness. Post-production events are necessary to prolong the life of AM components. This process is extremely difficult, especially for AM components with complex geometry that want to be used in the form of a network as they are manufactured. As a result, one of the primary advantages of AM, it loses the ability to produce complex geometry that traditional manufacturing cannot do after surface treatment. The surface hardness of the surface of the product produced by the AM process may be affected by the type of equipment used, dust size, process parameters and the direction of the device. Therefore, the fatigue of the components and their relationship to the surface coating should be fully known. DLD techniques, due to the use of thicker covers, cause a lot of dust, fragments and layers to be present, and therefore produce rough surfaces, which are typically found in L-DBF techniques. This increases the surface hardness of AM components, increased hatching pitch, layer thickness and/or dust size [3, 4, 10, 14, 19, 20, 38, 40].

12.4.5 FRACTURE TOUGHNESS PROPERTIES OF AM MATERIALS

Breaking strength is the capacity of a material to endure breaking. In studies on regulating the crack strength benefits of metal AM components, considering the crack life anisotropy in both SLM-mounted and EBM-mounted Ti-6Al-4 V, due to the fracture toughness, the effect of anisotropy on how breaks propagate – It has been found that breaks occur along columnar grains in equally oriented samples, while breaks occur at grain column boundaries in upward oriented samples. In defining the post-heat treatment approaches, understanding the microstructures built for the unique metal AM frames to ensure superior fracture toughness is critical [4, 9, 14, 27, 35].

12.4.6 FATIGUE STRENGTH IN AM MATERIALS

The fatigue strength of metal AM components is determined by static mechanical properties of materials, which are identical to the microstructure of various metals. At the same time, natural properties such as surface robustness and material failures weaken the fatigue performance of parts manufactured by AM. Research on the tensile

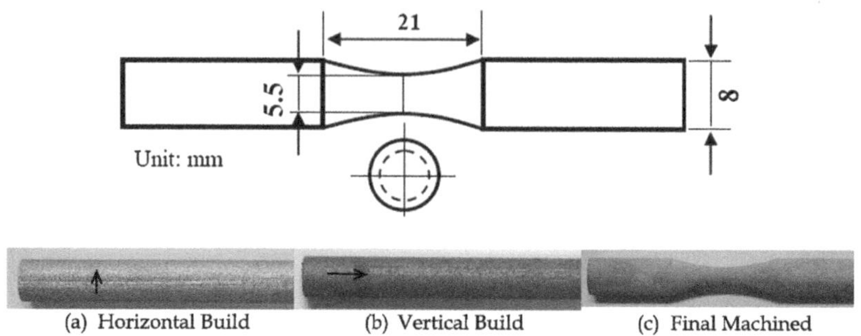

FIGURE 12.9 Fatigue test samples; (a) Built horizontally (b) Built vertically (c) Final machined [41]

behavior of various metals and alloys is done using the AM method (Figure 12.9). In the study where the wear properties of the additive Ti-6Al-4 V were examined due to its potential use in aerospace and biomedical applications, it is revealed that the lower the wear rate, the better the fatigue resistance of PBF-EB and DMLS. Polishing, for example, can help improve fatigue properties. According to another study, they concluded that the factors that make it difficult to evaluate tribological properties are the dispersion of research data, porosity and lack of adhesion on the layer. Similar to AM processes, higher wear and access to real data results in the use of hot isostatic presses to improve and intensify these defects. In the study, the fatigue strength of AM-based metals such as aluminum alloys, Ti-6Al-4 V and steels undergoing different surface and heat treatment conditions was compared. The examination of AM metals and alloys may thus be done using principles from mechanical behaviour, and the physio-mechanical properties of AM raw materials are equivalent to those of conventional goods like static and fatigue resistance [9, 10, 11, 14, 27, 41].

12.5 STUDIES ON MECHANICAL PROPERTIES

Different mechanical tests are applied to the materials or parts produced with AM, taking into consideration the purpose of its use. Thus, their suitability for the intended use is determined. These mechanical tests determine whether they are suitable for the intended use or not. In mechanical tests, the parts produced with AM are generally evaluated by applying tests such as hardness, fracture toughness, tensile, impact strength, and fatigue [3, 14, 39]. Summaries of some studies in the literature regarding mechanical tests are given in Table 12.2.

12.6 CONCLUSION

Depending on the rapid progress of technology, AM technologies are also developing daily. However, due to the building nature of AM technologies, many defects occur in manufactured parts. Therefore, it is necessary to research the ideal print settings and ideal mechanical qualities of the items that will be made using AM technologies. In the literature, there are many studies on optimization of machining parameters of parts

TABLE 12.2
Overview of mechanical tests

	Material	UTS (MPa) *	YS (MPa)*	Elongation (%)*	Ref.
Summary of tensile properties	Ti-6Al-4V	1040	947	14.6	[42]
	Ti-6Al-4V	943	887	18.5	[43]
	Ti-6Al-4V ELI	942	824	13.4	[34]
	Ti-6Al-4V ELI	994	905	16.5	[44]
	Ti-6Al-4V ELI	1051	988	14.4	[45]
	Ti-6Al-4V	1068	994	14.5	[46]
	Ti-6Al-4V	1222	1058	4.9	[47]

Ultimate Tensile Strength: UTS, Yield Strength: YS, Ref.: Reference

	Material	Microhardness (Hardness Vickers (HV))*	Ref.
Summary of hardness properties	Ti-6Al-4V	460	[48]
	Ti-6Al-4V	347	[49]
	Ti-6Al-4V ELI	320	[34]
	SS316L	185	[50]
	IN718	242	[50]
	Al-8.5Fe-1.3V-1.7Si	154	[51]
	Ti-6Al-4V	360	[52]

	Material	Fracture toughness (MPa $\sqrt{m}$)*	Ref.
Summary of fracture toughness properties	Ti-6Al-4V	111	[53]
	Ti-6Al-4V	75	[54]
	Ti-6Al-4V	67	[55]
	Ti-6Al-4V	29	[35]
	Ti-6Al-4V	42	[35]
	Al-12Si	47	[56]
	Al-12Si	22	[56]

*Note: approximate values have been taken.

built with different AM technologies. However, these studies are not a continuation of each other, they are independent from each other and there are quite large differences between the parameters. As a natural consequence of this situation, it becomes very difficult to make comprehensive evaluations. On the other hand, it is vital to carry out mechanical tests to realize the suitability of the products built with AM technologies for their use. As a matter of fact, when the studies in the literature are examined, it is seen that there are differences in this subject. In addition, it has been seen that there are almost no studies in the literature on standard samples used for different mechanical tests. Moreover, it is mentioned that there are extensive manufacturing defects in AM parts. The reason for this is that parts are produced without optimizing the manufacturing processes. As a natural consequence of all these situations, both the properties and the performance of the products built with AM methods are quite different from each other. Future-oriented analysis calls for high-quality optimization of all AM techniques as well as accurate application of ASTM and ISO standards. Thanks to the fulfilment of these conditions, a logarithmic increase will occur in the variety of products manufactured by AM methods.

REFERENCES

[1]   Chua, C. K., and Leong, K. F., *3D Printing and additive manufacturing: Principles and applications (with companion media pack)-of rapid prototyping*: World Scientific Publishing Company, 2014.

[2]   Kok, Y., Tan, X. P., Wang, P., Nai, M., Loh, N. H., Liu, E., and Tor, S. B., "Anisotropy and heterogeneity of microstructure and mechanical properties in metal additive manufacturing: A critical review," *Materials & Design,* vol. 139, pp. 565–586, 2018.

[3]   Godec, D., Breški, T., Katalenić, M., Nordin, A., Diegel, O., Kristav, P., Motte, D., and Tavčar, J., Applications of AM, in: D. Godec, J. Gonzalez-Gutierrez, A. Nordin, E. Pei, J. Ureña Alcázar, *A guide to additive manufacturing,* Springer Nature Switzerland AG, pp. 1–344, 2022.

[4]   Frazier, W. E., "Metal additive manufacturing: A review," *Journal of Materials Engineering and performance,* vol. 23, pp. 1917–1928, 2014.

[5]   Patil, N., Pal, D., and Stucker, B., "A new finite element solver using numerical eigen modes for fast simulation of additive manufacturing processes," 2013.

[6]   Aboulkhair, N. T., Simonelli, M., Parry, L., Ashcroft, I., Tuck, C., and Hague, R., "3D printing of Aluminium alloys: Additive Manufacturing of Aluminium alloys using selective laser melting," *Progress in materials science,* vol. 106, pp. 100578, 2019.

[7]   Shiyas, K., and Ramanujam, R., "A review on post processing techniques of additively manufactured metal parts for improving the material properties," *Materials Today: Proceedings,* vol. 46, pp. 1429–1436, 2021.

[8]   Galy, C., Le Guen, E., Lacoste, E., and Arvieu, C., "Main defects observed in aluminum alloy parts produced by SLM: From causes to consequences," *Additive Manufacturing,* vol. 22, pp. 165–175, 2018.

[9]   Hegab, H., Khanna, N., Monib, N., and Salem, A., "Design for sustainable additive manufacturing: A review," *Sustainable Materials and Technologies,* vol. 35, pp. e00576, 2023.

[10] Ngo, T. D., Kashani, A., Imbalzano, G., Nguyen, K. T., and Hui, D., "Additive manufacturing (3D printing): A review of materials, methods, applications and challenges," *Composites Part B: Engineering,* vol. 143, pp. 172–196, 2018.

[11] Gibson, I., Rosen, D., and Stucker, B., *Additive manufacturing technologies: Rapid prototyping to direct digital manufacturing.* Springer, 2014.

[12] Sturm, L. D., Williams, C. B., Camelio, J. A., White, J., and Parker, R., "Cyber-physical vulnerabilities in additive manufacturing systems: A case study attack on the. STL file with human subjects," *Journal of Manufacturing Systems,* vol. 44, pp. 154–164, 2017.

[13] Ding, D., Shen, C., Pan, Z., Cuiuri, D., Li, H., Larkin, N., and van Duin, S., "Towards an automated robotic arc-welding-based additive manufacturing system from CAD to finished part," *Computer-Aided Design,* vol. 73, pp. 66–75, 2016.

[14] Khan, M. A., and Jappes, J. W., *Innovations in additive manufacturing.* Springer, 2022.

[15] Rypl, D., and Bittnar, Z., "Generation of computational surface meshes of STL models," *Journal of Computational and Applied Mathematics,* vol. 192, no. 1, pp. 148–151, 2006.

[16] López-Castro, J. D., Marchal, A., González, L., and Botana, J., "Topological optimization and manufacturing by Direct Metal Laser Sintering of an aeronautical part in 15-5PH stainless steel," *Procedia Manufacturing,* vol. 13, pp. 818–824, 2017.

[17] Halbrecht, A., Kinsbursky, M., Poranne, R., and Sterman, Y., "3D printed spacer fabrics," *Additive Manufacturing,* vol. 65, pp. 103436, 2023.

[18] Tiwary, V. K., Deshpande, A. S., and Rangaswamy, N., "Surface enhancement of FDM patterns to be used in rapid investment casting for making medical implants," *Rapid Prototyping Journal,* vol. 25, no. 5, pp. 904–914, 2019.

[19] Çevik, Ü., and Kam, M., "A review study on mechanical properties of obtained products by FDM method and metal/polymer composite filament production," *Journal of Nanomaterials,* vol. 2020, pp. 1–9, 2020.

[20] Kam, M., İpekçi, A., and Şengül, Ö., "Effect of FDM process parameters on the mechanical properties and production costs of 3D printed PowerABS samples," *International Journal of Analytical, Experimental and Finite Element Analysis, RAME Publishers,* vol. 7, no. 3, pp. 77–90, 2020.

[21] Kanagalingam, S., Dalton, C., Champneys, P., Boutefnouchet, T., Fernandez-Vicente, M., Shepherd, D. E., Wimpenny, D., and Thomas-Seale, L. E., "Detailed design for additive manufacturing and post processing of generatively designed high tibial osteotomy fixation plates," *Progress in Additive Manufacturing,* vol. 8, pp. 1–18, 2022.

[22] Kumbhar, N. N., and Mulay, A., "Post processing methods used to improve surface finish of products which are manufactured by additive manufacturing technologies: a review," *Journal of the Institution of Engineers (India): Series C,* vol. 99, pp. 481–487, 2018.

[23] Tiwary, V. K., Padmakumar, A., and Malik, V., "Adhesive bonding of similar/dissimilar three-dimensional printed parts (ABS/PLA) considering joint design, surface treatments, and adhesive types," *Proceedings of the Institution of Mechanical Engineers, Part C: Journal of Mechanical Engineering Science,* vol. 236, no. 16, pp. 8991–9002, 2022.

[24] Chohan, J. S., and Singh, R., "Pre and post processing techniques to improve surface characteristics of FDM parts: A state of art review and future applications," *Rapid Prototyping Journal,* vol. 23, pp. 495–515, 2017.

[25] Lalehpour, A., and Barari, A., "Post processing for fused deposition modeling parts with acetone vapour bath," *IFAC-PapersOnLine,* vol. 49, no. 31, pp. 42–48, 2016.

[26] Wang, X., Jiang, M., Zhou, Z., Gou, J., and Hui, D., "3D printing of polymer matrix composites: A review and prospective," *Composites Part B: Engineering,* vol. 110, pp. 442–458, 2017.

[27] Carlton, H. D., Haboub, A., Gallegos, G. F., Parkinson, D. Y., and MacDowell, A. A., "Damage evolution and failure mechanisms in additively manufactured stainless steel," *Materials Science and Engineering: A,* vol. 651, pp. 406–414, 2016.

[28] ISO/ASTM, *ISO/ASTM 52900: 2015, Additive Manufacturing–General Principles–Terminology*: ASTM F2792-10e1, West Conshohocken, PA 19428-2959, USA, 2015.

[29] Monzón, M., Ortega, Z., Martínez, A., and Ortega, F., "Standardization in additive manufacturing: Activities carried out by international organizations and projects," *International Journal of Advanced Manufacturing Technology,* vol. 76, pp. 1111–1121, 2015.

[30] Vilaro, T., Colin, C., and Bartout, J.-D., "As-fabricated and heat-treated microstructures of the Ti-6Al-4V alloy processed by selective laser melting," *Metallurgical and Materials Transactions A,* vol. 42, no. 10, pp. 3190–3199, 2011.

[31] Weingarten, C., Buchbinder, D., Pirch, N., Meiners, W., Wissenbach, K., and Poprawe, R., "Formation and reduction of hydrogen porosity during selective laser melting of AlSi10Mg," *Journal of Materials Processing Technology,* vol. 221, pp. 112–120, 2015.

[32] Wang, P., Tan, X., Nai, M. L. S., Tor, S. B., and Wei, J., "Spatial and geometrical-based characterization of microstructure and microhardness for an electron beam melted Ti–6Al–4V component," *Materials & Design,* vol. 95, pp. 287–295, 2016.

[33] Hrabe, N., and Quinn, T., "Effects of processing on microstructure and mechanical properties of a titanium alloy (Ti–6Al–4V) fabricated using electron beam melting (EBM), Part 2: Energy input, orientation, and location," *Materials Science and Engineering: A,* vol. 573, pp. 271–277, 2013.

[34] Tan, X., Kok, Y., Tan, Y. J., Descoins, M., Mangelinck, D., Tor, S. B., Leong, K. F., and Chua, C. K., "Graded microstructure and mechanical properties of additive manufactured Ti–6Al–4V via electron beam melting," *Acta Materialia,* vol. 97, pp. 1–16, 2015.

[35] Cain, V., Thijs, L., Van Humbeeck, J., Van Hooreweder, B., and Knutsen, R., "Crack propagation and fracture toughness of Ti6Al4V alloy produced by selective laser melting," *Additive Manufacturing,* vol. 5, pp. 68–76, 2015.

[36] ISO, E., "527-1. Plastics—Determination of tensile properties—Part 1: General principles," *International Organization of Standardization: Geneva, Switzerland,* 2012.

[37] Lazović, T. M., and Stojanović, M. S., "Preparation of specimens for standard tensile testing of plastic materials for FDM 3D printing," *Machines. Technologies. Materials.,* vol. 15, no. 5, pp. 205–208, 2021.

[38] Pernica, J., Sustr, M., Dostal, P., Brabec, M., and Dobrocky, D., "Tensile testing of 3D printed materials made by different temperature," *Manufacturing Technology,* vol. 21, no. 3, pp. 398–404, 2021.

[39] Sert, E., Hitzler, L., Hafenstein, S., Merkel, M., Werner, E., and Öchsner, A., "Tensile and compressive behaviour of additively manufactured AlSi10Mg samples," *Progress in Additive Manufacturing,* vol. 5, pp. 305–313, 2020.

[40] Tiwary, V. K., Ravi, N., Arunkumar, P., Shivakumar, S., Deshpande, A. S., and Malik, V. R., "Investigations on friction stir joining of 3D printed parts to overcome bed size limitation and enhance joint quality for unmanned aircraft systems," *Proceedings of the Institution of Mechanical Engineers, Part C: Journal of Mechanical Engineering Science,* vol. 234, no. 24, pp. 4857–4871, 2020.

[41] Walker, K., Liu, Q., and Brandt, M., "Evaluation of fatigue crack propagation behaviour in Ti-6Al-4V manufactured by selective laser melting," *International Journal of Fatigue,* vol. 104, pp. 302–308, 2017.

[42] Carroll, B. E., Palmer, T. A., and Beese, A. M., "Anisotropic tensile behavior of Ti–6Al–4V components fabricated with directed energy deposition additive manufacturing," *Acta Materialia,* vol. 87, pp. 309–320, 2015.

[43] Wang, P., Nai, M. L. S., Tan, X., Sin, W. J., Tor, S. B., and Wei, J., "Anisotropic mechanical properties in a big-sized Ti-6Al-4V plate fabricated by electron beam melting." In: TMS 2016 145th Annual Meeting & Exhibition. Springer, Cham. pp. 5–12. https://doi.org/10.1007/978-3-319-48254-5_1

[44] Lu, S. L., Tang, H., Ning, Y., Liu, N., StJohn, D., and Qian, M., "Microstructure and mechanical properties of long Ti-6Al-4V rods additively manufactured by selective electron beam melting out of a deep powder bed and the effect of subsequent hot isostatic pressing," *Metallurgical and Materials Transactions A,* vol. 46, pp. 3824–3834, 2015.

[45] Galarraga, H., Lados, D. A., Dehoff, R. R., Kirka, M. M., and Nandwana, P., "Effects of the microstructure and porosity on properties of Ti-6Al-4V ELI alloy fabricated by electron beam melting (EBM)," *Additive Manufacturing,* vol. 10, pp. 47–57, 2016.

[46] Sun, Y., Gulizia, S., Fraser, D., Oh, C., Lu, S., and Qian, M., "Layer additive production or manufacturing of thick sections of Ti-6Al-4V by selective electron beam melting (SEBM)," *Jom,* vol. 69, pp. 1836–1843, 2017.

[47] Palanivel, S., Dutt, A. K., Faierson, E., and Mishra, R., "Spatially dependent properties in a laser additive manufactured Ti–6Al–4V component," *Materials Science and Engineering: A,* vol. 654, pp. 39–52, 2016.

[48] Murr, L., Esquivel, E., Quinones, S., Gaytan, S., Lopez, M., Martinez, E., Medina, F., Hernandez, D., Martinez, E., and Martinez, J., "Microstructures and mechanical properties of electron beam-rapid manufactured Ti–6Al–4V biomedical prototypes compared to wrought Ti–6Al–4V," *Materials Characterization,* vol. 60, no. 2, pp. 96–105, 2009.

[49] Hrabe, N., and Quinn, T., "Effects of processing on microstructure and mechanical properties of a titanium alloy (Ti–6Al–4V) fabricated using electron beam melting (EBM), part 1: Distance from build plate and part size," *Materials Science and Engineering: A,* vol. 573, pp. 264–270, 2013.

[50] Hinojos, A., Mireles, J., Reichardt, A., Frigola, P., Hosemann, P., Murr, L. E., and Wicker, R. B., "Joining of Inconel 718 and 316 Stainless Steel using electron beam melting additive manufacturing technology," *Materials & Design,* vol. 94, pp. 17–27, 2016.

[51] Sun, S., Zheng, L., Peng, H., and Zhang, H., "Microstructure and mechanical properties of Al-Fe-V-Si aluminum alloy produced by electron beam melting," *Materials Science and Engineering: A,* vol. 659, pp. 207–214, 2016.

[52] Kasperovich, G., and Hausmann, J., "Improvement of fatigue resistance and ductility of TiAl6V4 processed by selective laser melting," *Journal of Materials Processing Technology,* vol. 220, pp. 202–214, 2015.

[53] Edwards, P., O'conner, A., and Ramulu, M., "Electron beam additive manufacturing of titanium components: properties and performance," *Journal of Manufacturing Science and Engineering,* vol. 135, no. 6, 2013.

[54] Seifi, M., Dahar, M., Aman, R., Harrysson, O., Beuth, J., and Lewandowski, J. J., "Evaluation of orientation dependence of fracture toughness and fatigue crack propagation behavior of as-deposited ARCAM EBM Ti-6Al-4V," *Jom,* vol. 67, pp. 597–607, 2015.

[55] Edwards, P., and Ramulu, M., "Effect of build direction on the fracture toughness and fatigue crack growth in selective laser melted Ti-6Al-4 V," *Fatigue & Fracture of Engineering Materials & Structures,* vol. 38, no. 10, pp. 1228–1236, 2015.

[56] Suryawanshi, J., Prashanth, K., Scudino, S., Eckert, J., Prakash, O., and Ramamurty, U., "Simultaneous enhancements of strength and toughness in an Al-12Si alloy synthesized using selective laser melting," *Acta Materialia,* vol. 115, pp. 285–294, 2016.

13 Exploring and Understanding the Possibilities of IoT in FDM-3D Printing

Vivek Kumar Tiwary, Arunkumar Padmakumar, Shreyans Bhandari and Vinayak R. Malik

13.1 INTRODUCTION

The phrase "Industry 4.0" refers to a number of modifications to the methods of production, most of which are propelled by the Information Technology (IT). These developments have broad organizational consequences in addition to their technological ones. Due to this, even in established businesses, a swing from a product to a service-oriented mentality is expected [1]. In the context of Industry 4.0, networked computers, intuitive machines, and smart materials interact with one another and their environment, communicate, and eventually make decisions with slight to no human involvement [2]. Industry 4.0 relies on the implementation and integration of a variety of simple and advances technologies, which include Internet of Things (IoT), Internet of Service (IoS), data analytics, cloud-based manufacturing, cloud computing, Radio Frequency Identification (RFID), Enterprise Resource Planning (ERP), Industrial sensors and controllers, Robots, Augmented and Virtual Reality (AVR), Additive Manufacturing (AM) and Artificial Intelligence (AI) [3]. Figure 13.1 depicts the important technological stakeholders of Industry 4.0. Beyond financial gains, Industry 4.0 can offer a variety of chances for social and environmental sustainability. The anticipated benefits of Industry 4.0 for sustainability aim to advance sustainable development generally by realizing increased resource effectiveness, waste decrement, and a more favorable working environment [4].

The Internet of Things (IoT) is a network of physical substances with electronics, allowing them to collect and share data. The physical objects may include machines, buildings, tools while the electronics includes software, sensors, actuators, networks etc. The IoT enables a more direct integration of the physical world with computer-based systems, which additionally results in enhanced effectiveness and accuracy [5]. Internet of Things relies on the technologies including Internet Protocol (IP), Barcode, Wireless Fidelity (Wi-Fi), ZigBee, Near Filed Communication (NFC), Bluetooth, Electronic Product Code (EPC), Actuators and Wireless Sensor Networks (WSN) [6].

DOI: 10.1201/9781032665351-16

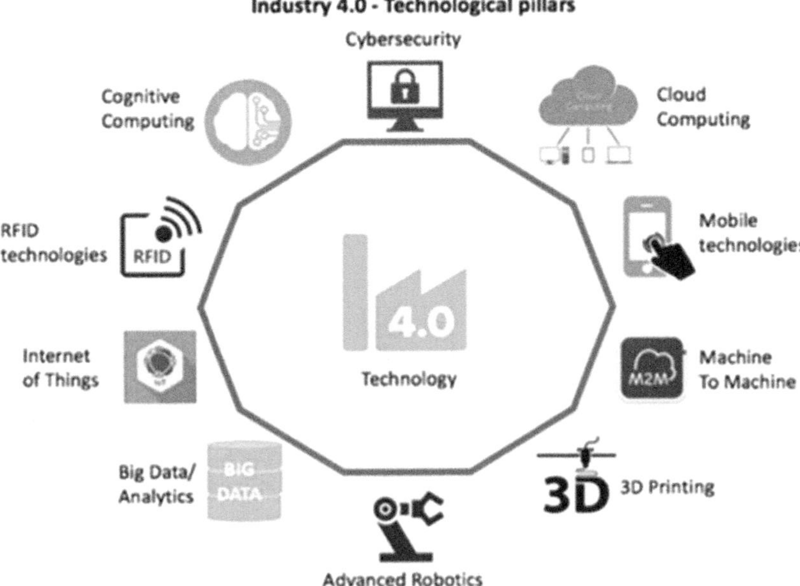

FIGURE 13.1 Important pillars of Industry 4.0 [12]

Figure 13.2 depicts the four-layered architecture design of IoT. This technology has already been implemented in agriculture, wearables, smart home and cities, industrial automation, quality control, healthcare, transportation, and automotive industries [7].

Additive Manufacturing (AM) (colloquially: 3D printing or Rapid Prototyping) is another vital technological pillar of Industry 4.0, which has extensively augmented as a disruptive technology of this age. Defined as a "layer-by-layer joining process", this technology is predicted to alter the scene of forthcoming manufacturing [8]. The main advantages of this technique over traditional manufacturing are its low buy-to-fly ratio, short cycle time, design freedom, customized production, reduced input waste, and comparatively low energy requirements [9]. Applications of 3D printing is not limited to but can be observed in aerospace, automotive, medical, consumer products, government, military and academic sectors [10], [11]. Figure 13.3 shows the working of a FDM-Fused Deposition Modelling 3D printer.

IoT and 3D printing are two emerging technologies that have the potential to revolutionize various industries. More and more 3D printer owners are looking for ways to operate their machines remotely. Combining IoT with 3D printing can unlock numerous advantages and create new opportunities including remote monitoring and control, automated material management, better quality control and maintenance, enhanced customization and personalization, sustainability and better waste reduction [15], [16]. Combining IoT with 3D printing has the likelihood to transform the traditional manufacturing process and unlock new possibilities. Considering this fact, a summary of the research progress achieved in the domain of IoT based 3D printing is presented in the present chapter. It also provides a step-by-step instruction to connect

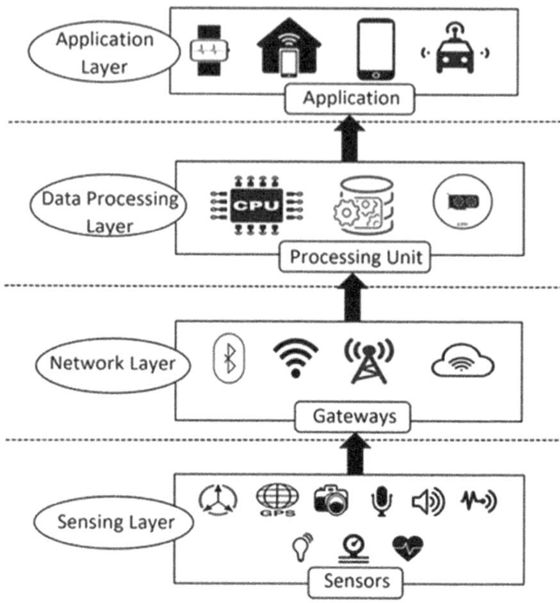

FIGURE 13.2   The four-layer architectural design of Internet of Things [7], [13]

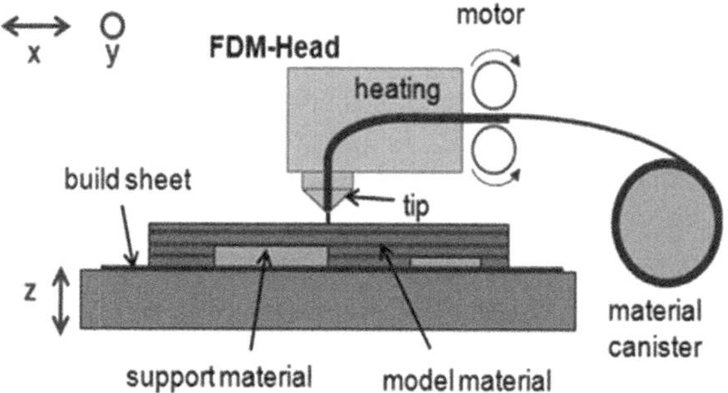

FIGURE 13.3   The process of FDM-3D printing [14]

FDM-3D printer to the Internet of Things (IoT) using OctoPrint. Additionally, camera integration for online tracking and demonstration of how to connect it to a Telegram bot for remote control and monitoring is also discussed.

13.2   IOT BASED 3D PRINTING—WHAT IT MEANS

IoT-based 3D printing denotes the integration of Internet of Things (IoT) technology and 3D printers. It combines the abilities of 3D printing with the connectivity and automation offered by IoT devices, resulting in a more advanced and efficient printing process. In traditional 3D printing, a computer-aided design (CAD) file is prepared and sent to the 3D printer for fabrication. With IoT-based 3D printing, the process becomes more automated and interconnected. Some key aspects of IoT-based 3D printing are as mentioned below:

i.   *Connectivity:* IoT devices (sensors and actuators) are assimilated into the 3D printer and connected to the internet. This enables remote monitoring and control of the printing process. Users can monitor the progress, adjust settings, and receive notifications through a web interface or mobile app.

ii. *Data collection and analysis:* IoT sensors embedded in the printer can collect real-time data about various parameters, such as temperature, humidity, filament usage, and print speed. This data can be analyzed to optimize the printing process, identify potential issues, and improve overall efficiency.

iii. *Predictive maintenance:* IoT-enabled 3D printers can utilize data analytics to predict and prevent maintenance issues. By monitoring parameters like printer temperature, vibration, and filament usage, the system can identify patterns that indicate potential failures or the need for maintenance. This helps in minimizing downtime and improving the longevity of the printer [17], [18].

iv. *Cloud connectivity and collaboration:* IoT-based 3D printers can be connected to cloud platforms, allowing users to access and share CAD files, track print jobs, and collaborate with others remotely. This feature enables seamless collaboration between multiple users and facilitates the sharing of 3D printing resources [19], [20].

v.   *Smart material management:* IoT-based systems can track the usage of printing materials and manage inventory levels. Sensors can be used to monitor the filament supply and automatically reorder when it reaches a certain threshold. This ensures a continuous printing process without the risk of running out of materials [21], [22].

IoT-based 3D printing enhances the functionality and efficiency of traditional 3D printers by incorporating connectivity, data analytics, remote monitoring, predictive maintenance, and collaboration capabilities. It offers increased control, automation, and optimization of the printing process, making it an exciting area of development in the field of additive manufacturing [23]. Figure 13.4 shows the workflow of current 3D printing services.

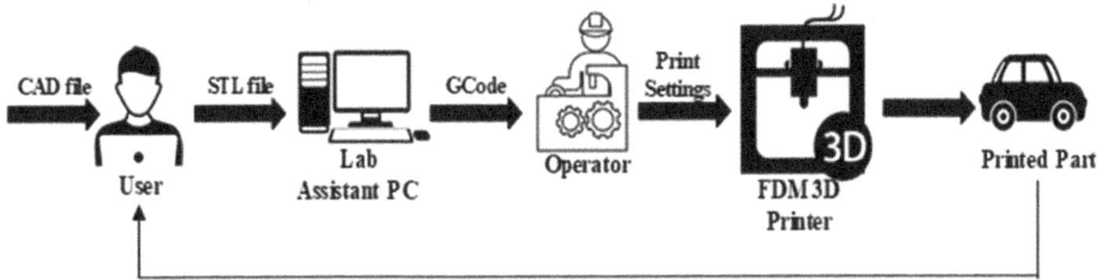

FIGURE 13.4   Workflow of current 3D printing service

13.3   IOT BASED 3D PRINTING—HOW TO IMPLEMENT

IoT (Internet of Things) based 3D printing can be implemented by integrating 3D printers with IoT technologies to create a networked and automated system. This includes a step-by-step approach, which has to be taken to achieve this. The following section presents a general overview of how the concept can be implemented:

Step 1 – IoT-enabled 3D Printers: Start by using 3D printers that are designed to be IoT-enabled. These printers are equipped with the necessary hardware such as sensors, actuators, and connectivity modules (e.g., Wi-Fi or Ethernet) to communicate with other devices and systems [24].

Step 2 – Sensor Integration: Install sensors within the 3D printer to gather data about various parameters such as temperature, humidity, print progress, filament usage, and machine status. These sensors can be integrated into the printer or added externally [25].

Step 3 – Connectivity: Connect the IoT-enabled 3D printers to a local network or the internet using Wi-Fi, Ethernet, or other suitable communication protocols. This connectivity allows the printers to communicate with other devices, services, and platforms [26].

Step 4 – Data Collection and Analysis: Collect the sensor data from the 3D printers and send it to a centralized system or cloud platform for further processing and analysis. This data can provide insights into printer performance, usage patterns, maintenance requirements, and quality control.

Step 5 – Remote Monitoring and Control: With IoT integration, you can remotely monitor and control the 3D printers from anywhere using a web-based dashboard or dedicated mobile applications. This allows you to track print progress, receive notifications, adjust print settings, and troubleshoot issues without physical access to the printer [27].

Step 6 – Automation and Optimization: Utilize the data collected from the 3D printers to automate certain tasks and optimize the printing process. For example, you can implement predictive maintenance based on sensor readings to schedule maintenance activities, or leverage machine learning algorithms to optimize print settings for better efficiency and quality.

Step 7 – Integration with Other Systems: Integrate the IoT-enabled 3D printers with other systems or applications, such as inventory management, production planning, or order fulfilment systems. This integration enables seamless coordination and automation of the entire workflow, from receiving orders to printing and shipping products [28].

Step 8 – Security and Privacy Considerations: Implement appropriate security measures to protect the IoT-enabled 3D printers and the data they generate. This may include authentication mechanisms, encryption, secure communication protocols, and regular software updates to address vulnerabilities.

Figure 13.5 represents the workflow for the proposed concept of IoT based FDM 3D printing. By implementing IoT in 3D printing, creation of a connected ecosystem

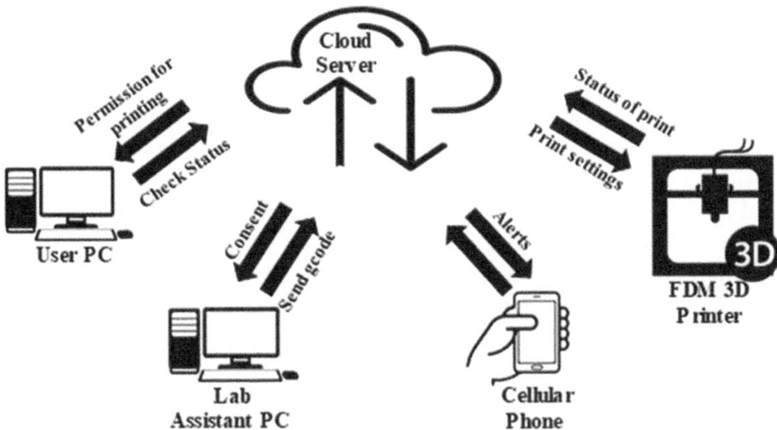

FIGURE 13.5   Workflow for cloud based FDM 3D printer

that enhances automation, monitoring, and control is possible leading to improved productivity, efficiency, and quality in the 3D printing process.

13.4   IOT BASED 3D PRINTING-CHALLENGES

While the amalgamation of IoT and 3D printing holds tremendous potential, there are several shortcomings and challenges that need to be addressed. Here are a few notable ones:

i.   *Data Security:* With IoT-enabled 3D printing, there is an increased risk of data breaches and intellectual property theft. 3D printing designs and files can be vulnerable to unauthorized access, potentially leading to the replication of copyrighted or patented objects. Robust security measures must be implemented to protect the digital assets and ensure the integrity of the manufacturing process [29].

ii.   *Quality Control:* 3D printing processes can be affected by various factors, such as material composition, printing parameters, and environmental conditions. When coupled with IoT, there is a need for real-time monitoring and feedback mechanisms to ensure consistent quality throughout the manufacturing process. Failure to address quality control issues can result in defective products and compromise safety.

iii.   *Limited Material Selection:* While 3D printing has made significant progress in terms of available materials, the range is still relatively limited compared to traditional manufacturing methods. IoT-driven 3D printing requires materials that can be used in conjunction with sensors and other embedded components. Expanding the range of compatible materials is crucial to unlock the full potential of IoT applications in 3D printing [30].

iv.   *Scalability and Speed:* Although 3D printing offers the advantage of on-demand manufacturing, the process is generally slower than conventional mass production methods. The integration of IoT can help optimize production schedules, monitor machine performance, and improve overall efficiency. However, achieving scalability and high-speed production remains a challenge, particularly for large-scale manufacturing applications [31].

v. *Cost Considerations:* 3D printing can be cost-effective for producing small batches, customized products, or complex geometries. However, the cost of materials, equipment, and maintenance can still be prohibitive for widespread adoption. Additionally, integrating IoT technology adds an additional layer of expenses, including sensors, connectivity infrastructure, and data management systems. Balancing cost-effectiveness with the benefits of IoT and additive manufacturing is essential for successful implementation [32].

vi. *Skill Gap:* The adoption of IoT-enabled 3D printing requires a skilled workforce capable of operating and maintaining the progressive technologies involved. The integration of IoT devices, data analytics, and connectivity solutions demands expertise in areas such as software development, data science, and cybersecurity. Bridging the skill gap and providing adequate training and education is crucial to fully leverage the potential of this technology combination [33].

Addressing these shortcomings will require ongoing research, technological advancements, and collaboration among various stakeholders, including manufacturers, software developers, regulatory bodies, and academic institutions. Despite the challenges, the combination of IoT and 3D printing holds tremendous promise for revolutionizing industries such as healthcare, aerospace, automotive, and consumer goods.

13.5 CASE STUDY: CREATION OF A IOT BASED 3D PRINTER

The following section describes a step by step procedure of how an FDM 3D printer can be successfully connected to the Internet of Things (IoT) using OctoPrint, together with a camera for online tracking and a Telegram bot for remote management and monitoring.

Equipment Required:

1. 3D printer (compatible with OctoPrint)
2. Raspberry Pi (preferably Raspberry Pi 4 Model B)
3. MicroSD card (minimum 16GB)
4. Power supply for Raspberry Pi
5. USB cable (compatible with your 3D printer)
6. Ethernet cable (or Wi-Fi dongle for wireless connectivity)
7. Computer/laptop with an internet connection
8. Webcam or Raspberry Pi Camera Module
9. Telegram account

Step 1: Preparing the Raspberry Pi

1. Download the latest version of OctoPi (a pre-configured OctoPrint image for Raspberry Pi) from the official OctoPrint website.
2. Use a tool like Etcher or Win32 Disk Imager to flash the OctoPi image onto the microSD card.
3. Insert the microSD card into the Raspberry Pi.
4. Connect the Raspberry Pi to your 3D printer using the USB cable.
5. Connect the power supply to the Raspberry Pi.

Step 2: Setting up OctoPrint

1. Connect the Raspberry Pi to your local network using an Ethernet cable or a Wi-Fi dongle.
2. Power on the Raspberry Pi and wait for it to boot up.
3. Open a web browser on your computer/laptop and enter the IP address of the Raspberry Pi followed by ":5000" (e.g., http:// 192.168.1.100:5000).
4. Follow the on-screen instructions to complete the OctoPrint setup wizard, including setting a username, password, and other preferences.
5. Once the setup is complete, you will be redirected to the OctoPrint interface.

Step 3: Configuring OctoPrint for IoT

1. From the OctoPrint interface, click on the wrench icon (Settings) in the top-right corner.
2. In the settings menu, select "Plugin Manager" from the sidebar.
3. Search for and install the "MQTT" plugin. This plugin enables MQTT (Message Queuing Telemetry Transport) integration.
4. After installation, click on the "MQTT" plugin to access its settings.
5. Enter your MQTT broker's address, port, and credentials (if applicable).

Step 4: Camera Integration

1. Connect a webcam to your Raspberry Pi or use the Raspberry Pi Camera Module.
2. Ensure that the camera is properly detected and configured on the Raspberry Pi.
3. In the OctoPrint interface, click on the wrench icon (Settings) and select "Webcam & Timelapse" from the sidebar.
4. Configure the webcam settings, including the stream URL or camera-specific settings for the Raspberry Pi Camera Module.
5. Save the settings and test the camera to ensure it is working correctly.

Step 5: Setting up a Telegram Bot

1. Install the Telegram app on your mobile device and create an account.
2. Open a web browser and visit the Telegram BotFather (https://telegram. me/botfather).
3. Follow the instructions provided by BotFather to create a new bot and obtain an API token.
4. Keep the API token handy for the next steps.

Step 6: Integrating Telegram Bot with OctoPrint

1. In the OctoPrint interface, click on the wrench icon (Settings) and select "Plugin Manager" from the sidebar.
2. Search for and install the "Telegram" plugin. This plugin enables integration with Telegram.
3. After installation, click on the "Telegram" plugin to access its settings.
4. Enter the Telegram Bot API token obtained from BotFather.
5. Configure other settings such as chat IDs, commands, and notifications as per your preference.
6. Save the settings and test the Telegram integration by sending commands or receiving notifications through the bot. Figure 13.6 shows the communication of 3D printers remotely using the telegram app while Figure 13.7 displays the complete layout of the proposed concept.

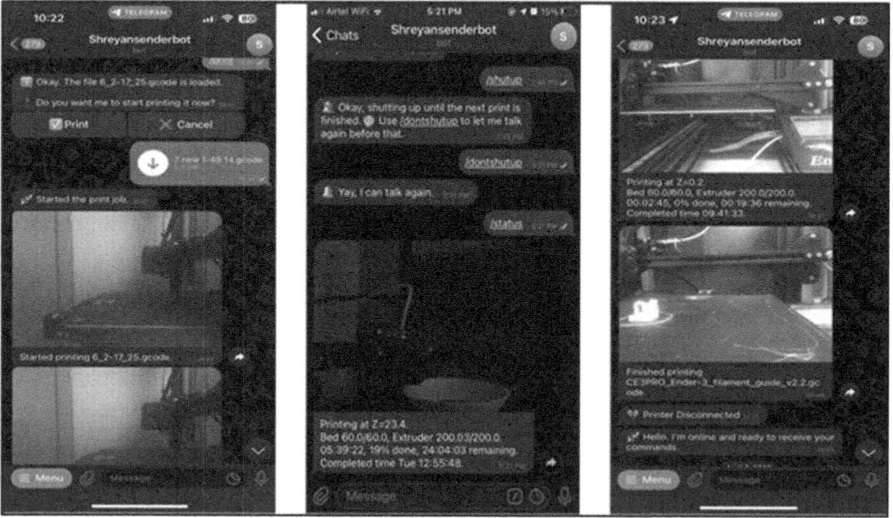

FIGURE 13.6   Print status display/communication on the cellular phone

FIGURE 13.7   The image of the complete setup

13.6   EXPECTED TRENDS IN IOT BASED 3D PRINTING:

13.6.1   INTEGRATION OF IoT, AI TOOLS AND 3D PRINTING

The integration of IoT (Internet of Things), AI tools like GPT (Generative Pre-trained Transformer), and 3D printing can open up numerous possibilities and applications across various industries [34]. Here are a few examples of how these technologies can work together:

1. *Smart Manufacturing:* IoT devices can monitor various aspects of the manufacturing process, such as machine performance, inventory levels, and quality control. AI tools like GPT can analyze the collected data to identify patterns, optimize production schedules, and predict maintenance requirements. 3D printing can be used to quickly produce prototypes or customized parts based on the insights gained from the AI analysis [35].
2. *Healthcare:* IoT-enabled medical devices can collect patient data in real-time, such as blood pressure, heart rate, and glucose levels. AI algorithms that process this data to detect anomalies, make predictions, and provide personalized healthcare recommendations. 3D printing can be used to create custom medical implants, prosthetics, or surgical models based on patient-specific data.
3. *Smart Homes:* IoT devices like smart thermostats, security cameras, and voice assistants can be interconnected to create an intelligent home ecosystem. AI tools like GPT can enhance the capabilities of voice assistants by enabling more natural language understanding and generation. 3D printing can be used to create personalized home decor items or customized fittings based on user preferences [36].
4. *Supply Chain Management:* IoT sensors can track and monitor goods throughout the supply chain, providing real-time visibility and data on factors like temperature, humidity, and location. AI algorithms can analyze this data to optimize logistics, predict demand, and identify potential bottlenecks. 3D printing can be used to produce spare parts or temporary fixes on-site, reducing downtime and improving efficiency.

5. *Education and Training:* IoT-enabled devices can create interactive learning environments, collecting data on student engagement, progress, and performance. AI tools like GPT can provide personalized tutoring, feedback, and content recommendations based on individual student needs. 3D printing can be used to create educational models or prototypes to enhance hands-on learning experiences.

These examples illustrate just a few ways in which the integration of IoT, AI tools like GPT, and 3D printing can revolutionize industries and create innovative solutions. The possibilities are vast, and as technology advances, we can expect even more transformative applications to emerge.

13.6.2   Integration of IoT, 3D Printing and Blockchain Technology

Integrating 3D printing, IoT (Internet of Things), and blockchain technology can have various applications and benefits. Here's a high-level overview of how you can integrate these technologies:

1. Understand the technologies:
   i. 3D Printing: 3D printing allows the creation of physical objects by layer to layer addition of materials based on a digital design.
   ii. IoT: The Internet of Things (IoT) is a network of connected physical substances that are able to share data.
   iii. Blockchain: Blockchain is a dispersed and decentralized ledger technology that ensures transparency, security, and immutability of data.
2. *Identify use cases:* Determine the specific use case where the integration of these technologies can provide value. For example:
   i. Supply chain management: Utilize 3D printing to create spare parts on-demand, track their usage and performance through IoT sensors, and store the related information on the blockchain for transparency.
   ii. Product authentication: Embed IoT sensors in 3D-printed products to monitor their usage, and record the product's history and authenticity on the blockchain.
   iii. Decentralized manufacturing: Use blockchain to establish a decentralized marketplace where 3D designs can be securely shared, traded, and protected.
3. Connect IoT devices and 3D printers:
   i. Establish communication protocols: Set up a communication framework that allows IoT devices and 3D printers to exchange information. This could involve using standard protocols like MQTT or HTTP.
   ii. Develop integration software: Create software that enables IoT devices to monitor and control 3D printers based on predefined conditions or triggers. This integration can be achieved through APIs or middleware.
4. Implement blockchain integration:
   i. Select a blockchain platform: Choose a suitable blockchain platform that aligns with your requirements, such as Ethereum, Hyperledger, or others.

    ii. Define data structure and smart contracts: Design the data structure to store relevant information, such as 3D design files, transaction history, and IoT sensor data. Develop smart contracts to automate and enforce business logic.

    iii. Integrate with IoT and 3D printing: Connect the IoT devices and 3D printers to the blockchain network. This can involve utilizing oracles or IoT gateways to securely transmit data to the blockchain.

5. Ensure security and privacy:

    i. Implement access control: Use cryptographic techniques to enforce access control mechanisms for data and transactions.

    ii. Data encryption: Employ encryption methods to secure sensitive data transmitted between devices and the blockchain.

    iii. Identity management: Establish a robust identity management system to authenticate and authorize participants accessing the system.

6. Test, iterate, and scale:

    i. Test the integrated system: Conduct thorough testing to ensure the functionality, interoperability, and security of the integrated solution.

    ii. Gather feedback and iterate: Collect feedback from users and stakeholders, and iterate on the solution to address any identified issues or improve its performance.

    iii. Scale the solution: Once the system is stable and validated, scale it to accommodate a larger number of users, devices, and transactions.

It is to be noted that integrating these technologies requires expertise in 3D printing, IoT, and blockchain development. Collaborating with professionals in each domain can help ensure a successful integration.

13.7 CONCLUSIONS

Industry 4.0 refers to a new industrial standard that is revolutionizing how manufacturing/production activities are controlled as a result of the convergence of physical and digital realms. The two important technological pillars of Industry 4.0, Internet of Things (IoT) and 3D printing, is expected to revolutionize the conventional manufacturing scenario. This chapter provides a summary of how these two important technologies can be integrated together for the benefit of the industries. The following important conclusions can be drawn from this chapter:

- IoT-based 3D printing is the blend of 3D printers and Internet of Things (IoT) technology creating more sophisticated and effective printing process by combining the capabilities of 3D printing with the connectivity and automation provided by IoT devices.
- A step-by-step procedure to monitor and control the 3D printers at their fingertips is described, implementation of the proposed technique would require a 3D printer that can be remotely controlled by setting up a Raspberry Pi, Camera, Octaprint and Telegram interface.

- However, the technique proposed faces a few challenges related to data security, quality control, cost, lack of required level of skill gap and scalability issues. It is expected that the ongoing research, technological advancements will overcome these barriers.
- The future trends in this technique could be to further integrate Artificial Intelligence (AI) tools like GPT, machine learning and blockchain technologies to the IoT based 3D printing.

As a universal remark, it can be perceived that the future of potential combination of IoT and 3D printing is very bright creating smart, interconnected printing systems offering enhanced functionality, efficiency and productivity. Further, as technology continues to evolve, new possibilities and applications will likely emerge at the intersection of IoT and 3D printing.

ACKNOWLEDGMENT

The authors would like to thank the Centre of Excellence on Industrial Microwave Heating Applications-Enerzi Microwave Systems Pvt. Ltd., Additive Manufacturing and Reverse Engineering Lab, for the research facilities and KLS Gogte Institute of Technology, Belagavi for the research encouragement.

REFERENCES

[1] H. Lasi, P. Fettke, H. G. Kemper, T. Feld, and M. Hoffmann, "Industry 4.0," *Bus. Inf. Syst. Eng.*, vol. 6, no. 4, pp. 239–242, 2014, doi: 10.1007/s12599-014-0334-4.

[2] M. Ghobakhloo, "Industry 4.0, digitization, and opportunities for sustainability," *J. Clean. Prod.*, vol. 252, 2020, doi: 10.1016/j.jclepro.2019.119869.

[3] W. Z. W. Yusoff and Maziah Ismail, "기사 (Article) 와 안내문 (Information)," *Eletronic Libr.*, vol. 34, no. 1, pp. 1–5, 2008.

[4] R. Brozzi, D. Forti, E. Rauch, and D. T. Matt, "The advantages of industry 4.0 applications for sustainability: Results from a sample of manufacturing companies," *Sustain.*, vol. 12, no. 9, pp. 1–19, 2020, doi: 10.3390/su12093647.

[5] J. Y. Khan, "Introduction to IoT Systems," *Internet of Things (IoT)*, no. January 2019, pp. 1–24, 2019, doi: 10.1201/9780429399084-1.

[6] S. Madakam, R. Ramaswamy, and S. Tripathi, "Internet of Things (IoT): A Literature Review," *J. Comput. Commun.*, vol. 3, no. 5, pp. 164–173, 2015, doi: 10.4236/jcc.2015.35021.

[7] S. Li, L. Da Xu, and S. Zhao, "The internet of things: A survey," *Inf. Syst. Front.*, vol. 17, no. 2, pp. 243–259, 2015, doi: 10.1007/s10796-014-9492-7.

[8] U. M. Dilberoglu, B. Gharehpapagh, U. Yaman, and M. Dolen, "The Role of Additive Manufacturing in the Era of Industry 4.0," *Procedia Manuf.*, vol. 11, no. June, pp. 545–554, 2017, doi: 10.1016/j.promfg.2017.07.148.

[9] Y. L. Yap, W. Toh, R. Koneru, R. Lin, K. I. Chan, H. Guang, W. Y. B. Chan, S. S. Teong, G. Zheng, T. Y. Ng, "Evaluation of structural epoxy and cyanoacrylate adhesives on jointed 3D printed polymeric materials," *Int. J. Adhes. Adhes.*, vol. 100, Jul. 2020, doi: 10.1016/j.ijadhadh.2020.102602.

[10] V. K. Tiwary, N. J. Ravi, P. Arunkumar, S. Shivakumar, A. S. Deshpande, and V. R. Malik, "Investigations on friction stir joining of 3D printed parts to overcome bed size limitation and enhance joint quality for unmanned aircraft systems," *Proc. Inst. Mech. Eng. Part C J. Mech. Eng. Sci.*, vol. 234, no. 24, pp. 4857–4871, Dec. 2020, doi: 10.1177/0954406220930049.

[11] V. K. Tiwary, A. Padmakumar, and V. R. Malik, "Investigations on microwave-assisted welding of MEX additive manufactured parts to overcome the bed size limitation," *J. Adv. Join. Process.*, vol. 7, no. February, p. 100141, 2023, doi: 10.1016/j.jajp.2023.100141.

[12] M. Saturno, V. Moura Pertel, F. Deschamps, and E. De Freitas Rocha Loures, "Proposal of an automation solutions architecture for Industry 4.0," *DEStech Trans. Eng. Technol. Res.*, no. icpr, 2018, doi: 10.12783/dtetr/icpr2017/17675.

[13] A. K. Sikder, G. Petracca, H. Aksu, T. Jaeger, and A. S. Uluagac, "A survey on sensor-based threats to Internet-of-Things (IoT) devices and applications," no. February, 2018, [Online]. Available: http://arxiv.org/abs/1802.02041

[14] D. Hod, A. Pandžić, I. Hajro, and P. Tasi, "Strength comparison of FDM 3D printed PLA made by different manufacturers," vol. 9, no. 3, pp. 966–970, 2020, doi: 10.18421/TEM93.

[15] V. K. Tiwary, A. Padmakumar, and V. R. Malik, "Investigations on FSW of nylon micro-particle enhanced 3D printed parts applied to a Clark-Y UAV wing," *Weld. Int.*, vol. 36, no. 8, pp. 474–488, 2022, doi: 10.1080/09507116.2022.2104141.

[16] V. K. Tiwary, A. Padmakumar, and V. Malik, "Adhesive bonding of similar / dissimilar three-dimensional printed parts (ABS / PLA) considering joint design, surface treatments, and adhesive types," vol. 236, no. 16, pp. 1–12, 2022, doi: 10.1177/09544062221089849.

[17] J.-S. J. Song and Y. Zhang, "Predictive 3D Printing with IoT," *SSRN Electron. J.*, no. Wikipedia, 2021, doi: 10.2139/ssrn.3895854.

[18] P. A. Bajakke, V. R. Malik, P. Mugali, and A. S. Deshpande, "Microwave Processing of Engineering Materials," in: Kumar, K., Babu, B.S., Davim, J.P. (eds), *Coatings. Materials Forming, Machining and Tribology*. Springer, Cham. pp. 31–55, 2021, https://doi.org/10.1007/978-3-030-62163-6_2.

[19] "What Happens When You Combine 3D Printing and IoT?," 2020. https://iotmktg.com/what-happens-when-you-combine-3d-printing-and-iot/ (accessed Jun. 19, 2023).

[20] V. Malik, N. K. Sanjeev, H. S. Hebbar, and S. V. Kailas, "Investigations on the effect of various tool pin profiles in friction stir welding using finite element simulations," *Procedia Eng.*, vol. 97, pp. 1060–1068, 2014, doi: 10.1016/j.proeng.2014.12.384.

[21] C. Crowder, "The role of 3D printing in IoT: Trends and new developments," 2021. www.iottechtrends.com/role-of-3d-printing-iot/ (accessed Jun. 19, 2023).

[22] P. A. Bajakke, S. C. Jambagi, V. R. Malik, and A. S. Deshpande, *Friction Stir Processing: An Emerging Surface Engineering Technique*, in: Gupta, K. (ed), *Surface Engineering of Modern Materials. Engineering Materials*. Springer, Cham. 2020. https://doi.org/10.1007/978-3-030-43232-4_1.

[23] Y. Wang, Y. Lin, R. Y. Zhong, and X. Xu, "IoT-enabled cloud-based additive manufacturing platform to support rapid product development," *Int. J. Prod. Res.*, vol. 57, no. 12, pp. 3975–3991, 2019, doi: 10.1080/00207543.2018.1516905.

[24] S. Li, E. Freije, and P. Yearling, "Monitoring 3D printer performance using internet of things (IoT) application," *ASEE Annu. Conf. Expo. Conf. Proc.*, vol. 2017-June, 2017, doi: 10.18260/1-2--28686.

[25] E. Prianto, H. S. Pramono, and Yuchofif, "IoT-Based 3D printer development for student competence improvement," *J. Phys. Conf. Ser.*, vol. 2111, no. 1, 2021, doi: 10.1088/1742-6596/2111/1/012002.

[26] Y. Shinde, R. Madaki, and S. Nadaf, "IoT based 3D Printer," *Int. Res. J. Eng. Technol.*, vol 06, no. 04, p. 1015, 2008, [Online]. Available: www.irjet.net

[27] T. Chen and Y. C. Wang, "An advanced IoT system for assisting ubiquitous manufacturing with 3D printing," *Int. J. Adv. Manuf. Technol.*, vol. 103, no. 5–8, pp. 1721–1733, 2019, doi: 10.1007/s00170-019-03691-5.

[28] S. K. Phang, N. Bin Ahmad, C. V. Aravind, and X. Chen, "Internet of Things based architecture for additive manufacturing interface," *Int. J. Grid Util. Comput.*, vol. 12, no. 5–6, pp. 460–468, 2021, doi: 10.1504/IJGUC.2021.120090.

[29] J. R. Mark Co and A. B. Culaba, "3D printing: Challenges and opportunities of an emerging disruptive technology," *2019 IEEE 11th Int. Conf. Humanoid, Nanotechnology, Inf. Technol. Commun. Control. Environ. Manag. HNICEM 2019*, pp. 1–6, 2019, doi: 10.1109/HNICEM48295.2019.9073427.

[30] L. Zhou, L. Zhang, L. Ren, and Y. Laili, "Matching and selection of distributed 3D printing services in cloud manufacturing," *Proc. IECON 2017 – 43rd Annu. Conf. IEEE Ind. Electron. Soc.*, vol. 2017-Janua, pp. 4728–4733, 2017, doi: 10.1109/IECON.2017.8216815.

[31] V. K. Tiwary, A. Padmakumar, and V. R. Malik, "An overview on joining/welding as post-processing technique to circumvent the build volume limitation of an FDM-3D printer," *Rapid Prototyping Journal*, vol. 27, no. 4, pp. 808–821, 2021. doi: 10.1108/RPJ-10-2020-0265.

[32] S. Fatima, A. Haleem, S. Bahl, M. Javaid, S. K. Mahla, and S. Singh, "Exploring the significant applications of Internet of Things (IoT) with 3D printing using advanced materials in medical field," *Mater. Today Proc.*, vol. 45, pp. 4844–4851, 2021, doi: 10.1016/j.matpr.2021.01.305.

[33] R. Grande, A. V. Peña, and C. U. Brancati, "JRC Technical Report. The impact of IoT and 3D printing on job quality and work organi s ation: a snapshot from Spain JRC Working Papers Series on Labour, education and Technology," 2021.

[34] K. W. Chun, H. Kim, and K. Lee, *A study on research trends of technologies for industry 4.0; 3D printing, artificial intelligence, big data, cloud computing, and internet of things*, vol. 518. Springer Singapore, 2019. doi: 10.1007/978-981-13-1328-8_51.

[35] J. Qin, Y. Liu, and R. Grosvenor, "A framework of energy consumption modelling for additive manufacturing using Internet of Things," *Procedia CIRP*, vol. 63, pp. 307–312, 2017, doi: 10.1016/j.procir.2017.02.036.

[36] Z. Xu and Q. Zhu, "Secure and resilient control of IoT-based 3D printers," *Model. Des. Secur. Internet Things*, no. September, pp. 383–405, 2020, doi: 10.1002/9781119593386.ch17.

14 Research Scope and Future Challenges in Post-Processing of FDM Parts

Vivek Kumar Tiwary, Arunkumar Padmakumar and Vinayak R. Malik

14.1 INTRODUCTION

The advent of additive manufacturing, particularly Fused Deposition Modeling (FDM), has revolutionized the way complex parts are designed and fabricated. FDM offers numerous advantages, including rapid prototyping, design flexibility, and reduced material wastage. At the same time, it is also seen that the printed parts suffer from bad surface quality, inferior dimensional accuracy, anisotropy, low mechanical strength as well as scaling up issues. Hence we often require post-processing to meet specific functional and aesthetic requirements [1].

The phrase "post-process" refers to the supplementary step performed on the unprocessed FDM samples. Post-processing encompasses a range of techniques aimed at enhancing the mechanical, surface, and overall performance of FDM parts. This interdisciplinary field presents a compelling research scope that delves into improving the quality, properties, and applications of these parts. However, it also presents a number of issues that must be resolved in order to effectively realize the prospect of FDM post-processing [2]. As we look ahead, understanding the research scope and confronting the future challenges in this domain is essential for advancing additive manufacturing and ensuring its seamless integration into various industries [3].

Going through the literatures, it could be observed that the majority of the research papers have tried to improve the properties of the 3D printed components by pre-processing as well as during the process techniques. Only a few literatures could be found addressing the post-processing issue related to FDM 3D printing. Considering this, the present chapter has been drafted to provide a first-hand experience of the research possibilities in the area of post-processing of FDM parts. Several post-processing techniques that could be employed on FDM printed parts including thermal annealing, resin infiltration, CNC machining, barrel finishing, electroplating are being introduced. Next, one of the most recent developments in additive manufacturing i.e.,

Hybrid Large Format Additive Manufacturing, which makes use of additive technique as the primary process and subtraction as the post-processing techniques is also introduced for the benefit of researchers. Lastly, the future challenges in the area of post-processing of FDM 3D printed parts is also discussed in the chapter.

After going through the chapter, it is anticipated that researchers and academics would view post-processing as a crucial stage in FDM technology and attempt to address its difficulties in order to make the technology more promising.

14.2   RESEARCH SCOPE IN POST-PROCESSING OF FDM PARTS

Post-processing of FDM parts is a critical aspect of additive manufacturing that aims to enhance the final quality, mechanical properties, and aesthetics of 3D printed objects. There are several areas within the scope of research for improving FDM part post-processing:

A)   Surface Finish Enhancement

Due to the finished part's rough geometrical textures, the usefulness of FDM components in many applications is still questionable. This is basically because of the layer-by-layer manufacturing approach resulting in a staircase effect in the fabricated components [4]. This effect is shown in the figure below. Different post-processing techniques including sanding, acetone vapor smoothening, filling and priming etc. has been already attempted by several researchers to overcome this issue.

The future research prospects include investigating methods to reduce layer lines, surface roughness, and visible defects on FDM parts [5]. Combining FDM with other manufacturing processes, such as CNC machining or resin casting, can allow for precise finishing of surfaces after 3D printing. Further, developing techniques for achieving smoother and more uniform surfaces without compromising part dimensions or accuracy can also be attempted. This is shown in Figure 14.1.

B)   Mechanical Property Improvement

Fused Deposition Modeling (FDM) has exceptional benefits in the rapid prototyping of thermoplastics which have been established in varied fields. Despite significant efforts to improve the FDM process the mechanical qualities of printed items are

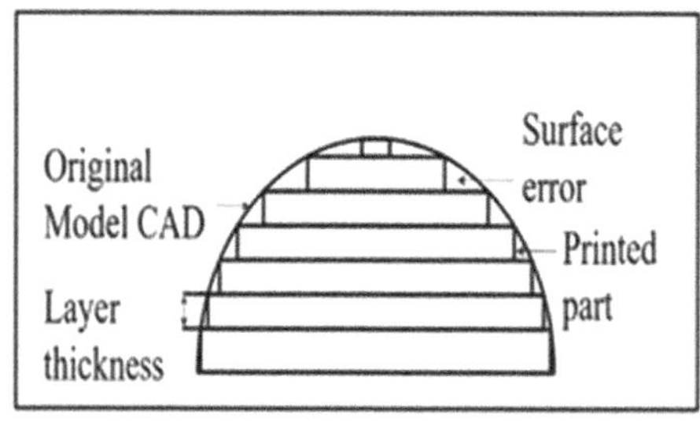

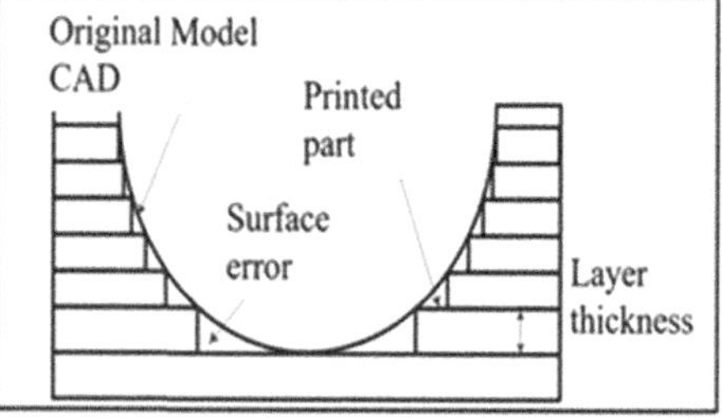

FIGURE 14.1   The staircase effect seen in FDM printed components [6]

constrained by the weak interlamination bonding and the weak performance of raw filaments employed, like PLA and ABS [7].

Various techniques including layer bonding and infill percentage optimization, better material choice, orientation optimization, topology optimization, post-processing and treatments have been attempted by researchers [8]. Future research work in this area could include developing research methods to enhance the mechanical properties of FDM parts, including tensile strength, impact resistance, and fatigue life. Secondly, explore treatments that can increase material interlayer adhesion and reduce anisotropy in mechanical properties.

C) Dimensional Accuracy and Tolerance Control

Although the benefits of 3D printed components are well known, due to the improper cooling of the plastic molten during printing, the printed parts suffer from inaccurate dimensions [9]. Different strategies, including regular calibration of the 3D printers, proper material selection for printing, providing a compensation in design, using advanced printing software's and using post-processing techniques can improve the dimensional accuracy of the part produced [10].

Future research works can include study of the techniques to minimize dimensional inaccuracies caused by thermal effects, material shrinkage, and other factors inherent to FDM. Further, developing strategies to improve part tolerances, especially for functional components that require precise dimensions can also be undertaken.

D) Joining and Assembly Methods

One of the major limitations that a FDM 3D printer suffers from is its limited bed size. This can be circumvented by employing joining and assembly methods [1]. Research in joining and assembly methods of 3D printed parts is an exciting and rapidly evolving field that addresses the challenges of creating functional and reliable structures from additive manufacturing processes. Some of the specific research areas within this domain may include Joining Dissimilar 3D Printed Materials, Hybrid Joining Methods, Bioinspired Joining, Design for Assembly, Heat and Pressure-based Joining Techniques etc. [11].

Future research works could include investigating ways to improve the bond strength and integrity of joints between FDM parts or FDM parts and components from other manufacturing processes. Further, exploring methods for designing and printing features that facilitate better assembly could also be tried out.

E) Multi-Material Printing and Graded Structures

Although 3D printing is revolutionizing manufacturing by transforming digital designs into tangible realities, laying the foundation for an era of boundless creativity and endless possibilities, right now it can print only one material at a time. Further, printing components with varying density is not possible with the existing 3D printers [12].

Considering this, future research could include an attempt to enable FDM printers to work with multiple materials or create parts with varying material properties within a single print. Secondly, research groups can study the mechanical, thermal, and electrical properties of multi-material FDM parts and their potential applications.

F) Support Removal Optimization

Overhanging objects always require a support structure to be printed. ASTM has defined that any object having an overhanging greater than 14° will require support structures to be 3D printed. Removal of the support structure after printing sometimes becomes a challenging task. Several techniques including smart support placement, reducing support density, support interface optimization, reducing contact points, using breakaway soluble supports, utilizing post-processing techniques are being attempted by researchers in this direction [13].

The future research directions could be to develop innovative methods for efficient and easy removal of support structures while minimizing damage to the main part. Secondly, investigating support materials that can be dissolved or detached more easily after printing [14]. This is shown in Figure 14.2.

G) Post-Processing Automation and Robotics

The technique of creating an apparatus, a process, or a system that operates automatically is known as automation. This concept can transform the post-processing of 3D printed parts, leading to improved efficiency, consistency, and overall quality of the final products. Automation concepts including robotics can be applied to post-processing techniques including support removal and sanding, surface smoothening,

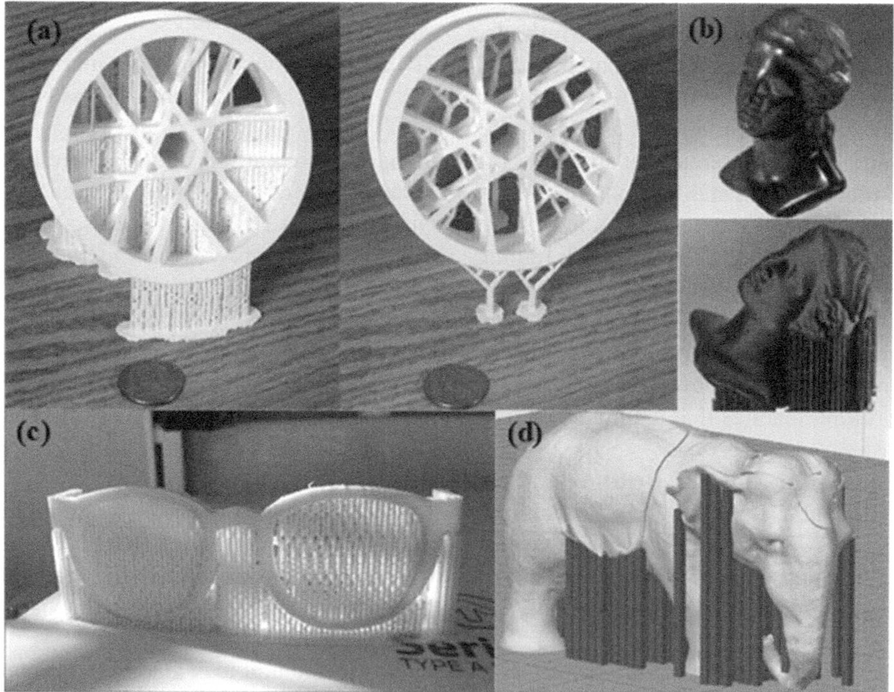

FIGURE 14.2 Support structures in overhanging 3D printed parts (a) Wheel (b) Statue (c) Spectacle frame (d) Elephant

painting and coating 3D printed parts, quality inspections, part assembly, texturing etc. [15].

However, it's important to tailor the automation approach to the specific needs of your organization, considering factors like the types of parts produced, materials used, required finishes, and production volumes [16].

Future research direction in this can include exploration of the use of automation, robotics, and computer vision to streamline and standardize post-processing tasks for FDM parts. Secondly, research can also be taken up in the area of developing algorithms for automatically detecting and addressing post-processing defects [17]. The figure below demonstrates how the concept of automation and robotics can be employed for 3D printing.

G) Surface Coating and Finishing Techniques

Surface coating and finishing techniques are crucial in 3D printing for several reasons trying to improve the aesthetics and visual appeal, functional enhancements, improving the texture and feel, to reduce the roughness and porosity, enhance the durability, have a uniform appearance and to protect the delicate 3D printed features [18]. Depending on the intended application, materials used, and desired outcomes, choosing the appropriate coating or finishing method is essential to achieving the best possible results. This is shown in Figure 14.3.

Research methods to apply coatings, paints, or finishes to FDM parts, improving their appearance, corrosion resistance, or other functional properties can be attempted. Further, exploring novel techniques for surface treatments that can bond effectively with FDM materials can also be tried out [19].

H) Environmental Sustainability

Environmental sustainability in the perspective of 3D printing refers to the accountable and well-organized use of resources, reduction of waste, and minimizing the negative environmental impacts associated with the technology. 3D printing, involves creating three-dimensional objects by depositing material on top of each other, often using digital designs. While 3D printing has the potential to revolutionize various industries by enabling on-demand production and customization, it also poses certain environmental challenges [20].

Designing products specifically for 3D printing can optimize their structural integrity and minimize the amount of material required. This approach, known as "design for additive manufacturing," can result in lighter and more efficient objects.

As 3D printing technology evolves, there's a need to address the potential for waste from discarded printers and components. Developing recycling programs and strategies for 3D printing equipment is crucial. Governments and industry bodies may establish regulations and standards to ensure environmentally responsible 3D printing practices. Compliance with these guidelines can help ensure sustainable operations [21].

Future research interests can include investigating environmentally friendly post-processing methods, such as biodegradable support materials or processes with reduced waste generation. Ongoing research into sustainable materials, energy-efficient processes, and waste reduction techniques can drive innovation and improvements in the environmental impact of 3D printing.

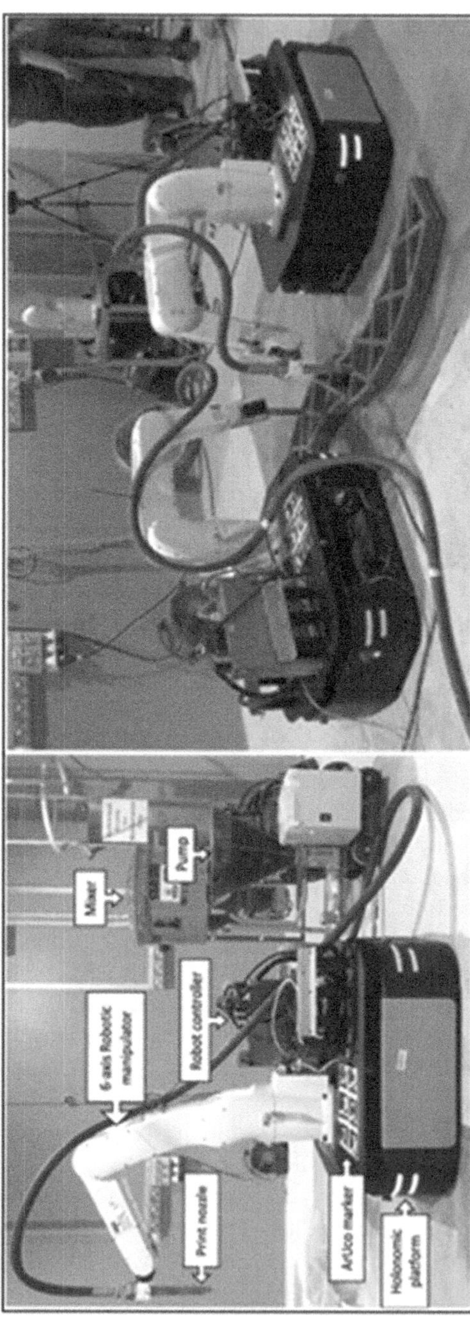

FIGURE 14.3 Automation and Robotics in 3D printing

I) Economic Analysis

Economic assessment and an evaluation of 3D printing is very crucial and an important phase to be observed. Economic analysis in the context of 3D printing involves evaluating the costs, benefits, and potential economic impacts associated with the adoption and integration of 3D printing technologies in various industries. As with any technology, 3D printing has both direct and indirect economic implications that need to be considered. The costs include initial investment costs, operating costs, materials costs, and labor costs [22].

Economic analysis in the context of 3D printing involves evaluating the costs, benefits, risks, and opportunities associated with the technology's adoption. It requires a comprehensive assessment of the various factors that contribute to the overall economic impact of integrating 3D printing into different industries and applications.

Research attempts can include evaluating the cost-effectiveness of different post-processing methods, considering factors like labor, time, equipment, and material usage. Secondly, research contributing to advancing the quality, versatility, and applicability of FDM-produced parts in various industries.

The ultimate goal is to establish reliable and efficient post-processing techniques that bridge the gap between the capabilities of FDM printers and the requirements of real-world applications.

14.3   POST-PROCESSING TECHNIQUES FOR FDM 3D PRINTED PARTS

14.3.1   Thermal Annealing Coupled with Isostatic Pressing

Anisotropy is a major issue observed in FDM-3D printed parts and occurring due to the creation of voids and non-uniform bonds between the printed layers. Due to this the strength in Z direction is usually lesser than that seen in an X and Y direction. To overcome this, thermal annealing coupled with isostatic pressing can be employed. In this process, the 3D printed samples are kept inside a vacuum bag, sealed with a tape, and then subjected to a vacuum using a pump. Next, the specimens are subjected to annealing heat treatment process which involves uniform heating at a rate of 5 °C min^{-1} them above Tg temperature and then cooling them slowly till room temperature is reached. With this it is observed that an enhanced mechanical behavior in all X-Y-Z directions and surface quality is attained [23].

14.3.2   RAISED TEMPERATURE FOLLOWED WITH RESIN INFILTRATION

Many mechanical and electrical component packages are made of transparent or translucent materials. To serve as a functional prototype, new translucent materials must be developed or existing opaque RP materials must be modified. To obtain this post-processing techniques involving raising the temperature of the printed samples is employed. This results in the increment of the transmissivity of the printed samples. However, this also results in the shrinkages in the specimens due to the air gaps. Hence to overcome this the samples can be further infiltrated with resins like cyano-acrylate and acrylic. With this a considerable increment in the transparency can be achieved for the FDM 3D printed parts [24].

14.3.3 Abrasive Flow Machining

Abrasive Flow machining is a machining technique used to finish and polish surfaces of 3D printed parts. The process involves flowing a semi-solid abrasive media through the internal passages and intricate features of the 3D printed component. The abrasive particles suspended in a viscoelastic carrier fluid helps in removal of material, burs and improves the surface finish. This process can be particularly useful for enhancing the accuracy, tolerances, and overall surface finish of 3D printed components [25].

14.3.4 CNC Machining

Finishing FDM parts through CNC machining involves using computer-controlled cutting tools to remove material from the printed part's surface. This process can help achieve tighter tolerances, smoother surfaces, and improved dimensional accuracy on the FDM parts. CNC machining can be employed to remove layer lines, excess material, and other imperfections from the 3D printed objects, resulting in a higher-quality final product with enhanced mechanical properties and aesthetics. It's a common method to refine FDM parts for applications that require higher precision and surface quality than what the initial 3D printing process can provide [26].

14.3.5 Barrel Finishing

Barrel finishing post-processing technique involves placing the printed parts along with abrasive media, water and sometimes finishing compounds into a rotating barrel or drum. As the barrel rotates, the parts and media inside rub against each other, causing friction and abrasion. This action helps to smooth out the surface of the FDM parts and remove layer lines, burrs, and other imperfections left behind by the 3D printed process.

Barrel finishing can achieve a more uniform and polished appearance on the parts, reducing the visible layer lines and creating a more aesthetically pleasing finish. The process can also help round sharp edges and corners, improve dimensional accuracy, and eliminate minor surface defects. However, it is important to note that barrel finishing might remove some material from the parts, so careful consideration should be given to the desired dimensions and tolerances of the final product [26], [27]. This is shown in Figure 14.4.

14.3.6 Electroplating

Through the use of an electric current, the process of electroplating is used to cover an object's surface with a thin layer of metal. This technique is commonly used to enhance the appearance, conductivity, and durability of objects. In the context of FDM (Fused Deposition Modeling) parts, which are produced through a 3D printing technique, electroplating can be employed to add a metallic layer to the printed plastic parts [28].

The first step involves preparing the FDM part that needs to be electroplated. The FDM part is usually made of plastic, and it needs to be properly cleaned to remove any

FIGURE 14.4   Barrel Finishing machine

dirt, oils, or residues from the surface. This cleaning is essential to ensure good adhesion of the metal layer. After cleaning, the FDM part's surface might need to be activated, which can include chemical treatments or roughening of the surface. The FDM part is then set up as the cathode (negative electrode) in an electroplating bath. The anode (positive electrode) is typically made of the metal that you want to deposit onto the FDM part. The electroplating bath contains a solution with metal ions of the desired coating metal. When an electric current is applied, metal ions from the solution start to deposit onto the surface of the FDM part. This deposition happens because the electric current causes the metal ions to be reduced at the surface of the cathode (the FDM part), leading to the formation of a thin layer of metal. Once the desired thickness of the metal layer is achieved, the plated FDM part is carefully removed from the electroplating bath [19].

Electroplating FDM parts offers several benefits, such as giving the appearance of solid metal parts without the added weight, improving the part's mechanical properties, and providing electrical conductivity when using metals like copper or nickel [29]. The main drawbacks of this technique are the higher expense of the procedure and restricted number of applications.

14.4   LFAM SYSTEMS—A TECHNOLOGY TO EVADE THE BED SIZE LIMITATION

Large Format Additive Manufacturing (LFAM) is a type of 3D printing or additive manufacturing technology that focuses on producing large-scale objects or structures. The term Large Format Additive Manufacturing is generally used for equipment capable of printing over $1m^3$ printing volume. In traditional 3D printing, objects are built layer by layer using materials such as plastic, metal, or composite materials while in LFAM, polymer pellets are used [30]. The bed size limitation of the desktop 3D printers can also be circumvented by this technique.

Large Format Additive Manufacturing takes this concept to a larger scale, enabling the creation of objects that are much bigger than what conventional 3D printers can produce. Additionally, LFAM typically uses a feedstock system based on pellets that melts polymeric pellets and extrudes them "on the move," creating fused filament that typically has a diameter greater than 2.5 mm. A preceding additional filament-extruding step is not necessary during the pellet-based AM process, which can lower costs by a factor of 10 and improve production speeds by up to 200 times. Additionally, the variety of materials available in polymer filaments is limited, while all industrial polymers are available as pellets, which promotes the use of pellets [31], [32].

Key features of Large Format Additive Manufacturing include:

1. *Scale*: LFAM systems are designed to handle larger build volumes, often exceeding those of standard 3D printers. This makes them suitable for creating large parts or components that might be used in industries like construction, aerospace, automotive, naval industries, and architecture.
2. *Speed*: LFAM technologies often incorporate techniques to increase printing speed. This can involve using larger print nozzles or extruders, optimizing layer adhesion, and improving material deposition rates.
3. *Materials*: While LFAM can work with various materials, including polymer pellets, metals, ceramics, and composites, the choice of materials can be more limited compared to traditional 3D printing. This is due to the challenges of managing the properties of larger volumes of material and maintaining consistent quality [33].
4. *Applications*: Large Format Additive Manufacturing has a range of applications. It can be used for creating prototypes, molds and tooling, architectural structures, sculptures, industrial parts, and more. Some companies are even exploring its potential for building large-scale housing or infrastructure components [34], [35].
5. *Equipment*: LFAM systems may require specialized equipment and infrastructure to handle the larger scale. This can include larger print beds, more robust gantries or robotic arms, and more powerful extrusion systems. Additionally, the cost is still expensive, and the majority of businesses who produce and employ LFAM systems are based in the United States. However, fascinating projects are currently being carried out in Sweden, Italy, Spain, and Spain [36].
6. *Challenges*: LFAM comes with its own set of challenges, including maintaining structural integrity over large volumes, managing heat distribution, avoiding warping or distortion, and ensuring material consistency throughout the printing process. Internal thermal stresses are created in parts created by the LFAM equipment by temperature gradients that happen during the material deposition process. Maintaining the material behavior during the long periods of prints is another challenge which requires attention. The images below represent some of the defects like distortion and warping seen in LFAM [37].

Overall, Large Format Additive Manufacturing represents an advancement in additive manufacturing technology that enables the creation of larger and potentially

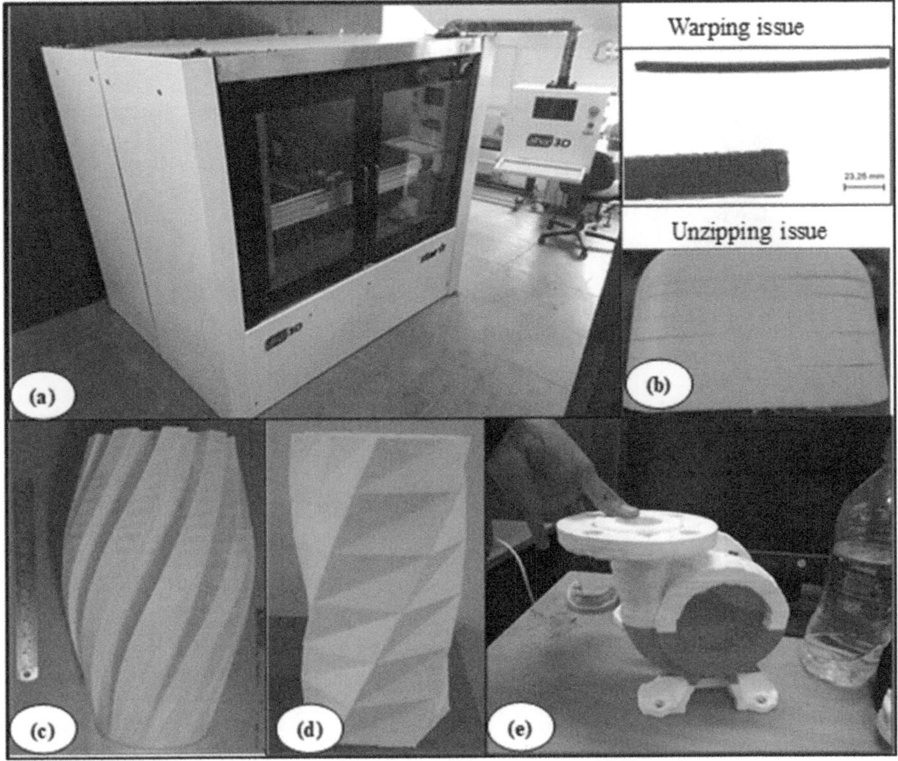

FIGURE 14.5 (a) LFAM with 1.2 cubic meter build volume (b) Challenges in LFAM (c) Vase 1 (d) Vase 2 (e) A centrifugal pump fabricated in the LFAM machine

more complex objects. It has the potential to revolutionize industries that require large-scale production of parts and structures, but it also requires careful engineering and material considerations to achieve successful outcomes. One of the most recent advances going on in this area is the hybrid LFAM. A description of Hybrid LFAM is attempted in the next section. This is shown in Figure 14.5.

14.5 HYBRID LARGE FORMAT ADDITIVE MANUFACTURING (LFAM) SYSTEM

Hybrid LFAM uses the "Near Net Shape" approach to part production where the part is first printed at high speed slightly larger than needed, then trimmed to the final size and shape. This is the quickest and most effective way to create huge structures using 3D technology. Printing and trimming can both be done with LFAM on the same equipment. The vast majority of thermoplastic composite materials, including high temperature materials that are appropriate for molds and tooling and must function at high temperatures, may be processed into parts by LFAM. This is shown in Figure 14.6.

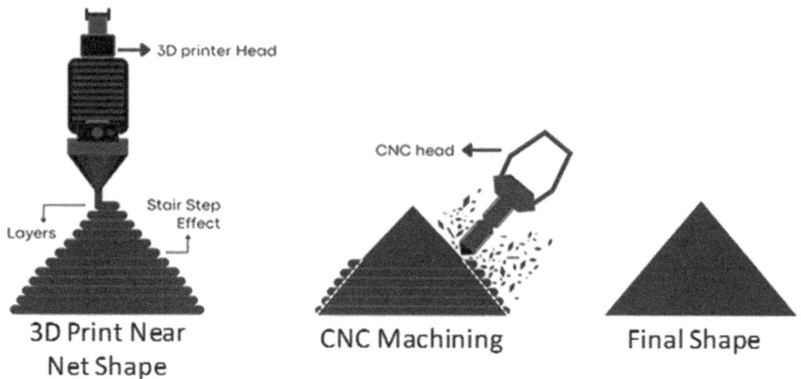

FIGURE 14.6   The Hybrid Large Format Additive Manufacturing concept [38]

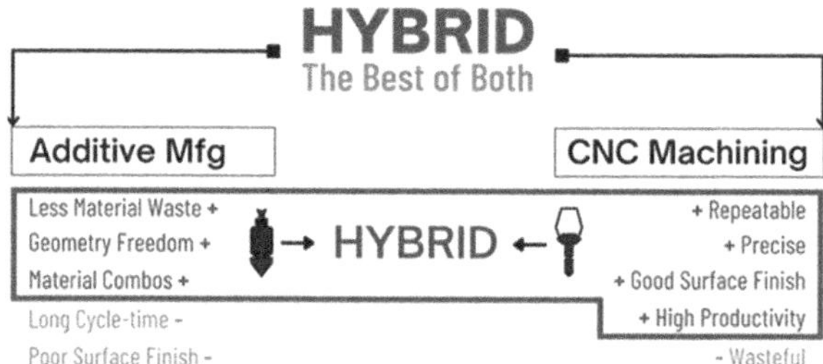

FIGURE 14.7   The hybrid approach in Additive Manufacturing [38]

The printing technology employed by LFAM creates parts that are virtually void-free, solid, totally fused, and vacuum-tight. For serious industrial production, LFAM is designed. It is a fully-fledged industrial additive manufacturing system designed for the production of large-scale components, not a lab, evaluation, or demonstration machine [39].

Hybrid LFAM is a new concept that has three things different compared to a desktop 3D printer. First, it has a provision for pellet inputs, secondly, the printing area is large (larger than 1 cubic meter) and thirdly, both additive as well as sub-tractive operations are performed simultaneously. This is shown in Figure 14.7.

Only LFAM cannot deliver the finish and tolerances which CNC Machining offers. However, traditionally, sections have been produced by machining an enlarged blank and eliminating material until the desired net form was obtained. More material is frequently removed than is kept. With near-net-shape additive manufacturing, a product is printed almost exactly to final dimensions before being cut [40]. Less material is removed, which leads to quicker processing, cheaper costs, and better material utilization. It is the perfect solution for producing particularly huge pieces when other manufacturing techniques might not be feasible. Even additional resources and time might be saved by printing a working mould following a computer design rather than first manufacturing a master, provided the right material is used. This direct digital, additive manufacturing method is noticeably faster and noticeably less expensive for industrial tooling. The advantage of this hybrid mode of additive manufacturing is displayed in the figure below.

Applications of hybrid LFAM include Tooling such as mold and patterns, composite patterns and molds, packaging industry for custom packing, recreational vehicles, concrete molds for construction, jigs and fixture, electric vehicles, Unmanned Aerial Vehicle (UAV), furniture, interior designing, wall panels, defense, aerospace, automotive, boat, railway, tool-rooms.

14.6 FUTURE CHALLENGES IN POST-PROCESSING OF FDM PARTS

As additive manufacturing, including Fused Deposition Modeling (FDM), continues to evolve and gain widespread adoption, there are several future challenges that researchers and industry professionals will need to address in the post-processing of FDM parts:

1. Automated Post-Processing Solutions: Developing automated and robotic systems that can handle various post-processing tasks, such as support removal, surface finishing, and inspection, is a challenge. Integrating automation into the post-processing workflow can increase efficiency and reduce manual labor.
2. In-Process Monitoring and Quality Control: Ensuring consistent part quality is challenging due to the layer-by-layer nature of FDM. Developing in-process monitoring and quality control techniques that can detect defects or deviations during printing and post-processing can lead to more reliable final parts.
3. Multi-Material and Multi-Component Printing: As FDM technology advances to accommodate multiple materials and components within a single print, the post-processing of such complex structures becomes challenging. Developing methods to handle dissimilar materials and optimize post-processing for these designs is crucial.
4. Mechanical Property Enhancement: Improving the mechanical properties of FDM parts through post-processing while maintaining design integrity remains a challenge. Addressing anisotropic properties and material interlayer bonding is essential for producing parts with consistent and isotropic performance [41].
5. Surface Finish and Aesthetics: Achieving high-quality surface finishes on FDM parts is challenging due to the layer-by-layer nature of printing. Developing post-processing techniques that result in smooth, uniform, and aesthetically pleasing surfaces without altering part dimensions is an ongoing challenge.
6. Material Compatibility: Post-processing methods should be compatible with a wide range of FDM materials. As new materials are developed, ensuring that post-processing techniques can be applied uniformly across different materials is crucial for versatility.

7. Scaling Up for Large Parts: Post-processing techniques developed for small to medium-sized parts might not be directly applicable to larger parts. Scaling up post-processing methods while maintaining quality and efficiency presents challenges in terms of equipment, process control, and logistics.

8. Support Removal Optimization: Efficient and easy removal of support structures remains a challenge, especially for intricate geometries and hard-to-reach areas. Developing support materials that are easy to remove without damaging the main part is an ongoing focus.

9. Environmental Impact: Addressing the environmental impact of post-processing methods, such as chemical usage and waste generation, is a growing concern. Developing sustainable and eco-friendly post-processing solutions is essential for long-term viability [42].

10. Integration with CAD and Design Software: Seamless integration of post-processing considerations into the design and CAD software is a challenge. Enabling designers to account for post-processing requirements and constraints during the design phase can lead to more efficient and optimized parts.

11. Process Standardization: Developing standardized post-processing procedures and guidelines is crucial for ensuring consistent quality across different FDM systems and industries. Establishing industry-wide standards for post-processing can lead to greater reliability and easier adoption.

12. Training and Skill Development: As new post-processing techniques and equipment are developed, training the workforce to effectively use and implement these methods becomes a challenge. Bridging the skills gap and providing adequate training is essential for successful implementation.

Addressing these challenges will require collaboration between researchers, engineers, manufacturers, and industry stakeholders. As FDM technology continues to mature, overcoming these hurdles will contribute to unlocking the full potential of additive manufacturing and its applications in various industries.

14.7 CONCLUSIONS

FDM 3D printing unlocks the power of imagination by transforming digital designs into tangible realities, layer by layer. However, some of its inherent limitations like bad surface quality, low mechanical strengths, inaccurate dimensional accuracies, anisotropy etc. are limiting its further acceptability by the business tycoons. Post processing techniques referring to the aftermath processes applied to the FDM 3D printed parts can be a rational solution to the aforesaid issue. Unfortunately, much of the research in the area of 3D printing is only concentrated on pre-processing while post-processing of FDM 3D printed parts has taken a backseat.

The chapter tries to bridge this gap by discussing the research prospects and future challenges in the post-processing domain of 3D printing. Various post-processing techniques including thermal annealing, resin infiltration, CNC machining, barrel finishing, and abrasive machining are discussed. Lastly, an important development in the area of 3D printing, hybrid Large Format Additive Manufacturing technique that employs the concept of both additive and subtractive manufacturing as a post-processing technique is also discussed in the chapter.

After reading the chapter, it is anticipated that researchers and academics would view post-processing as a crucial stage in FDM technology and try to address its difficulties in order to make the technology more promising to work on.

ACKNOWLEDGMENTS

The authors would like to acknowledge the Centre of Excellence on Industrial Microwave Heating Applications-Enerzi Microwave Systems Pvt. Ltd., Additive Manufacturing and Reverse Engineering Lab, for the research facilities and KLS Gogte Institute of Technology, Belagavi for the research encouragement.

REFERENCES

[1] V. K. Tiwary, A. Padmakumar, and V. R. Malik, "An overview on joining/welding as post-processing technique to circumvent the build volume limitation of an FDM-3D printer," *Rapid Prototyp. J.*, vol. 27, no. 4, pp. 808–821, 2021, doi: 10.1108/RPJ-10-2020-0265.

[2] W. Kuczko, F. Górski, R. Wichniarek, and P. Buń, "Influence of post-processing on accuracy of FDM products," *Adv. Sci. Technol. Res. J.*, vol. 11, no. 2, pp. 172–179, 2017, doi: 10.12913/22998624/70996.

[3] A. Lalehpour and A. Barari, "Post processing for Fused Deposition Modeling Parts with Acetone Vapour Bath," *IFAC-PapersOnLine*, vol. 49, no. 31, pp. 42–48, 2016, doi: 10.1016/j.ifacol.2016.12.159.

[4] R. Singh, R. Kumar, L. Feo, and F. Fraternali, "Friction welding of dissimilar plastic/polymer materials with metal powder reinforcement for engineering applications," *Compos. Part B Eng.*, vol. 101, pp. 77–86, 2016, doi: 10.1016/j.compositesb.2016.06.082.

[5] A. Mathew, S. Ram Kishore, A. T. Tomy, M. Sugavaneswaran, S. G. Scholz, A. Elkaseer, V. H. Wilson and A. J. Rajan, "Vapour polishing of fused deposition modelling (FDM) parts: a critical review of different techniques, and subsequent surface finish and mechanical properties of the post-processed 3D-printed parts," *Prog. Addit. Manuf.*, vol. 8, pp.1161–1178, 2023, doi: 10.1007/s40964-022-00391-7.

[6] J. I. Aguilar-Duque, C. O. Balderrama-Armendáriz, C. A. Puente-Montejano, A. S. Ontiveros-Zepeda, and J. L. García-Alcaraz, "Genetic algorithm for the reduction printing time and dimensional precision improvement on 3D components printed by Fused Filament Fabrication," *Int. J. Adv. Manuf. Technol.*, vol. 115, no. 11–12, pp. 3965–3981, 2021, doi: 10.1007/s00170-021-07314-w.

[7] P. Wang, B. Zou, S. Ding, L. Li, and C. Huang, "Effects of FDM-3D printing parameters on mechanical properties and microstructure of CF/PEEK and GF/PEEK," *Chinese J. Aeronaut.*, vol. 34, no. 9, pp. 236–246, 2021, doi: 10.1016/j.cja.2020.05.040.

[8] D. Syrlybayev, B. Zharylkassyn, A. Seisekulova, M. Akhmetov, A. Perveen, and D. Talamona, "Optimisation of strength properties of FDM printed parts—A," *Polymers (Basel).*, vol. 13, pp. 1–35, 2021.

[9]   M. M. Hanon, L. Zsidai, and Q. Ma, "Accuracy investigation of 3D printed PLA with various process parameters and different colors," *Mater. Today Proc.*, vol. 42, pp. 3089–3096, 2021, doi: 10.1016/j.matpr.2020.12.1246.

[10] K. Tiwari and S. Kumar, "Analysis of the factors affecting the dimensional accuracy of 3D printed products," *Mater. Today Proc.*, vol. 5, no. 9, pp. 18674–18680, 2018, doi: 10.1016/j.matpr.2018.06.213.

[11] V. K. Tiwary, N. J. Ravi, P. Arunkumar, S. Shivakumar, A. S. Deshpande, and V. R. Malik, "Investigations on friction stir joining of 3D printed parts to overcome bed size limitation and enhance joint quality for unmanned aircraft systems," *Proc. Inst. Mech. Eng. Part C J. Mech. Eng. Sci.*, vol. 234, no. 24, pp. 4857–4871, Dec. 2020, doi: 10.1177/0954406220930049.

[12] I. F. Ituarte, N. Boddeti, V. Hassani, M. L. Dunn, and D. W. Rosen, "Design and additive manufacture of functionally graded structures based on digital materials," *Addit. Manuf.*, vol. 30, no. August, p. 100839, 2019, doi: 10.1016/j.addma.2019.100839.

[13] N. H. Mohd Yusoff, C. H. Chong, Y. K. Wan, K. H. Cheah, and V. L. Wong, "Optimization strategies and emerging application of functionalized 3D-printed materials in water treatment: A review," *J. Water Process Eng.*, vol. 51, no. December 2022, p. 103410, 2023, doi: 10.1016/j.jwpe.2022.103410.

[14] J. Jiang, J. Stringer, and X. Xu, "Support optimization for flat features via path planning in additive manufacturing," *3D Print. Addit. Manuf.*, vol. 6, no. 3, pp. 171–179, 2019, doi: 10.1089/3dp.2017.0124.

[15] R. Bogue, "3D printing: The dawn of a new era in manufacturing?," *Assem. Autom.*, vol. 33, no. 4, pp. 307–311, 2013, doi: 10.1108/AA-06-2013-055.

[16] X. Zhang, M. Li, J. H. Lim, Y. Weng, Y. W. D. Tay, H. Pham, Q.-C. Pham , "Large-scale 3D printing by a team of mobile robots," *Autom. Constr.*, vol. 95, no. July, pp. 98–106, 2018, doi: 10.1016/j.autcon.2018.08.004.

[17] F. Craveiro, S. Nazarian, H. Bartolo, P. J. Bartolo, and J. Pinto Duarte, "An automated system for 3D printing functionally graded concrete-based materials," *Addit. Manuf.*, vol. 33, no. January, p. 101146, 2020, doi: 10.1016/j.addma.2020.101146.

[18] J. Žigon, M. Kariž, and M. Pavlič, "Surface finishing of 3d-printed polymers with selected coatings," *Polymers (Basel).*, vol. 12, no. 12, pp. 1–14, 2020, doi: 10.3390/polym12122797.

[19] A. H. M. Haidiezul, A. F. Aiman, and B. Bakar, "Surface finish effects using coating method on 3D printing (FDM) parts," *IOP Conf. Ser. Mater. Sci. Eng.*, vol. 318, no. 1, pp. 1–8, 2018, doi: 10.1088/1757-899X/318/1/012065.

[20] Z. Liu, Q. Jiang, Y. Zhang, T. Li, and H. C. Zhang, "Sustainability of 3D printing: A critical review and recommendations," *ASME 2016 11th Int. Manuf. Sci. Eng. Conf. MSEC 2016*, vol. 2, pp. 1–8, 2016, doi: 10.1115/MSEC2016-8618.

[21] M. Gebler, A. J. M. Schoot Uiterkamp, and C. Visser, "A global sustainability perspective on 3D printing technologies," *Energy Policy*, vol. 74, no. C, pp. 158–167, 2014, doi: 10.1016/j.enpol.2014.08.033.

[22] B. Peeters, N. Kiratli, and J. Semeijn, "A barrier analysis for distributed recycling of 3D printing waste: Taking the maker movement perspective," *J. Clean. Prod.*, vol. 241, p. 118313, 2019, doi: 10.1016/j.jclepro.2019.118313.

[23] A. Chueca de Bruijn, G. Gómez-Gras, L. Fernández-Ruano, L. Farràs-Tasias, and M. A. Pérez, "Optimization of a combined thermal annealing and isostatic pressing process for mechanical and surface enhancement of Ultem FDM parts using Doehlert experimental designs," *J. Manuf. Process.*, vol. 85, no. December 2022, pp. 1096–1115, 2023, doi: 10.1016/j.jmapro.2022.12.027.

[24] S. H. Ahn, C. S. Lee, and W. Jeong, "Development of translucent FDM parts by post-processing," *Rapid Prototyp. J.*, vol. 10, no. 4, pp. 218–224, 2004, doi: 10.1108/13552540410551333.

[25] J. Sambharia and H. S. Mali, "Recent developments in abrasive flow finishing process: A review of current research and future prospects," *Proc. Inst. Mech. Eng. Part B J. Eng. Manuf.*, vol. 233, no. 2, pp. 388–399, 2019, doi: 10.1177/0954405417731466.

[26] A. Boschetto and L. Bottini, "Surface improvement of fused deposition modeling parts by barrel finishing," *Rapid Prototyp. J.*, vol. 21, no. 6, pp. 686–696, 2015, doi: 10.1108/RPJ-10-2013-0105.

[27] M. Prakash, A. S. Deshpande, P. A. Bajakke, V. R. Malik, "Microwave Processing of Engineering Materials," *Mater. Forming, Mach. Tribol.*, 2021, doi: https://link.springer.com/chapter/10.1007/978-3-030-62163-6_2.

[28] K. Angel, H. H. Tsang, S. S. Bedair, G. L. Smith, and N. Lazarus, "Selective electroplating of 3D printed parts," *Addit. Manuf.*, vol. 20, no. January, pp. 164–172, 2018, doi: 10.1016/j.addma.2018.01.006.

[29] E. Vaněčková, M. Bouša, R. Sokolová, P. Moreno-García, P. Broekmann, V. Shestivska, J. Rathouský, M. Gál, T. Sebechlebská, and V. Kolivoška, "Copper electroplating of 3D printed composite electrodes," *J. Electroanal. Chem.*, vol. 858, p. 113763, 2020, doi: 10.1016/j.jelechem.2019.113763.

[30] D. Moreno Nieto, V. Casal López, and S. I. Molina, "Large-format polymeric pellet-based additive manufacturing for the naval industry," *Addit. Manuf.*, vol. 23, no. July, pp. 79–85, 2018, doi: 10.1016/j.addma.2018.07.012.

[31] D. Moreno Nieto and S. I. Molina, "Large-format fused deposition additive manufacturing: A review," *Rapid Prototyp. J.*, vol. 26, no. 5, pp. 793–799, 2020, doi: 10.1108/RPJ-05-2018-0126.

[32] V. Malik, N. K. Sanjeev, H. S. Hebbar, and S. V. Kailas, "Investigations on the effect of various tool pin profiles in friction stir welding using finite element simulations," *Procedia Eng.*, vol. 97, pp. 1060–1068, 2014, doi: 10.1016/j.proeng.2014.12.384.

[33] E. Meraz Trejo, X. Jimenez, K. M. M. Billah, J. Seppala, R. Wicker, and D. Espalin, "Compressive deformation analysis of large area pellet-fed material extrusion 3D printed parts in relation to in situ thermal imaging," *Addit. Manuf.*, vol. 33, no. January, p. 101099, 2020, doi: 10.1016/j.addma.2020.101099.

[34] C. M. S. Vicente, M. Sardinha, L. Reis, A. Ribeiro, and M. Leite, "Large-format additive manufacturing of polymer extrusion-based deposition systems: review and applications," *Prog. Addit. Manuf.*, no. 0123456789, 2023, doi: 10.1007/s40964-023-00397-9.

[35] V. Malik and S. V. Kailas, "Plasticine modeling of material mixing in friction stir welding," *J. Mater. Process. Technol.*, vol. 258, no. July 2017, pp. 80–88, 2018, doi: 10.1016/j.jmatprotec.2018.03.008.

[36] F. Pignatelli and G. Percoco, "An application- and market-oriented review on large format additive manufacturing, focusing on polymer pellet-based 3D printing," *Prog. Addit. Manuf.*, vol. 7, no. 6, pp. 1363–1377, 2022, doi: 10.1007/s40964-022-00309-3.

[37] V. Malik, N. . Sanjeev, H. S. Hebbar, and S. V. Kailas, "Time efficient simulations of plunge and dwell phase of FSW and its significance in FSSW," *Procedia Mater. Sci.*, vol. 5, pp. 630–639, 2014, doi: 10.1016/j.mspro.2014.07.309.

[38] "Hybrid Large Format Additive Manufacturing (LFAM) System." www.3deltasys.com/ (accessed Aug. 11, 2023).

[39] Phillips, "Phillips Additive Hybrid." www.phillipscorp.com/hybrid/ (accessed Aug. 11, 2023).

[40]   V.-T. heart of additive Manufacturing, "CMS presents Kreator hybrid LFAM system." www.voxelmatters.com/cms-presents-kreator-hybrid-lfam-system/ (accessed Aug. 11, 2023).

[41]   V. K. Tiwary, A. Padmakumar, and V. R. Malik, "Investigations on FSW of nylon micro-particle enhanced 3D printed parts applied to a Clark-Y UAV wing," *Weld. Int.*, vol. 36, no. 8, pp. 474–488, 2022, doi: 10.1080/09507116.2022.2104141.

[42]   A. L. Woern, D. J. Byard, R. B. Oakley, M. J. Fiedler, S. L. Snabes, and J. M. Pearce, "Fused particle fabrication 3-D printing: Recycled materials' optimization and mechanical properties," *Materials (Basel).*, vol. 11, no. 8, pp. 1–19, 2018, doi: 10.3390/ma11081413.

Index

Note: Page numbers in *italic* refer to figures, page numbers in **bold** refer to tables.